Effective Computation in Physics

Anthony Scopatz and Kathryn D. Huff

Beijing · Boston · Farnham · Sebastopol · Tokyo   O'REILLY®

Effective Computation in Physics

by Anthony Scopatz and Kathryn D. Huff

Copyright © 2015 Anthony Scopatz and Kathryn D. Huff. All rights reserved.

Printed in the United States of America.

Published by O'Reilly Media, Inc., 1005 Gravenstein Highway North, Sebastopol, CA 95472.

O'Reilly books may be purchased for educational, business, or sales promotional use. Online editions are also available for most titles (*http://safaribooksonline.com*). For more information, contact our corporate/institutional sales department: 800-998-9938 or *corporate@oreilly.com*.

Editor: Meghan Blanchette
Production Editor: Nicole Shelby
Copyeditor: Rachel Head
Proofreader: Rachel Monaghan

Indexer: Judy McConville
Interior Designer: David Futato
Cover Designer: Ellie Volckhausen
Illustrator: Rebecca Demarest

June 2015:   First Edition

Revision History for the First Edition
2015-06-09:   First Release

See *http://oreilly.com/catalog/errata.csp?isbn=9781491901533* for release details.

The O'Reilly logo is a registered trademark of O'Reilly Media, Inc. *Effective Computation in Physics*, the cover image, and related trade dress are trademarks of O'Reilly Media, Inc.

While the publisher and the authors have used good faith efforts to ensure that the information and instructions contained in this work are accurate, the publisher and the authors disclaim all responsibility for errors or omissions, including without limitation responsibility for damages resulting from the use of or reliance on this work. Use of the information and instructions contained in this work is at your own risk. If any code samples or other technology this work contains or describes is subject to open source licenses or the intellectual property rights of others, it is your responsibility to ensure that your use thereof complies with such licenses and/or rights.

978-1-491-90153-3

[LSI]

To THW and friends: gonuke, animal1, kmo, redbeard, spidr, slayer, nicopresto, wolfman, blackbeard, johnnyb, jdangerx, punkish, radio, crbates, 3rdbit, fastmath, and others, this one is for you.

Table of Contents

Foreword... xv

Preface.. xvii

Part I. Getting Started

1. Introduction to the Command Line................................. 1
    Navigating the Shell   1
        The Shell Is a Programming Language   2
        Paths and pwd   3
        Home Directory (~)   5
        Listing the Contents (ls)   6
        Changing Directories (cd)   7
        File Inspection (head and tail)   10
    Manipulating Files and Directories   11
        Creating Files (nano, emacs, vi, cat, >, and touch)   11
        Copying and Renaming Files (cp and mv)   17
        Making Directories (mkdir)   18
        Deleting Files and Directories (rm)   18
        Flags and Wildcards   20
    Getting Help   21
        Reading the Manual (man)   21
        Finding the Right Hammer (apropos)   24
        Combining Utilities with Redirection and Pipes (>, >>, and |)   25
    Permissions and Sharing   26
        Seeing Permissions (ls -l)   26
        Setting Ownership (chown)   28

v

Setting Permissions (chmod)	29
Creating Links (ln)	29
Connecting to Other Computers (ssh and scp)	30
The Environment	31
Saving Environment Variables (.bashrc)	33
Running Programs (PATH)	34
Nicknaming Commands (alias)	36
Scripting with Bash	36
Command Line Wrap-up	38

2. Programming Blastoff with Python... 39
Running Python	40
Comments	41
Variables	42
Special Variables	44
Boolean Values	45
None Is Not Zero!	45
NotImplemented Is Not None!	45
Operators	46
Strings	49
String Indexing	50
String Concatenation	53
String Literals	54
String Methods	55
Modules	57
Importing Modules	58
Importing Variables from a Module	58
Aliasing Imports	59
Aliasing Variables on Import	59
Packages	60
The Standard Library and the Python Ecosystem	62
Python Wrap-up	63

3. Essential Containers... 65
Lists	66
Tuples	70
Sets	71
Dictionaries	73
Containers Wrap-up	75

4. Flow Control and Logic... 77
Conditionals	77

if-else Statements	80
if-elif-else Statements	81
if-else Expression	82
Exceptions	82
Raising Exceptions	84
Loops	85
while Loops	86
for Loops	88
Comprehensions	90
Flow Control and Logic Wrap-up	93

5. Operating with Functions.. 95
Functions in Python	96
Keyword Arguments	99
Variable Number of Arguments	101
Multiple Return Values	103
Scope	104
Recursion	107
Lambdas	108
Generators	109
Decorators	112
Function Wrap-up	116

6. Classes and Objects.. 117
Object Orientation	118
Objects	119
Classes	123
Class Variables	124
Instance Variables	126
Constructors	127
Methods	129
Static Methods	132
Duck Typing	133
Polymorphism	135
Decorators and Metaclasses	139
Object Orientation Wrap-up	141

Part II. Getting It Done

7. Analysis and Visualization... 145
Preparing Data	145

Experimental Data	149
Simulation Data	150
Metadata	151
Loading Data	151
NumPy	152
PyTables	153
Pandas	153
Blaze	155
Cleaning and Munging Data	155
Missing Data	158
Analysis	159
Model-Driven Analysis	160
Data-Driven Analysis	162
Visualization	162
Visualization Tools	164
Gnuplot	164
matplotlib	167
Bokeh	172
Inkscape	174
Analysis and Visualization Wrap-up	175

8. Regular Expressions. 177

Messy Magnetism	178
Metacharacters on the Command Line	179
Listing Files with Simple Patterns	180
Globally Finding Filenames with Patterns (find)	182
grep, sed, and awk	187
Finding Patterns in Files (grep)	188
Finding and Replacing Patterns in Files (sed)	190
Finding and Replacing a Complex Pattern	192
sed Extras	193
Manipulating Columns of Data (awk)	195
Python Regular Expressions	197
Regular Expressions Wrap-up	199

9. NumPy: Thinking in Arrays. 201

Arrays	202
dtypes	204
Slicing and Views	208
Arithmetic and Broadcasting	211
Fancy Indexing	215
Masking	217

	Structured Arrays	220
	Universal Functions	223
	Other Valuable Functions	226
	NumPy Wrap-up	228

10. Storing Data: Files and HDF5. 229

	Files in Python	230
	An Aside About Computer Architecture	235
	Big Ideas in HDF5	237
	File Manipulations	239
	Hierarchy Layout	242
	Chunking	245
	In-Core and Out-of-Core Operations	249
	In-Core	249
	Out-of-Core	250
	Querying	252
	Compression	252
	HDF5 Utilities	254
	Storing Data Wrap-up	255

11. Important Data Structures in Physics. 257

	Hash Tables	258
	Resizing	259
	Collisions	261
	Data Frames	263
	Series	264
	The Data Frame Structure	266
	B-Trees	269
	K-D Trees	272
	Data Structures Wrap-up	277

12. Performing in Parallel. 279

	Scale and Scalability	280
	Problem Classification	282
	Example: N-Body Problem	284
	No Parallelism	285
	Threads	290
	Multiprocessing	296
	MPI	300
	Parallelism Wrap-up	307

13. Deploying Software. 309
Deploying the Software Itself   311
  pip   312
  Conda   316
  Virtual Machines   319
  Docker   321
Deploying to the Cloud   325
Deploying to Supercomputers   327
Deployment Wrap-up   329

Part III. Getting It Right

14. Building Pipelines and Software. 333
make   334
  Running make   337
  Makefiles   337
  Targets   338
  Special Targets   340
Building and Installing Software   341
  Configuration of the Makefile   343
  Compilation   345
Installation   346
Building Software and Pipelines Wrap-up   346

15. Local Version Control. 349
What Is Version Control?   349
  The Lab Notebook of Computational Physics   350
  Version Control Tool Types   351
Getting Started with Git   352
  Installing Git   352
  Getting Help (git --help)   352
  Control the Behavior of Git (git config)   354
Local Version Control with Git   355
  Creating a Local Repository (git init)   355
  Staging Files (git add)   357
  Checking the Status of Your Local Copy (git status)   357
  Saving a Snapshot (git commit)   358
  git log: Viewing the History   361
  Viewing the Differences (git diff)   362
  Unstaging or Reverting a File (git reset)   363
  Discard Revisions (git revert)   364

    Listing, Creating, and Deleting Branches (git branch)   365
    Switching Between Branches (git checkout)   366
    Merging Branches (git merge)   367
    Dealing with Conflicts   369
Version Conrol Wrap-Up   369

16. Remote Version Control. 371
Repository Hosting (github.com)   371
Creating a Repository on GitHub   373
Declaring a Remote (git remote)   373
Sending Commits to Remote Repositories (git push)   374
Downloading a Repository (git clone)   375
Fetching the Contents of a Remote (git fetch)   379
Merging the Contents of a Remote (git merge)   380
Pull = Fetch and Merge (git pull)   380
Conflicts   381
Resolving Conflicts   382
Remote Version Control Wrap-up   384

17. Debugging. 385
Encountering a Bug   386
Print Statements   387
Interactive Debugging   389
Debugging in Python (pdb)   390
    Setting the Trace   391
    Stepping Forward   392
    Querying Variables   393
    Setting the State   393
    Running Functions and Methods   394
    Continuing the Execution   394
    Breakpoints   395
Profiling   396
    Viewing the Profile with pstats   396
    Viewing the Profile Graphically   397
    Line Profiling with Kernprof   400
Linting   401
Debugging Wrap-up   402

18. Testing. 403
Why Do We Test?   404
When Should We Test?   405
Where Should We Write Tests?   405

What and How to Test?	406
Running Tests	409
Edge Cases	409
Corner Cases	410
Unit Tests	412
Integration Tests	414
Regression Tests	416
Test Generators	417
Test Coverage	418
Test-Driven Development	419
Testing Wrap-up	422

Part IV. Getting It Out There

19. Documentation... 427
Why Prioritize Documentation?	427
Documentation Is Very Valuable	428
Documentation Is Easier Than You Think	429
Types of Documentation	429
Theory Manuals	430
User and Developer Guides	431
Readme Files	431
Comments	432
Self-Documenting Code	434
Docstrings	435
Automation	436
Sphinx	437
Documentation Wrap-up	440

20. Publication... 441
Document Processing	441
Separation of Content from Formatting	442
Tracking Changes	443
Text Editors	443
Markup Languages	444
LaTeX	445
Bibliographies	456
Publication Wrap-up	459

21. Collaboration.. 461
Ticketing Systems	462

      Workflow Overview   462
      Creating an Issue   464
      Assigning an Issue   466
      Discussing an Issue   467
      Closing an Issue   468
    Pull Requests and Code Reviews   468
      Submitting a Pull Request   469
      Reviewing a Pull Request   469
      Merging a Pull Request   470
    Collaboration Wrap-up   470

22. Licenses, Ownership, and Copyright. 471
    What Is Copyrightable?   472
    Right of First Publication   473
    What Is the Public Domain?   473
    Choosing a Software License   474
    Berkeley Software Distribution (BSD) License   475
    GNU General Public License (GPL)   477
    Creative Commons (CC)   479
    Other Licenses   480
    Changing the License   482
    Copyright Is Not Everything   483
    Licensing Wrap-up   485

23. Further Musings on Computational Physics. 487
    Where to Go from Here   487

Glossary. 493

Bibliography. 499

Index. 503

Foreword

Right now, somewhere, a grad student is struggling to make sense of some badly formatted data in a bunch of folders called *final*, *final_revised*, and *final_updated*. Nearby, her supervisor has just spent four hours trying to reconstruct the figures in a paper she wrote six months ago so that she can respond to Reviewer Number Two. Down the hall, the lab intern is pointing and clicking in a GUI to run an analysis program for the thirty-fifth of two hundred input files. He won't realize that he used the wrong alpha for all of them until Thursday...

This isn't science: it's what scientists do when they don't have the equivalent of basic lab skills for scientific computing. They spend hours, days, or even weeks doing things that the computer could do for them, or trying to figure out what they or their colleagues did last time when the computer could tell them. What's worse, they usually have no idea when they're done how reliable their results are.

Starting with their work at the Hacker Within, a grassroots group at the University of Wisconsin that they helped found, Katy and Anthony have shown that none of this pain is necessary. A few basic tools like the command shell and version control, and a few basic techniques like writing modular code, can save scientists hours or days of work per week today, and simultaneously make it easier for others (including their future selves) to reproduce and build on their work tomorrow.

This book won't make you a great programmer—not on its own—but it will make you a *better* programmer. It will teach you how to do everyday tasks without feeling like you're wading through mud, and give you the background knowledge you need to make effective use of the thousands of tutorials and Q&A forums now available on the Web. I really wish I had written it, but if I had, I couldn't have done a better job than Anthony and Katy. I hope you enjoy it as much as I have.

—*Gregory V. Wilson*

Preface

Welcome to *Effective Computation in Physics*. By reading this book, you will learn the essential software skills that are needed by anyone in a physics-based field. From astrophysics to nuclear engineering, this book will take you from not knowing how to make a computer add two variables together to being the software development guru on your team.

Physics and computation have a long history together. In many ways, computers and modern physics have co-evolved. Only cryptography can really claim the same timeline with computers as physics. Yet in spite of this shared growth, physicists are not the premier software developers that you would expect. Physicists tend to suffer from two deadly assumptions:

1. Software development and software engineering are easy.
2. Simply by knowing physics, someone knows how to write code.

While it is true that some skills are transferable—for example, being able to reason about abstract symbols is important to both—the fundamental concerns, needs, interests, and mechanisms for deriving the truth of physics and computation are often distinct.

For physicists, computers are just another tool in the toolbox. Computation plays a role in physics that is not unlike the role of mathematics. You can understand physical concepts without a computer, but knowing how to speak the language(s) of computers makes practicing physics much easier. Furthermore, a physical computer is not unlike a slide rule or a photon detector or an oscilloscope. It is an experimental device that can help inform the science at hand when set up properly. Because computers are much more complicated and configurable than any previous experimental device, however, they require more patience, care, and understanding to properly set up.

More and more physicists are being asked to be software developers as part of their work or research. This book aims to make growing as a software developer as easy as possible. In the long run, this will enable you to be more productive as a physicist.

On the other end of the spectrum, computational modeling and simulation have begun to play an important part in physics. When experiments are too big or expensive to perform in statistically significant numbers, or when theoretical parameters need to be clamped down, simulation science fills a vital role. Simulations help tell experimenters where to look and can validate a theory before it ever hits a bench. Simulation is becoming a middle path for physicists everywhere, separate from theory and experiment. Many simulation scientists like to think of themselves as being more theoretical. In truth, though, the methods that are used in simulations are more similar to experimentalism.

What Is This Book?

All modern physicists, no matter how experimental, rely on a computer in some part of their scientific workflow. Some researchers only use computers as word processing devices. Others may employ computers that tirelessly collect data and churn analyses through the night, outpacing most other members of their research teams. This book introduces ways to harness computers to accomplish and automate nearly any aspect of research, and should be used as a guide during each phase of research.

Reading this book is a great way to learn about computational physics from all angles. It will help you to gain and hone software development skills that will be invaluable in the context of your work as a physicist. To the best of our knowledge, another book like this does not exist. This is not a physics textbook. This book is not the only way to learn about Python and other programming concepts. This book is about what happens when those two worlds inelastically collide. This book is about computational physics. You are in for a treat!

Who This Book Is For

This book is for anyone in a physics-based field who must do some programming as a result of their job or one of their interests. We specifically cast a wide net with the term "physics-based field." We take this term to mean any of the following fields: physics, astronomy, astrophysics, geology, geophysics, climate science, applied math, biophysics, nuclear engineering, mechanical engineering, material science, electrical engineering, and more. For the remainder of this book, when the term *physics* is used it refers to this broader sense of physics and engineering. It does not simply refer to the single area of study that shares that name.

Even though this book is presented in the Python programming language, the concepts apply to a wide variety of programming languages, both modern and historical.

Python was chosen here because it is easy and intuitive to use in a wide variety of situations. While you are trying to learn concepts in computational physics, Python gets out of your way. You can take the skills that you learn here and apply them equally well in other programming contexts.

Who This Book Is Not For

While anyone is welcome to read this book and learn, it is targeted at people in physics who need to learn computational skills. The examples will draw from a working knowledge of physics concepts. If you primarily work as a linguist or anthropologist, this book is probably not for you. No knowledge of computers or programming is assumed. If you have already been working as a software developer for several years, this book will help you only minimally.

Case Study on How to Use This Book: Radioactive Decay Constants

To demonstrate, let's take the example of a team of physicists using a new detector to measure the decay constants of radium isotopes at higher precision. The physicists will need to access data that holds the currently accepted values. They may also want to write a small program that gives the expected activity of each isotope as a function of time. Next, the scientists will collect experimental data from the detector, store the raw output, compare it to the expected values, and publish a paper on the differences. Since the heroes of this story value the tenets of science and are respectful of their colleagues, they'll have been certain to test all of their analyses and to carefully document each part of the process along the way. Their colleagues, after all, will need to repeat this process for the thousands of other isotopes in the table of nuclides.

Accessing Data and Libraries

To access a library that holds nuclear data such as currently accepted nuclear decay constants, λ_i, for each isotope i, our heroes may have to install the ENSDF (*http://www.nndc.bnl.gov/ensdf/*) database into their filesystem. Insights about the shell (Chapter 1) and systems for building software (Chapter 14) will be necessary in this simple endeavor.

Creating a Simple Program

The expected activity for an isotope as a function of time is very simple ($A_i = N_i e^{-\lambda_i t}$). No matter how simple the equation, though, no one wants to solve it by hand (or by copying and pasting in Excel) for every 10^{-10} second of the experiment. For this step, Chapter 2 provides a guide for creating a simple function in the Python programming language. For more sophisticated mathematical models, object orientation (Chapter 6), numerical Python (Chapter 9), and data structures (Chapter 11) may be needed.

Automating Data Collection

A mature experiment is one that requires no human intervention. Said another way, a happy physicist sleeps at home while the experiment is running unaided all night back at the lab. The skills gained in Chapter 1 and Chapter 2 can help to automate data collection from an experiment. Methods for storing that data can be learned in Chapter 10, which covers HDF5.

Analyzing and Plotting the Data

Once the currently accepted values are known and the experimental data has been collected, the next step of the experiment is to compare the two datasets. Along with lessons learned from Chapter 1 and Chapter 2, this step will be aided by a familiarity with sophisticated tools for analysis and visualization (Chapter 7). For very complex data analysis, parallelism (the basics of which are discussed in Chapter 12) can speed up the work by employing many processors at once.

Keeping Track of Changes

Because this is science, reproducibility is paramount. To make sure that they can repeat their results, unwind their analysis to previous versions, and replicate their plots, all previous versions of the scientists' code and data should be under version control. This tool may be the most essential one in this book. The basics of version control can be found in Chapter 15, and the use of version control within a collaboration is discussed in Chapter 16.

Testing the Code

In addition to being reproducible, the theory, data collection, analysis, and plots must be correct. Accordingly, Chapter 17 will cover the basics of how to debug software and how to interpret error messages. Even after debugging, the fear of unnoticed software bugs (and subsequent catastrophic paper retractions) compels our hero to test the code that's been written for this project. Language-independent principles for testing code will be covered in Chapter 18, along with specific tools for testing Python code.

Documenting the Code

All along, our physicists should have been documenting their computing processes and methods. With the tools introduced in Chapter 19, creating a user manual for code doesn't have to be its own project. That chapter will demonstrate how a clickable, Internet-publishable manual can be generated in an automated fashion based on comments in the code itself. Even if documentation is left to the end of a project, Chapter 19 can still help forward-thinking physicists to curate their work for posterity. The chapters on licenses (Chapter 22) and collaboration (Chapter 21) will also be helpful when it's time to share that well-documented code.

Publishing

Once the software is complete, correct, and documented, our physicists can then move on to the all-important writing phase. Sharing their work in a peer-reviewed publication is the ultimate reward of this successful research program. When the data is in and the plots are generated, the real challenge has often only begun, however. Luckily, there are tools that help authors be more efficient when writing scientific documents. These tools will be introduced in Chapter 20.

What to Do While Reading This Book

You learn by doing. We want you to learn, so we expect you to follow along with the examples. The examples here are practical, not theoretical. In the chapters on Python, you should fire up a Python session (don't worry, we'll show you how). Try the code out for yourself. Try out your own variants of what is presented in the book. Writing out the code yourself makes the software and the physics real.

If you run into problems, try to solve them by thinking about what went wrong. Googling the error messages you see is a huge help. The question and answer website Stack Overflow (*http://stackoverflow.com/*) is your new friend. If you find yourself truly stuck, feel free to contact us. This book can only give you a finite amount of content to study. However, with your goals and imagination, you will be able to practice computational physics until the end of time.

Furthermore, if there are chapters or sections whose topics you already feel comfortable with or that you don't see as being directly relevant to your work, feel free to skip them! You can always come back to a section if you do not understand something or you need a refresher. We have inserted many back and forward references to topics throughout the course of the text, so don't worry if you have skipped something that ends up being important later. We've tried to tie everything together so that you can know what is happening, while it is happening. This book is one part personal odyssey and one part reference manual. Please use it in both ways.

Conventions Used in This Book

The following typographical conventions are used in this book:

Italic
: Indicates new terms, URLs, email addresses, filenames, and file extensions.

`Constant width`
: Used for program listings, as well as within paragraphs to refer to program elements such as variable or function names, databases, data types, environment variables, statements, and keywords.

> This element signifies a tip or suggestion.

> This element signifies a general note.

> This element indicates a warning or caution.

This book also makes use of a fair number of "code callouts." This is where the coding examples are annotated with numbers in circles. For example:

```
print("This is code that you should type.")  ❶
```

❶ This is used to annotate something special about the software you are writing.

These are useful for drawing your attention to specific parts of the code and to explain what is happening on a step-by-step basis. You should not type the circled numbers, as they are not part of the code itself.

Using Code Examples

Supplemental material (code examples, exercises, etc.) is available for download at *https://github.com/physics-codes/examples*.

This book is here to help you get your job done. In general, if example code is offered with this book, you may use it in your programs and documentation. You do not need to contact us for permission unless you're reproducing a significant portion of the code. For example, writing a program that uses several chunks of code from this book does not require permission. Selling or distributing a CD-ROM of examples from O'Reilly books does require permission. Answering a question by citing this book and quoting example code does not require permission. Incorporating a significant amount of example code from this book into your product's documentation does require permission.

We appreciate, but do not require, attribution. An attribution usually includes the title, author, publisher, and ISBN. For example: "*Effective Computation in Physics* by Anthony Scopatz and Kathryn D. Huff (O'Reilly). Copyright 2015 Anthony Scopatz and Kathryn D. Huff, 978-1-491-90153-3."

If you feel your use of code examples falls outside fair use or the permission given above, feel free to contact us at *permissions@oreilly.com*.

Installation and Setup

This book will teach you to use and master many different software projects. That means that you will have to have a lot of software packages on your computer to follow along. Luckily, the process of installing the packages has recently become much easier and more consistent. We will be using the conda package manager for all of our installation needs.

Step 1: Download and Install Miniconda (or Anaconda)

If you have not done so already, please download and install Miniconda. Alternatively, you can install Anaconda (*http://bit.ly/dl-anaconda*). Miniconda is a stripped-down version of Anaconda, so if you already have either of these, you don't need the other. Miniconda is a Python distribution that comes with Conda, which we will then use to install everything else we need. The Conda website (*http://bit.ly/install-conda*) will help you download the Miniconda version that is right for your system. Linux, Mac OS X, and Windows builds are available for 32- and 64-bit architectures. You do not need administrator privileges on your computer to install Miniconda. We recommend that you install the Python 3 version, although all of the examples in this book should work with Python 2 as well.

If you are on Windows, we recommend using Anaconda because it allievates some of the other package installation troubles. However, on Windows you can install Miniconda simply by double-clicking on the executable and following the instructions in the installation wizard.

> **Special Windows Instructions Without Anaconda: msysGit and Git Bash**
>
> If you are on Windows and are not using Anaconda, please download and install msysGit, which you can find on GitHub (*http://msysgit.github.io*). This will provide you with the version control system called Git as well as the bash shell, both of these, which we will discuss at length. Neither is automatically available on Windows or through Miniconda. The default install settings should be good enough for our purposes here.

If you are on Linux or Mac OS X, first open your Terminal application. If you do not know where your Terminal lives, use your operating system's search functionality to find it. Once you have an open terminal, type in the following after the dollar sign ($). Note that you may have to change the version number in the filename (the Miniconda-3.7.0-Linux-x86_64.sh part) to match the file that you downloaded:

```
# On Linux, use the following to install Miniconda:
$ bash ~/Downloads/Miniconda-3.7.0-Linux-x86_64.sh

# On Mac OS X, use the following to install Miniconda:
$ bash ~/Downloads/Miniconda3-3.7.0-MacOSX-x86_64.sh
```

Here, we have downloaded Miniconda into our default download directory, *~/Downloads*. The file we downloaded was the 64-bit version; if you're using the 32-bit version you will have to adjust the filename accordingly.

On Linux, Mac OS X, and Windows, when the installer asks you if you would like to automatically change or update the *.bashrc* file or the system `PATH`, say yes. That will make it so that Miniconda is automatically in your environment and will ease further installation. Otherwise, all of the other default installation options should be good enough.

Step 2: Install the Packages

Now that you have Conda installed, you can install the packages that you'll need for this book. On Windows, open up the command prompt, *cmd.exe*. On Linux and Mac OS X, open up a terminal. You may need to open up a new terminal window for the installation of Miniconda to take effect. Now, no matter what your operating system is, type the following command:

```
$ conda install --yes numpy scipy ipython ipython-notebook matplotlib pandas \
    pytables nose setuptools sphinx mpi4py
```

This may take a few minutes to download. After this, you are ready to go!

Safari® Books Online

Safari Books Online is an on-demand digital library that delivers expert content in both book and video form from the world's leading authors in technology and business.

Technology professionals, software developers, web designers, and business and creative professionals use Safari Books Online as their primary resource for research, problem solving, learning, and certification training.

Safari Books Online offers a range of plans and pricing for enterprise, government, education, and individuals.

Members have access to thousands of books, training videos, and prepublication manuscripts in one fully searchable database from publishers like O'Reilly Media, Prentice Hall Professional, Addison-Wesley Professional, Microsoft Press, Sams, Que, Peachpit Press, Focal Press, Cisco Press, John Wiley & Sons, Syngress, Morgan Kaufmann, IBM Redbooks, Packt, Adobe Press, FT Press, Apress, Manning, New Riders, McGraw-Hill, Jones & Bartlett, Course Technology, and hundreds more. For more information about Safari Books Online, please visit us online.

How to Contact Us

Please address comments and questions concerning this book to the publisher:

O'Reilly Media, Inc.
1005 Gravenstein Highway North
Sebastopol, CA 95472
800-998-9938 (in the United States or Canada)
707-829-0515 (international or local)
707-829-0104 (fax)

We have a web page for this book, where we list errata, examples, and any additional information. You can access this page at *http://bit.ly/effective-comp*.

To comment or ask technical questions about this book, send email to *bookquestions@oreilly.com*.

For more information about our books, courses, conferences, and news, see our website at *http://www.oreilly.com*.

Find us on Facebook: *http://facebook.com/oreilly*

Follow us on Twitter: *http://twitter.com/oreillymedia*

Watch us on YouTube: *http://www.youtube.com/oreillymedia*

Acknowledgments

This work owes a resounding thanks to Greg Wilson and to Software Carpentry (*http://software-carpentry.org/*). The work you have done has changed the conversation surrounding computational science. You have set the stage for this book to even exist. The plethora of contributions to the community cannot be understated.

Equally, we must thank Paul P.H. Wilson and The Hacker Within (*http://thehackerwithin.github.io/*) for continuing to inspire us throughout the years. Independent of age and affiliation, you have always challenged us to learn from each other and unlock what was already there.

Stephen Scopatz and Bruce Rowe also deserve the special thanks afforded only to parents and professors. Without them helping connect key synapses at the right time, this book would never have been proposed.

The African Institute for Mathematical Sciences deserves special recognition for demonstrating the immense value of scientific computing, even to those of us who have been in the field for years. Your work inspired this book, and we hope that we can give back to your students by writing it.

We also owe thanks to our reviewers for keeping us honest: Jennifer Klay, Daniel Wooten, Michael Sarahan, and Denia Djokić.

To baristas all across the world, in innumerable cafés, we salute you.

PART I
Getting Started

CHAPTER 1

Introduction to the Command Line

The command line, or *shell*, provides a powerful, transparent interface between the user and the internals of a computer. At least on a Linux or Unix computer, the command line provides total access to the files and processes defining the state of the computer—including the files and processes of the operating system.

Also, many numerical tools for physics can only be installed and run through this interface. So, while this transparent interface could inspire the curiosity of a physicist all on its own, it is much more likely that you picked up this book because there is something you need to accomplish that only the command line will be capable of. While the command line may conjure images of *The Matrix*, do not let it intimidate you. Let's take the red pill.

Navigating the Shell

You can access the shell by opening a terminal emulator ("terminal" for short) on a Linux or Unix computer. On a Windows computer, the Git Bash program is equivalent. Launching the terminal opens an interactive shell program, which is where you will run your executable programs. The shell provides an interface, called the *command-line interface*, that can be used to run commands and navigate through the filesystem(s) to which your computer is connected. This command line is also sometimes called the *prompt*, and in this book it will be denoted with a dollar sign ($) that points to where your cursor is ready to enter input. It should look something like Figure 1-1.

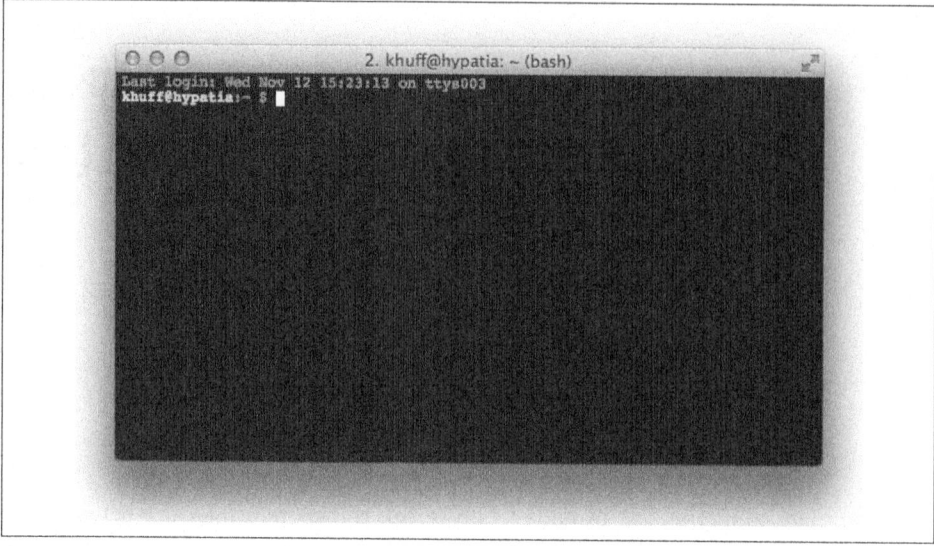

Figure 1-1. A terminal instance

This program is powerful and transparent, and provides total access to the files and processes on a computer. But what *is* the shell, exactly?

The Shell Is a Programming Language

The shell is a programming language that is run by the terminal. Like other programming languages, the shell:

- Can collect many operations into single entities
- Requires input
- Produces output
- Has variables and state
- Uses irritating syntax
- Uses special characters

Additionally, as with programming languages, there are more shells than you'll really care to learn. Among shells, bash is most widely used, so that is what we'll use in this discussion. The csh, tcsh, and ksh shell types are also popular. Features of various shells are listed in Table 1-1.

Table 1-1. Shell types

Shell	Name	Description
sh	Bourne shell	Popular, ubiquitous shell developed in 1977, still guaranteed on all Unixes
csh	C shell	Improves on sh
ksh	Korn shell	Backward-compatible with sh, but extends and borrows from other shells
bash	Bourne again shell	Free software replacement for sh, much evolved
tcsh	Tenex C shell	Updated and extended C shell

Exercise: Open a Terminal

1. Search your computer's programs to find one called Terminal. On a Windows computer, remember to use Git Bash as your bash terminal.
2. Open an instance of that program. You're in the shell!

The power of the shell resides in its transparency. By providing direct access to the entire filesystem, the shell can be used to accomplish nearly any task. Tasks such as finding files, manipulating them, installing libraries, and running programs begin with an understanding of paths and locations in the terminal.

Paths and pwd

The space where your files are—your file space—is made up of many nested directories (folders). In Unix parlance, the location of each directory (and each file inside them) is given by a "path." These can be either absolute paths or relative paths.

Paths are *absolute* if they begin at the top of the filesystem directory tree. The very top of the filesystem directory tree is called the root directory. The path to the root directory is /. Therefore, absolute paths start with /.

In many UNIX and Linux systems, the root directory contains directories like *bin* and *lib*. The absolute paths to the *bin* and *lib* directories are then */bin* and */lib*, respectively. A diagram of an example directory tree, along with some notion of paths, can be seen in Figure 1-2.

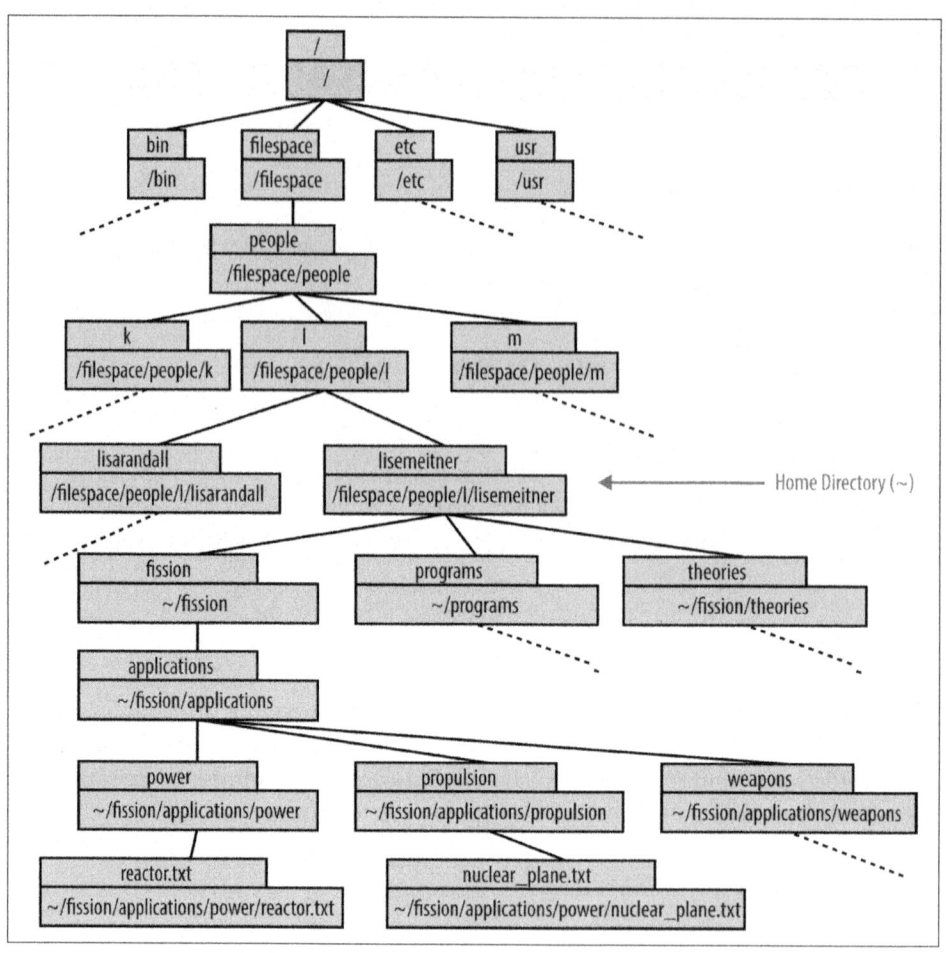

Figure 1-2. An example directory tree

> The / syntax is used at the beginning of a path to indicate the top-level directory. It is also used to separate the names of directories in a path, as seen in Figure 1-2.

Paths can, instead, be *relative* to your current working directory. The current working directory is denoted with one dot (.), while the directory immediately above it (its "parent") is denoted with two dots (..). Relative paths therefore often start with a dot or two.

As we have learned, absolute paths describe a file space location relative to the root directory. Any path that describes a location relative to the current working directory instead is a relative path. Bringing these together, note that you can always print out the full, absolute path of the directory you're currently working in with the command `pwd` (print working directory).

Bash was not available in the 1930s, when Lise Meitner was developing a theoretical framework for neutron-induced fission. However, had Bash been available, Prof. Meitner's research computer might have contained a set of directories holding files about her theory of fission as well as ideas about its application (see Figure 1-2). Let's take a look at how Lise would have navigated through this directory structure.

> You can work along with Lise while you read this book. The directory tree she will be working with in this chapter is available in a repository on GitHub (*https://github.com/physics-codes/examples*). Read the instructions at that site to download the files.

When she is working, Lise enters commands at the command prompt. In the following example, we can see that the command prompt gives an abbreviated path name before the dollar sign (this is sometimes a greater-than sign or other symbol). That path is *~/fission*, because *fission* is the directory that Lise is currently working in:

```
~/fission $
```

When she types `pwd` at the command prompt, the shell returns (on the following line) the full path to her current working directory:

```
~/fission $ pwd
/filespace/people/l/lisemeitner/fission/
```

When we compare the absolute path and the abbreviated prompt, it seems that the prompt replaces all the directories up to and including *lisemeitner* with a single character, the tilde (~). In the next section, we'll see why.

Home Directory (~)

The shell starts your session from a special directory called your *home directory*. The tilde (~) character can be used as a shortcut to your home directory. Thus, when you log in, you probably see the command prompt telling you you're in your home directory:

```
~ $
```

These prompts are not universal. Sometimes, the prompt shows the username and the name of the computer as well:

```
<user>@<machine>:~ $
```

For Prof. Meitner, who held a research position at the prestigious Kaiser Wilhelm Institute, this might appear as:

Navigating the Shell | 5

```
meitner@kaiser-wilhelm-cluster:~ $
```

Returning to the previous example, let us compare:

```
~/fission
```

to:

```
/filespace/people/l/lisemeitner/fission
```

It seems that the tilde has entirely replaced the home directory path (*/filespace/people/l/lisemeitner*). Indeed, the tilde is an abbreviation for the home directory path —that is, the sequence of characters (also known as a string) beginning with the root directory (*/*). Because the path is defined relative to the absolute top of the directory tree, this:

```
~/fission
```

and this:

```
/filespace/people/l/lisemeitner/fission
```

are both absolute paths.

Exercise: Find Home

1. Open the Terminal.
2. Type `pwd` at the command prompt and press Enter to see the absolute path to your home directory.

Now that she knows where she is in the filesystem, curious Lise is interested in what she'll find there. To list the contents of a directory, she'll need the `ls` command.

Listing the Contents (ls)

The `ls` command allows the user to print out a list of all the files and subdirectories in a directory.

Exercise: List the Contents of a Directory

1. Open the Terminal.
2. Type `ls` at the command prompt and press Enter to see the contents of your home directory.

From the *fission* directory in Professor Meitner's home directory, `ls` results in the following list of its contents:

```
~/fission $ ls ❶
applications/  heat-production.txt  neutron-release.txt  ❷
```

❶ In the *fission* directory within her home directory, Lise types `ls` and then presses Enter.

❷ The shell responds by listing the contents of the current directory.

When she lists the contents, she sees that there are two files and one subdirectory. In the shell, directories may be rendered in a different color than files or may be indicated with a forward slash (/) at the end of their name, as in the preceding example.

Lise can also provide an argument to the `ls` command. To list the contents of the *applications* directory without entering it, she can execute:

```
~/fission $ ls applications ❶
power/  propulsion/  weapons/  ❷
```

❶ Lise lists the contents of the *applications* directory without leaving the *fission* directory.

❷ The shell responds by listing the three directories contained in the *applications* directory.

The `ls` command can inform Lise about the contents of directories in her filesystem. However, to actually navigate to any of these directories, Lise will need the command `cd`.

Changing Directories (cd)

Lise can change directories with the `cd` command. When she types only those letters, the `cd` command assumes she wants to go to her home directory, so that's where it takes her:

```
~/fission $ cd ❶
~ $
```

❶ Change directories to the default location, the home directory!

As you can see in this example, executing the `cd` command with no arguments results in a new prompt. The prompt reflects the new current working directory, home (~). To double-check, `pwd` can be executed and the home directory will be printed as an absolute path:

```
~ $ pwd ❶
/filespace/people/l/lisemeitner  ❷
```

❶ Print the working directory.

❷ The shell responds by providing the absolute path to the current working directory.

However, the `cd` command can also be customized with an argument, a parameter that follows the command to help dictate its behavior:

```
~/fission $ cd [path]
```

If Lise adds a space followed by the path of another directory, the shell navigates to that directory. The argument can be either an absolute path or a relative path.

> **Angle and Square Bracket Conventions**
>
> Using `<angle brackets>` is a common convention for terms that must be included and for which a real value must be substituted. You should not type in the less-than (<) and greater-than (>) symbols themselves. Thus, if you see `cd <argument>`, you should type in something like `cd mydir`. The `[square brackets]` convention denotes optional terms that may be present. Likewise, if they do exist, do not type in the `[` or `]`. Double square brackets (`[[ ]]`) are used to denote optional arguments that are themselves dependent on the existence of other `[optional]` arguments.

In the following example, Lise uses an absolute path to navigate to a sub-subdirectory. This changes the current working directory, which is visible in the prompt that appears on the next line:

```
~ $ cd /filespace/people/l/lisemeitner/fission   ❶
~/fission $   ❷
```

❶ Lise uses the full, absolute path to the *fission* directory. This means, "change directories into the root directory, then the *filespace* directory, then the *people* directory, and so on until you get to the *fission* directory." She then presses Enter.

❷ She is now in the directory *~/fission*. The prompt has changed accordingly.

Of course, that is a lot to type. We learned earlier that the shorthand ~ means "the absolute path to the home directory." So, it can be used to shorten the absolute path, which comes in handy here, where that very long path can be replaced with `~/fission`:

```
~/ $ cd ~/fission   ❶
~/fission $
```

❶ The tilde represents the home directory, so the long absolute path can be shortened, accomplishing the same result.

Another succinct way to provide an argument to cd is with a relative path. A relative path describes the location of a directory *relative* to the location of the current directory. If the directory where Lise wants to move is inside her current directory, she can drop everything up to and including the current directory's name. Thus, from the *fission* directory, the path to the *applications* directory is simply its name:

```
~/fission $ cd applications    ❶
~/fission/applications $
```

❶ The *applications* directory must be present in the current directory for this command to succeed.

If a directory does not exist, bash will not be able to change into that location and will report an error message, as seen here. Notice that bash stays in the original directory, as you might expect:

```
~/fission $ cd biology
-bash: cd: biology: No such file or directory
~/fission $
```

Another useful convention to be aware of when forming relative paths is that the current directory can be represented by a single dot (.). So, executing cd ./power is identical to executing cd power:

```
~/fission/applications/ $ cd ./power    ❶
~/fission/applications/power/ $
```

❶ Change directories into this directory, then into the *power* directory.

Similarly, the parent of the current directory's parent is represented by two dots (..). So, if Lise decides to move back up one level, back into the *applications* directory, this is the syntax she could use:

```
~/fission/applications/power/ $ cd ..
~/fission/applications/ $
```

Using the two-dots syntax allows relative paths to point anywhere, not just at subdirectories of your current directory. For example, the relative path ../../../ means three directories *above* the current directory.

Exercise: Change Directories

1. Open the Terminal.
2. Type `cd ..` at the command prompt and press Enter to move from your home directory to the directory above it.
3. Move back into your home directory using a relative path.
4. If you have downloaded Lise's directory tree from the book's GitHub repository (*https://github.com/physics-codes/examples*), can you navigate to that directory using what you know about `ls`, `cd`, and `pwd`?

A summary of a few of these path-generating shortcuts is listed in Table 1-2.

Table 1-2. Path shortcuts

Syntax	Meaning
/	The root, or top-level, directory of the filesystem (also used for separating the names of directories in paths)
~	The home directory
.	This directory
..	The parent directory of this directory
../..	The parent directory of the parent directory of this directory

While seeing the names of files and directories is helpful, the content of the files is usually the reason to navigate to them. Thankfully, the shell provides myriad tools for this purpose. In the next section, we'll learn how to inspect that content once we've found a file of interest.

File Inspection (head and tail)

When dealing with input and output files for scientific computing programs, you often only need to see the beginning or end of the file (for instance, to check some important input parameter or see if your run completed successfully). The command head prints the first 10 lines of the given file:

```
~/fission/applications/power $ head reactor.txt

# Fission Power Idea
```

```
The heat from the fission reaction could be used to heat fluids. In
the same way that coal power starts with the production heat which
turns water to steam and spins a turbine, so too nuclear fission
might heat fluid that pushes a turbine. If somehow there were a way to
have many fissions in one small space, the heat from those fissions
could be used to heat quite a lot of water.
```

As you might expect, the `tail` command prints the last 10:

```
~/fission/applications/power $ head reactor.txt
```

```
the same way that coal power starts with the production heat which
turns water to steam and spins a turbine, so too nuclear fission
might heat fluid that pushes a turbine. If somehow there were a way to
have many fissions in one small space, the heat from those fissions
could be used to heat quite a lot of water.

Of course, it would take quite a lot of fissions.

Perhaps Professors Rutherford, Curie, or Fermi have some ideas on this
topic.
```

Exercise: Inspect a File

1. Open a terminal program on your computer.
2. Navigate to a text file.
3. Use head and tail to print the first and last lines to the terminal.

This ability to print the first and last lines of a file to the terminal output comes in handy when inspecting files. Once you know how to do this, the next tasks are often creating, editing, and moving files.

Manipulating Files and Directories

In addition to simply finding files and directories, the shell can be used to act on them in simple ways (e.g., copying, moving, deleting) and in more complex ways (e.g., merging, comparing, editing). We'll explore these tasks in more detail in the following sections.

Creating Files (nano, emacs, vi, cat, >, and touch)

Creating files can be done in a few ways:

- With a graphical user interface (GUI) outside the terminal (like Notepad, Eclipse, or the IPython Notebook)
- With the touch command
- From the command line with cat and redirection (>)
- With a sophisticated text editor inside the terminal, like *nano*, *emacs*, or *vi*

Each has its own place in a programming workflow.

GUIs for file creation

Readers of this book will have encountered, at some point, a graphical user interface for file creation. For example, Microsoft Paint creates *.bmp* files and word processors create *.doc* files. Even though they were not created in the terminal, those files are (usually) visible in the filesystem and can be manipulated in the terminal. Possible uses in the terminal are limited, though, because those file types are not plain text. They have *binary* data in them that is not readable by a human and must be interpreted through a GUI.

Source code, on the other hand, is written in plain-text files. Those files, depending on the conventions of the language, have various filename extensions. For example:

- *.cc* indicates C++
- *.f90* indicates Fortran90
- *.py* indicates Python
- *.sh* indicates bash

Despite having various extensions, source code files are plain-text files and should not be created in a GUI (like Microsoft Word) unless it is intended for the creation of plain-text files. When creating and editing these source code files in their language of choice, software developers often use interactive development environments (IDEs), specialized GUIs that assist with the syntax of certain languages and produce plain-text code files. Depending on the code that you are developing, you may decide to use such an IDE. For example, MATLAB is the appropriate tool for creating *.m* files, and the IPython Notebook is appropriate for creating *.ipynb* files.

Some people achieve enormous efficiency gains from IDEs, while others prefer tools that can be used for any text file without leaving the terminal. The latter type of text editor is an essential tool for many computational scientists—their hammer for every nail.

Creating an empty file (touch)

A simple, empty text file, however, can be created with a mere "touch" in the terminal. The touch command, followed by a filename, will create an empty file with that name.

Suppose Lise wants to create a file to act as a placeholder for a new idea for a nuclear fission application, like providing heat sources for remote locations such as Siberia. She can create that file with the touch command:

    ~/fission/applications $ touch remote_heat.txt

If the file already exists, the touch command does no damage. All files have metadata, and touch simply updates the file's metadata with a new "most recently edited" timestamp. If the file does not already exist, it is created.

> Note how the *remote_heat.txt* file's name uses an underscore instead of a space. This is because spaces in filenames are error-prone on the command line. Since the command line uses spaces to separate arguments from one another, filenames with spaces can confuse the syntax. Try to avoid filenames with spaces. If you can't avoid them, note that the escape character (\) can be used to alert the shell about a space. A filename with spaces would then be referred to as *my\ file\ with\ spaces\ in\ its\ name.txt*.

While the creation of empty files can be useful sometimes, computational scientists who write code do so by adding text to code source files. For that, they need text editors.

The simplest text editor (cat and >)

The simplest possible way, on the command line, to add text to a file without leaving the terminal is to use a program called cat and the shell syntax >, which is called redirection.

The cat command is meant to help concatenate files together. Given a filename as its argument, cat will print the full contents of the file to the terminal window. To output all content in *reactor.txt*, Lise could use cat as follows:

    ~fission/applications/power $ cat reactor.txt

    # Fission Power Idea

    The heat from the fission reaction could be used to heat fluids. In
    the same way that coal power starts with the production heat which
    turns water to steam and spins a turbine, so too nuclear fission
    might heat fluid that pushes a turbine. If somehow there were a way to
    have many fissions in one small space, the heat from those fissions

```
          could be used to heat quite a lot of water.

          Of course, it would take quite a lot of fissions.

          Perhaps Professors Rutherford, Curie, or Fermi have some ideas on this topic.
```

This quality of `cat` can be combined with redirection to push the output of one file into another. Redirection, as its name suggests, *redirects* output. The greater-than symbol, >, is the syntax for redirection. The arrow collects any output from the command preceding it and redirects that output into whatever file or program follows it. If you specify the name of an existing file, its contents will be overwritten. If the file does not already exist, it will be created. For example, the following syntax pushes the contents of *reactor.txt* into a new file called *reactor_copy.txt*:

```
~fission/applications/power $ cat reactor.txt > reactor_copy.txt
```

Without any files to operate on, `cat` accepts input from the command prompt.

Killing or Interrupting Programs

In the exercise above, you needed to use Ctrl-d to escape the `cat` program. This is not uncommon. Sometimes you'll run a program and then think better of it, or, even more likely, you'll run it incorrectly and need to stop its execution. Ctrl-c will usually accomplish this for noninteractive programs. Interactive programs (like `less`) typically define some other keystroke for killing or exiting the program. Ctrl-d will normally do the trick in these cases.

As an example of a never-terminating program, let's use the `yes` program. If you call yes, the terminal will print y ad infinitum. You can use Ctrl-c to make it stop.

```
~/fission/supercritical $ yes
y
y
y
y
y
y
y
y
y
Ctrl-c
```

Exercise: Learn About a Command

1. Open a terminal.
2. Type cat and press Enter. The cursor will move to a blank line.
3. Try typing some text. Note how every time you press Enter, a copy of your text is repeated.
4. To exit, type Ctrl-d. That is, hold down the Control key and press the lowercase *d* key at the same time.

Used this way, cat reads any text typed into the prompt and emits it back out. This quality, combined with redirection, allows you to push text into a file without leaving the command line. Therefore, to insert text from the prompt into the *remote_heat.txt* file, the following syntax can be used:

```
~fission/applications/power $ cat > remote_heat.txt
```

After you press Enter, the cursor will move to a blank line. At that point, any text typed in will be inserted into *remote_heat.txt*. To finish adding text and exit cat, type Ctrl-d.

Be careful. If the file you redirect into is not empty, its contents will be erased before it adds what you're writing.

Using cat this way is the simplest possible way to add text to a file. However, since cat doesn't allow the user to go backward in a file for editing, it isn't a very powerful text editor. It would be incredibly difficult, after all, to type each file perfectly the first time. Thankfully, a number of more powerful text editors exist that can be used for much more effective text editing.

More powerful text editors (nano, emacs, and vim)

A more efficient way to create and edit files is with a text editor. Text editors are programs that allow the user to create, open, edit, and close plain-text files. Many text editors exist. *nano* is a simple text editor that is recommended for first-time users. The most common text editors in programming circles are *emacs* and *vim*; these provide more powerful features at the cost of a sharper learning curve.

Typing the name of the text editor opens it. If the text editor's name is followed by the name of an existing file, that file is opened with the text editor. If the text editor's name is followed by the name of a nonexistent file, then the file is created and opened.

To use the *nano* text editor to open or create the *remote_heat.txt* file, Lise Meitner would use the command:

```
~fission/applications/power $ nano remote_heat.txt
```

Figure 1-3 shows the *nano* text editor interface that will open in the terminal. Note that the bottom of the interface indicates the key commands for saving, exiting, and performing other tasks.

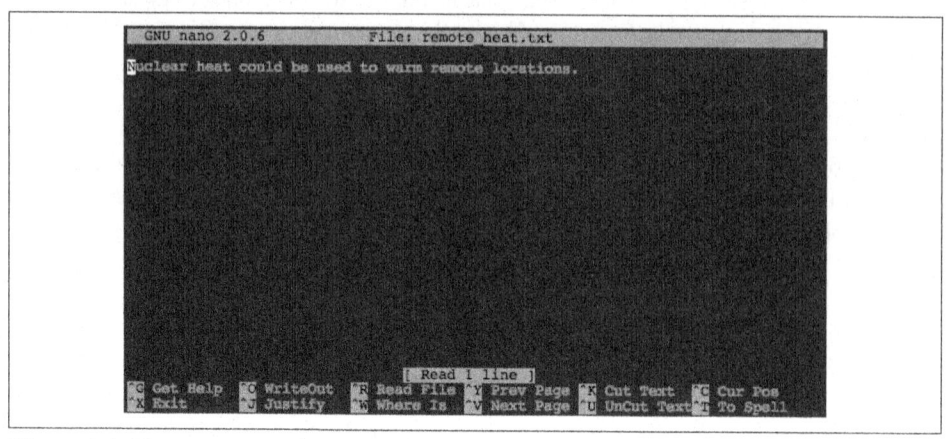

Figure 1-3. The nano text editor

If Lise wanted to use the *vim* text editor, she could use either the command `vim` or the command `vi` on the command line to open it in the same way. On most modern Unix or Linux computers, `vi` is a short name for `vim` (*vim* is *vi, improved*). To use *emacs*, she would use the `emacs` command.

Choose an Editor, Not a Side

A somewhat religious war has raged for decades in certain circles on the topic of which text editor is superior. The main armies on this battlefield are those that herald *emacs* and those that herald *vim*. In this realm, the authors encourage the reader to maintain an attitude of radical acceptance. In the same way that personal choices in lifestyle should be respected unconditionally, so too should be the choice of text editor. While the selection of a text editor can powerfully affect one's working efficiency and enjoyment while programming, the choice is neither permanent nor an indication of character.

Because they are so powerful, many text editors have a steep learning curve. The many commands and key bindings in a powerful text editor require practice to master. For this reason, readers new to text editors should consider starting with *nano*, a low-powered text editor with a shallower learning curve.

Exercise: Open nano

1. Open the Terminal.
2. Execute the command nano.
3. Add some text to the file.
4. Use the instructions at the bottom of the window to name and save the file, then exit *nano*.

Copying and Renaming Files (cp and mv)

Now that we've explored how to create files, let's start learning how to move and change them. To make a copy of a file, use the cp command. The cp command has the syntax cp <source> <destination>. The first required argument is the source file (the one you want to make a copy of), as a relative or absolute path. The second is the destination file (the new copy itself), as a relative or absolute path:

```
~/fission/applications/power $ ls
reactors.txt
~/fission/applications/power $ cp reactors.txt heaters.txt
~/fission/applications/power $ ls
reactors.txt  heaters.txt
```

However, if the destination is in another directory, the named directory must already exist. Otherwise, the cp command will respond with an error:

```
~/fission/applications/power $ cp ./reactors.txt ./electricity/power-plant.txt
cp: cannot create regular file `./electricity/power-plant.txt':
No such file or directory
```

If Lise doesn't need to keep the original file during a copy, she can use mv (move), which renames the file instead of copying it. The command evokes "move" because if the new name is a path in another directory, the file is effectively moved there.

Suppose that when browsing through her ideas, Lise notices an idea for a nuclear plane in the *propulsion* directory:

```
~/fission/applications/propulsion $ ls
nuclear_plane.txt
```

It really was not such a good idea, actually. A nuclear plane would probably be too heavy to ever fly. She decides to rename the idea, as a warning to others. It should be called *bad_idea.txt*. The mv command accepts two arguments: the original file path followed by the new file path. She renames *nuclear_plane.txt* to *bad_idea.txt*:

```
~/fission/applications/propulsion $ mv nuclear_plane.txt bad_idea.txt   ❶
~/fission/applications/propulsion $ ls   ❷
bad_idea.txt   ❸
~/fission/applications/propulsion $ mv ./bad_idea.txt ../   ❹
```

Manipulating Files and Directories | 17

```
~/fission/applications/propulsion $ ls ..    ❺
bad_idea.txt power/ propulsion/ weapons/    ❻
```

❶ Move (rename) *nuclear_plane.txt* to *bad_idea.txt*.

❷ Show the resulting contents of the directory.

❸ Indeed, the file is now called *bad_idea.txt*.

❹ Now, try moving *bad_idea.txt* to the *applications* directory.

❺ List the contents of the *applications* directory to see the result.

❻ The renamed file is now located in the *applications* directory above the *propulsion* directory.

Once all of her files have been properly named, Lise may need new directories to reorganize them. For this, she'll need the `mkdir` command.

Making Directories (mkdir)

You can make new directories with the `mkdir` (make directory) command. Using our usual path conventions, you can make them anywhere, not just in your current working directory. When considering a new class of theories about the nucleus, Lise might decide to create a directory called *nuclear* in the *theories* directory. The `mkdir` command creates a new directory at the specified path:

```
~/theories $ mkdir nuclear
```

The path can be relative or absolute. In order to create a new directory within the new *nuclear* directory, she can specify a longer path that delves a few levels deep:

```
~/theories $ mkdir ./nuclear/fission
```

Note, however, that the rule about not putting a file in a nonexistent directory applies to new directories too:

```
~/theories/nuclear $ mkdir ./nuclear/fission/uranium/neutron-induced
mkdir: cannot create directory `./nuclear/uranium/neutron-induced':
No such file or directory
```

Making directories like this on the command line speeds up the process of organization and reduces the overhead involved. Of course, sometimes you may make a file or directory by mistake. To fix this, you'll need the `rm` command.

Deleting Files and Directories (rm)

Files and directories can be deleted using the `rm` (remove) command. Recall that there was a bad idea in the *applications* directory:

```
~/fission/applications $ ls
bad_idea.txt power/ propulsion/ weapons/
```

After some time, Lise might want to delete that bad idea file entirely. To do so, she can use the `rm` command. Given a path to a file, `rm` deletes it:

```
~/fission/applications $ rm bad_idea.txt
```

Once it's removed, she can check again for its presence with the `ls` command. As you can see, it has disappeared:

```
~/fission/applications $ ls
power/ propulsion/ weapons/
```

Note that once a file is removed, it is gone forever. There is no safety net, no trash can, and no recycling bin. Once you delete something with `rm`, it is truly gone.

Be very careful when using `rm`. It is permanent. With `rm`, recall the adage "Measure twice, cut once." Before using `rm`, consciously consider whether you really want to remove the file.

Since propulsion with nuclear heat, in general, seems unlikely given the weight, Lise may decide to delete the *propulsion* directory entirely. However, if she just provides the path to the directory, the `rm` command returns an error, as shown here:

```
~/fission/applications $ rm propulsion
rm: propulsion: is a directory
```

This error is a safety feature of `rm`. To delete directories, it is necessary to use the `-r` (recursive) flag. Flags such as `-r` modify the behavior of a command and are common in the shell. This flag tells `rm` to descend into the directory and execute the command all the way down the tree, deleting all files and folders below *propulsion*:

```
~/fission/applications $ rm -r propulsion
```

This requirement prevents you from deleting entire branches of a directory tree without confirming that you do, in fact, want the shell to descend into all subdirectories of the given directory and delete them, as well as their contents.

On some platforms, just to be safe, the `rm` command requests confirmation at each new subdirectory it encounters. Before it deletes a subdirectory, it will ask: "rm: descend into directory 'subdirectoryname'?" Type y or n to confirm "yes" or "no," respectively. This can be avoided if an f (for force) is added to the flags. The command to force removal of a directory and all its subdirectories is `rm -rf <directory name>`.

While `rm -rf` can be used carefully to great effect, never execute `rm -rf *`. Unscrupulous mischief-makers may recommend this, but it will have catastrophic consequences. Do not fall for this tomfoolery.

The next section will cover some examples of more flags available to commands in the shell.

Exercise: Make and Remove Files and Directories

1. Open the Terminal.
2. Use `mkdir` to create a directory with a few empty subdirectories.
3. Use `touch` to create five empty files in those directories, and use `ls` to inspect your work.
4. With one command (hint: it will have to be recursive), remove the whole directory. Do you need to use the force flag to avoid typing y repeatedly?

Flags and Wildcards

Flags are often important when using these file and directory manipulation commands. For instance, you can `mv` a directory without any flags. However, copying a directory without the recursive flag fails. Let's look at an example. Since all applications generating power start by generating heat, a new directory called *heat* could start as a duplicate of the *power* directory:

```
~/fission/applications $ cp power/ heat/
cp: omitting directory `power/'
```

The copy command, not accepting a directory as a valid copy target, throws the error "cp: omitting directory *directoryname*". To copy the directory and its contents with `cp`, the `-r` (recursive) flag is necessary:

```
~/fission/applications $ cp -r power/ heat/
```

An alternative to copying, moving, or removing entire directories is to use a wildcard character to match more than one file at once. In the bash shell, the asterisk (*) is a wildcard character. We'll talk about this more in Chapter 8; for now, just note that the asterisk means, approximately, *match everything*.

In the following example, all the files in the directory are matched by the asterisk. Those files are *all* copied into the destination path:

```
~ $ cp beatles/* brits/
~ $ cp zeppelin/* brits/
~ $ cp beatles/john* johns/
~ $ cp zeppelin/john* johns/
~ $ ls brits
george  jimmy  john  john_paul  paul  ringo  robert
~ $ ls johns
john  john_paul
```

But notice that we've overwritten a "john" during the second copy into each directory. To help avoid making such mistakes, you can use `-i` to run the command interactively; the shell will then ask you to confirm any operations it thinks seem suspicious:

```
~ $ cp beatles/john* johns/.
~ $ cp -i beatles/john* johns/.
cp: overwrite `johns/./john'? y
```

In a sense, `-i` is the opposite of `-f`, which forces any operations that the shell might otherwise warn you about:

```
~ $  mv zeppelin/john deceased/.
~ $  mv beatles/john deceased/.
mv: overwrite `deceased/./john'? n
~ $  mv -f beatles/john deceased/.
```

In this section, we have covered a few flags commonly available to commands on the command line. However, we have only scratched the surface. Most available commands possess many customized behaviors. Since there are far too many to memorize, the following section discusses how to get help and find more information about commands.

Getting Help

Now that you have become familiar with the basics, you can freely explore the terminal. The most important thing to know before venturing forth, however, is how to get help.

Reading the Manual (man)

The program `man` (manual) is an interface to online reference manuals. If you pass the name of a command or program to `man` as an argument, it will open the help file for that command or program. To determine what flags and options are available to the `ls` command, then, typing `man ls` would provide the instructions for its use. Since `man` is itself a program, we can type `man man` to get the instructions for using `man`:

```
~ $ man man

NAME
       man - an interface to the on-line reference manuals
```

```
SYNOPSIS
       man  [-c|-w|-tZ]  [-H[browser]]  [-T[device]]  [-adhu7V]
       [-i|-I] [-m system[,...]] [-L locale] [-p  string]  [-C
       file]  [-M  path]  [-P pager] [-r prompt] [-S list] [-e
       extension] [[section] page ...] ...
       man  -l  [-7]  [-tZ]  [-H[browser]]  [-T[device]]   [-p
       string] [-P pager] [-r prompt] file ...
       man -k [apropos options] regexp ...
       man -f [whatis options] page ...

DESCRIPTION
       man  is  the  systems manual pager. Each page argument
       given to man is normally the name of a program, utility
       or  function.  The manual page associated with each of
       these arguments is then found and displayed. A section,
       if  provided, will direct man to look only in that sec
       tion of the manual.  The default action is to search in
       all  of the available sections, following a pre-defined
       order and to show only the first page  found,  even  if
       page exists in several sections.


   <snip>
```

What follows `man` in the `SYNOPSIS` is a listing of the optional and required arguments, options, and variables.

Arguments, options, and variables

In these `man` pages, you'll see that there are different ways to pass information to the command-line programs and commands you need to use. We've seen the first one: *arguments*. An argument simply gets added after the command. You can add multiple arguments if the command expects that behavior. We've added single arguments when we've changed into a specific directory (e.g., `cd ..`). We also used two arguments at once with `cp` (e.g., `cp <source> <destination>`). We also saw, for example, that the `ls` command with the single argument . lists the contents of the current directory:

```
~/weaponry $ ls .
fear  ruthless_efficiency  surprise
```

We've also seen *options*, also called *flags* or *switches* (e.g., the recursive flag, `-r`). These tell the program to run in some predefined way. Options are usually specified with a minus sign (-) in front of them. For instance, if we run `man ls` and scroll down, we see that the `-r` option lists directory contents in reverse order. That is:

```
~/weaponry $ ls -r .
surprise  ruthless_efficiency  fear
```

> Be careful—flags (like -r) don't necessarily have the same meaning for every command. For many commands, -r indicates *recursive* behavior, but for ls, it prints the directory contents in *reverse* order.

Variables can be used to pass in specific kinds of information and are usually specified with a double minus sign (--, typically pronounced "minus minus" or "dash dash"). Further perusal of the ls man page indicates that a variable called sort can be set to certain values to sort directory contents in various ways. To provide a value to sort, we use an equals sign (=). For instance, --sort=time sorts directory contents by file modification time, with the most recent file first:

```
~/weaponry $ ls --sort=time .
fear   surprise   ruthless_efficiency
```

All of the arguments, options, and variables for a command are detailed in the man page for that command. To see how they are used, you will need to scroll down in the man page document to where they are explained. To scroll down, it's helpful to know how to use less.

Moving around in less

man opens the help documents in a program called less, which you can use to look at other text files as well (just call less [filename]). There's lots to learn about less (use man less to get an overview), but the most important things to know are as follows:

- Use the up and down arrows to scroll up and down.
- Use Page Up and Page Down (or the space bar) to move up or down by an entire page.
- Use a forward slash (/) followed by a search term and then Enter to search for a particular word. The letter n (next) toggles through each occurrence.
- Use h to display help inside less—this displays all the possible commands that less understands.
- Use q to quit.

less is modeled on an earlier program called more. However, more has fewer features, and you probably shouldn't bother with it. So, always remember: *less is more.*

Exercise: Use the man Pages with less

1. Open the Terminal.
2. Use the `man` command and the preceding notes on `less` to learn about the commands covered already in this chapter (e.g., `mkdir`, `touch`, `mv`, `cp`, etc.).

Of course, before you can use `man` and `less` to find information about available commands, you must know what commands are available. For that, we need a command called apropos.

Finding the Right Hammer (apropos)

The bash shell has so many built-in programs, practically no one has all of their names memorized. Since the `man` page is only helpful if you know the name of the command you're looking for, you need some tool to determine what that command is. Thankfully, this tool exists. You can search the `man` pages for keywords with a command called apropos. Let's say you want to know what text editors are available. You might search for the string "text editor":

```
~ $ apropos "text editor"   ❶
ed(1), red(1)       - text editor  ❷
vim(1)              - Vi IMproved, a programmers text editor  ❸
```

❶ To search for an installed command based on a keyword string, use `apropos`.

❷ `ed` and `red` show up together, because their full description is "text editor."

❸ `vim` appears next, with its longer description. Other installed editors will not appear if the exact phrase "text editor" does not appear in their `man` pages. What happens if you try `apropos editor`?

An optimistic physicist, Lise might have been curious enough to query physics-related commands. Unfortunately, she might be disappointed to find there aren't many:

```
~ $ apropos physics
physics: nothing appropriate
```

> **Exercise: Find and Learn About a Command**
>
> 1. Open the Terminal.
> 2. Search your computer for commands by using `apropos` and a keyword.
> 3. Take some time to explore the `man` page of a command we've discussed or of another command or program you know of. Learn about a couple of new arguments or options and try them out. Practice killing or interrupting programs if necessary.

Now that this chapter has touched on the various commands for running processes and manipulating files, let's see how those commands can be combined into powerful pipelines using redirection and pipes.

Combining Utilities with Redirection and Pipes (>, >>, and |)

The power of the shell lies in the ability to combine these simple utilities into more complex algorithms very quickly. A key element of this is the ability to send the output from one command into a file or to pass it directly to another program.

To send the output of a command into a file, rather than printing it to the screen as usual, redirection is needed. A text or data stream generated by the command on the lefthand side of the arrow is sent (redirected) into the file named on the righthand side. One arrow (>) will create a new file or overwrite the contents of an existing one with the stream provided by the lefthand side. However, two arrows (>>) will append the stream to the end of an existing file, rather than overwriting it. If Lise wants to create a new file containing only the first line of another, she can combine the `head` command and the redirection method to achieve this in one line:

```
~/fission/applications/power $ head -1 reactor.txt > reactor_title.txt
```

Now, the content of *reactor_title.txt* is simply:

```
# Fission Power Idea
```

To chain programs together, the pipe (|) command can be used in a similar fashion. The output of one program can be used as the input of another. For example, to print the middle lines of a file to the screen, `head` and `tail` can be combined. To print only line 11 from the *reactor.txt* file, Lise can use `head`, `tail`, and a pipe:

```
~/fission/applications/power $ head -1 reactor.txt | tail -1
Of course, it would take quite a lot of fissions.
```

With these methods, any program that reads lines of text as input and produces lines of text as output can be combined with any other program that does the same.

Now that you've seen how the many simple commands available in the shell can be combined into ad hoc pipelines, the incredible combinatoric algorithmic power of the shell is at your fingertips—but only if you have the right permissions.

Permissions and Sharing

Permissions are a subtle but important part of using and sharing files and using commands on Unix and Linux systems. This topic tends to confuse people, but the basic gist is that different people can be given different types of access to a given file, program, or computer.

At the highest level, the filesystem is only available to users with a user account on that computer. Based on these permissions, some commands allow users to connect to other computers or send files. For example:

- `ssh [user@host]` connects to another computer.
- `scp [file] [user@host]:path` copies files from one computer to another.

Those commands only work if the user issuing them has *permission* to log into the filesystem. Otherwise, he will not be able to access the file system at all.

Once they have accessed a computer's filesystem, however, different types of users may have different types of access to the various files on that system. The "different types of people" are the individual *user* (u) who owns the file, the *group* (g) who's been granted special access to it, and all *others* (o). The "different types of access" are permission to *read* (r), *write* to (w), or *execute* (x) a file or directory.

This section will introduce three commands that allow you to manage the permissions of your files:

- `ls -l [file]` displays, among other things, the permissions for that file.
- `chown [-R] [[user]][:group] target1 [[target2 ..]]` changes the individual user and group ownership of the target(s), recursively if `-R` is used and one or more targets are directories.
- `chmod [options] mode[,mode] target1 [[target2 ...]]` changes or sets the permissions for the given target(s) to the given mode(s).

The first of these, `ls -l`, is the most fundamental. It helps us find out what permission settings apply to files and directories.

Seeing Permissions (ls -l)

We learned earlier in this chapter that `ls` lists the contents of a directory. When you explored the `man` page for `ls`, perhaps you saw information about the `-l` flag, which lists the directory contents in the "long format." This format includes information about permissions.

Namely, if we run `ls -l` in a directory in the filesystem, the first thing we see is a code with 10 permission digits, or "bits." In her *fission* directory, Lise might see the following "long form" listing. The first 10 bits describe the permissions for the directory contents (both files and directories):

```
~/fission $ ls -l
drwxrwxr-x  5 lisemeitner  expmt  170 May 30 15:08 applications
-rw-rw-r--  1 lisemeitner  expmt   80 May 30 15:08 heat-generation.txt
-rw-rw-r--  1 lisemeitner  expmt   80 May 30 15:08 neutron-production.txt
```

The first bit displays as a d if the target we're looking at is a directory, an l if it's a link, and generally - otherwise. Note that the first bit for the *applications* directory is a d, for this reason.

To see the permissions on just one file, the `ls -l` command can be followed by the filename:

```
~/fission $ ls -l heat-generation.txt
-rw-rw-r--  1 lisemeitner  expmt   80 May 30 15:08 heat-generation.txt
```

In this example, only the permissions of the desired file are shown. In the output, we see one dash followed by three sets of three bits for the *heat-generation.txt* file (-rw-rw-r--). Let's take a look at what this means:

- The first bit is a dash, -, because it is not a directory.
- The next three bits indicate that the user owner (lisemeitner) can read (r) or write (w) this file, but not execute it (-).
- The following three bits indicate the same permissions (rw-) for the group owner (expmt).
- The final three bits (r--) indicate read (r) but not write or execute permissions for everyone else.

All together, then, Lise (lisemeitner) and her Experiment research group (expmt) can read or change the file. They cannot run the file as an executable. Finally, other users on the network can only read it (they can never write to or run the file).

Said another way, the three sets of three bits indicate permissions for the user owner, group owner, and others (in that order), indicating whether they have read (r), write (w), or execute (x) privileges for that file.

The `ls man` page provides additional details on the rest of the information in this display, but for our purposes the other relevant entries here are the two names that follow the permission bits. The first indicates that the user `lisemeitner` is the individual owner of this file. The second says that the group `expmt` is the group owner of the file.

> **Exercise: View the Permissions of Your Files**
>
> 1. Open a terminal.
> 2. Execute `ls -l` on the command line. What can you learn about your files?
> 3. Change directories to the / directory (try `cd /`). What are the permissions in this directory? What happens if you try to create an empty file (with `touch <filename>`) in this directory?

In addition to just observing permissions, making changes to permissions on a file system is also important.

Setting Ownership (chown)

It is often helpful to open file permissions up to one's colleagues on a filesystem. Suppose Lise, at the Kaiser Wilhelm Institute, wants to give all members of the institute permission to read and write to one of her files, *heat-generation.txt*. If those users are all part of a group called `kwi`, then she can give them those permissions by changing the group ownership of the file. She can handle this task with `chown`:

```
~/fission $ chown :kwi heat-generation.txt
~/fission $ ls -l heat-generation.txt
-rw-rw-r--  1 lisemeitner  kwi  80 May 30 15:08 heat-generation.txt
```

> **Exercise: Change Ownership of a File**
>
> 1. Open a terminal.
> 2. Execute the `groups` command to determine the groups that you are a part of.
> 3. Use `chown` to change the ownership of a file to one of the groups you are a part of.
> 4. Repeat step 3, but change the group ownership back to what it was before.

However, just changing the permissions of the file is not quite sufficient, because directories that are not executable by a given user can't be navigated into, and directories that aren't readable by a given user can't be printed with `ls`. So, she must also make sure that members of this group can navigate to the file. The next section will show how this can be done.

Setting Permissions (chmod)

Lise must make sure her colleagues can visit and read the dictionary containing the file. Such permissions can be changed by using chmod, which changes the file *mode*. Since this is a directory, it must be done in recursive mode. If she knows her home directory can be visited by members of the kwi group, then she can set the permissions on the entire directory tree under *~/fission* with two commands. The first is again chown. It sets the *fission* directory's group owner (recursively) to be kwi:

    ~ $ chown -R :kwi fission/

Next, Lise changes the file mode with chmod. The chmod syntax is chmod [options] <mode> <path>. She specifies the recursive option, -R, then the mode to change the group permissions, adding (+) reading and execution permissions with g+rx:

    ~ $ chmod -R g+rx fission/

Many other modes are available to the chmod command. The mode entry g+rx means we *add* the read and execution bits to the group's permissions for the file. Can you guess the syntax for subtracting the group's read permissions? The manual page for chmod goes into exceptional detail about the ways to specify file permissions. Go there for special applications.

Physicists using large scientific computing systems rely heavily on permissions to securely and robustly share data files and programs with multiple users. All of these permissions tools are helpful with organizing files. Another tool available for organizing files across filesystems is the *symbolic link*.

Creating Links (ln)

The ln command allows a user to create a hard or symbolic link to a file or program. This effectively creates more than one reference pointing to where the contents of the file are stored. This section will focus on symbolic links rather than hard links.

Symbolic links are useful for providing access to large, shared resources on a networked filesystem. Rather than storing multiple copies of large datasets in multiple locations, an effective physicist can create symbolic links instead. This saves hard drive space, since a symbolic link only takes up a few bytes. Also, it saves time. Since the links can be placed in easy-to-find locations, colleagues will spend less time searching deeply nested subdirectories for desired programs and data files.

For our purposes, symbolic links (created with ln -s) are the safest and most useful. Let's say, for instance, that Lise has compiled a program that suggests a random pair of isotopes likely to result from a uranium fission. Her colleagues have a hard time remembering whether the program is called *fission_fragments* or just *fragments*. When they try *fission_fragments*, bash responds with a warning—the command is not a valid path:

```
~/programs/fission $ ./fission_fragments
./fission_fragments: Command not found.
```

One solution is to add a symbolic link. A new link at the incorrect filename, pointing to the correct filename, can be created with the syntax ln -s <source_path> <link_path>:

```
~/programs/fission $ ln -s fragments fission_fragments
```

With that complete, a new symbolic link has been created. It can be viewed with ls -l, and appears to be just like any other file except for the arrow showing it is just a pointer to the *fragments* program:

```
~/programs/fission $ ls -l    ❶
-rwxrwxr-x  1 lisemeitner  staff  20 Nov 13 19:02 fragments    ❷
lrwxrwxr-x  1 lisemeitner  staff   5 Nov 13 19:03 fission_fragments -> fragments
```

❶ Input: Execute the "list in long form" command on the command line.

❷ Output: the file listing now shows both the *fragments* file and, on the next line, the *fission fragments* file, with an arrow indicating that it is a symbolic link to the fragments executable. Note also that the first of the 10 permission bits for that file is an l for "link."

Now, with this symbolic link in the directory, Lise's colleagues can use either name with the same success. Furthermore, recall that a dot (.) stands for the current directory and that slashes (/) separate directory and file names. Therefore, *./myfile* refers to *myfile* in the current directory. When running a program in the current directory, you must include the dot-slash. As you can see from the following, this works equally well on symbolic links as it does on normal files:

```
~/programs/fission$ ./fission_fragments
140Xe 94Sr
```

Symbolic links are useful for providing access to large, shared resources, rather than storing multiple copies in multiple hard-to-reach locations. Another common way physicists gain access to large, shared resources is by accessing them on remote machines. We'll discuss the power of connecting to other computers in the next section.

Connecting to Other Computers (ssh and scp)

This powerful feature of the command line, providing access to networked and remote filesystems, is key to high-performance computing tasks. Since most large high-performance or high-throughput computing resources can only be accessed by SSH (Secure SHell) or similar protocols through the command line, truly high-powered computer systems simply are not accessible without use of the shell.

If you have the right credentials, you can even get access to another machine through the shell. You can do this using a program called *ssh*. For instance, for the user `grace` to log on to a networked computer `mk1`, she would use the `ssh` command with an argument specifying her username and the computer name, connected by the @ symbol:

```
~ $ ssh grace@mk1
```

Or, if `mk1` is a computer located on the remote network domain *harvard.edu*, Grace can connect to that computer from her home computer with the full location of the computer in its domain:

```
~ $ ssh grace@mk1.harvard.edu
```

Once logged into the computer, Grace has access to the files and directories in the remote filesystem and can interact with them just as she does locally.

She can use the `scp` (secure copy) command to copy files and directories from one computer to another. It has the syntax `scp <source_file> [[user@]host]:<destination>`. So, to copy a *notes.txt* file from her local computer to the *COBOL* directory on the *mk1.harvard.edu* filesystem, she would execute:

```
~ $ scp ./notes.txt grace@mk1.harvard.edu:~/COBOL/notes.txt
```

> Both `ssh` and `scp` require a valid username and password on the remote machine.

When she connects to another computer, Grace has access to its filesystem. On that system, there are not only different files, but also a different *environment*. We'll look at the environment and how it is configured next.

The Environment

In addition to providing commands, a filesystem hierarchy, and a syntax for navigation, the bash shell defines a computing environment. This computing environment can be customized using environment variables. We can investigate our environment with a program called *echo*. The `echo` command prints arguments to the terminal. In the case of a string argument, the string is printed verbatim:

```
~ $ echo "Hello World"
Hello World
```

In the case of environment variables, however, echo performs *expansion*, printing the values of the variables rather than just their names. You invoke these variables on the command line by prepending a $ to the variable name.

When, in 1959, she began to design the first machine-independent programming language (COBOL), Grace Hopper did not have bash. Bash, after all, could never have come into existence without her breakthrough. Hypothetically, though, if she had had bash, her environment might have behaved liked this:

```
~ $ echo $USERNAME  ❶
grace
~ $ echo $PWD
/filespace/people/g/grace  ❷
```

❶ Echo the value of the USERNAME environment variable. On certain platforms, this variable is called USER.

❷ The computer stores the working directory in the environment variable PWD; the command pwd is simply a shortcut for echo $PWD.

Shell variables are replaced with their values when executed. In bash, you can create your own variables and change existing variables with the export command:

```
~ $ export GraceHopper="Amazing Grace"
```

Variables are case-sensitive. For this reason, the following command will successfully echo the assigned string:

```
~ $ echo $GraceHopper
Amazing Grace
```

However, none of the following will succeed:

```
~ $ echo GraceHopper
~ $ echo GRACEHOPPER
~ $ echo $GRACEHOPPER
```

Table 1-3 lists some of the most common and important shell variables. These variables often become essential for defining the computer's behavior when the user compiles programs and builds libraries from the command line.

Table 1-3. Common environment variables

Variable name	Meaning
USER	User name
PATH	List of absolute paths that are searched for executables
PWD	Current directory (short for print working directory)

Variable name	Meaning
EDITOR	Default text editor
GROUP	Groups the user belongs to
HOME	Home directory
~	Same as HOME
DISPLAY	Used in forwarding graphics over a network connection
LD_LIBRARY_PATH	Like PATH, but for precompiled libraries
FC	Fortran compiler
CC	C compiler

Environment variables can be used to store information about the environment and to provide a shorthand for long but useful strings such as absolute paths. To see all of the environment variables that are active in your terminal session, use the `env` command. Rear Admiral Hopper might see something like:

```
~/fission $ env
SHELL=/bin/bash
USER=grace
EDITOR=vi
LD_LIBRARY_PATH=/opt/local/lib:/usr/local
PATH=/opt/local/lib:/filespace/people/g/grace/anaconda/bin:/opt/local/bin
PWD=/filespace/people/g/grace/languages
LANG=en_US.utf8
PWD=/filespace/people/g/grace
LOGNAME=grace
OLDPWD=/filespace/people/g/grace/languages/COBOL
```

To make an environment variable definition active every time you open a new terminal, you must add it to a file in your home directory. This file must be called *.bashrc*.

Saving Environment Variables (.bashrc)

A number of files in a bash shell store the environment variables that are active in each terminal session. They are plain-text files containing bash commands. These commands are executed every time a terminal window is opened. Thus, any environment variables set with `export` commands in those files are active for every new terminal session.

To configure and customize your environment, environment variables can be added or edited in *~/.bashrc*, the main user-level bash configuration file. The `export` commands we executed in the terminal before added new environment variables for a single terminal session. To add or change an environment variable for every session, we use *.bashrc*.

> The leading . in *.bashrc* makes the file a hidden file.

User-specific configuration exists in many files. In addition to the *.bashrc* file, you may see others, such as *.bash_profile* or, on newer Mac OS machines, *.profile*. Do any of those exist on your computer? If so, open that file and confirm that it contains the text `source ~/.bashrc`.

Exercise: Configure Your Shell with .bashrc

1. Use your text editor to open the *.bashrc* file in your home directory. If no such file exists, create it.
2. Add an `export` command to set a variable called `DATA` equal to the location of some data on your filesystem.
3. Open a new terminal window and query the `DATA` variable with echo.
4. What is the result of `cd $DATA`? Can you imagine ways this behavior could make navigating your files easier?

A new terminal instance will automatically reflect changes to the *.bashrc* file. However, the `source` command can be used to make changes to *.bashrc* take effect immediately, in the current session:

```
~ $ source .bashrc
```

The bash shell can be customized enormously with commands executed in the *.bashrc* file. This customization is ordinarily used to specify important paths and default behaviors, making the shell much more efficient and powerful. The most important variable in your *.bashrc* file is the PATH.

Running Programs (PATH)

Based on the environment, the shell knows where to find the commands and programs you use at the command line. Unless you modify your environment, you can't run just any old program on your computer from any directory. If you want to run a program in a nonstandard location, you have to tell the shell exactly where that program is by invoking it with an absolute or relative Unix path.

For instance, in Chapter 14, we will learn to build a program. However, after we've done so, we can still only run that program if we tell the shell exactly where it is. With the programs we have seen so far, the name of the command is sufficient. However, because bash only searches certain locations for available commands, the fragments command will not be found:

```
~/programs/fission $ fragments ❶
fragments: Command not found. ❷
```

❶ We attempt to run the *fragments* program.

❷ The shell's response indicates that it cannot find the named program (because it is not in the PATH).

Indeed, even in the proper directory, you must indicate the full path to the program by adding the leading dot-slash before the computer understands what program to run:

```
~/programs/fission $ ./fragments
136Cs 99Tc
```

In order for the computer to find the *fragments* program without us typing the full path, the PATH environment variable must contain the directory holding the program. Without the full path, the bash shell only executes commands found when searching the directories specified in the PATH environment variable. To add this folder to the PATH environment variable, Lise can execute the following command:

```
~/programs $ export PATH=$PATH:/filespace/people/l/lisemeitner/programs/fission
```

The first part of this command uses `export` to set the PATH variable. Everything on the righthand side of the equals sign will become the new PATH. The first element of that path is the old PATH variable value. The second element, after the colon, is the new directory to add to the list of those already in the PATH. It will be searched last.

Exercise: Customize Your PATH

1. In the terminal, use `echo` to determine the current value of the PATH environment variable. Why do you think these directories are in the PATH?

2. Use `export` to add your current directory to the end of the list. Don't forget to include the previous value of PATH.

3. Use `echo` once again to determine the new value.

Can you think of a way that the `PWD` environment variable could be used to shorten the preceding command? In addition to shortening commands and paths by setting environment variables, configuration files are an excellent place to permanently give shorter nicknames to other commands. In the next section, we'll see how to do this with the `alias` command.

Nicknaming Commands (alias)

In the same way that you can create variables for shortening long strings (like $DATA, the path to your data), you can create shorthand aliases for commands. `alias` does a simple replacement of the first argument by the second. If, for example, you like colors to distinguish the files and directories in your shell session, you'll always want to use the `--color` variable when calling `ls`. However, `ls --color` is quite a lot to type. It is preferable to reset the meaning of `ls` so that it behaves like `ls --color`. The `alias` command allows you to do just that. To replace `ls` with `ls --color`, you would type:

```
alias ls 'ls --color'
```

Once that command has been executed in the terminal, typing `ls` is equivalent to typing `ls --color`. Just like an environment variable, to make an alias active every time you open a new terminal, you must add this definition to your *.bashrc* file, which is executed upon login.

> To keep their *.bashrc* files cleaner and more readable, many individuals choose to keep all of their aliases in a separate hidden file (called something like *.bash_aliases*). To load it, they include the following line in their *.bashrc* files:
>
> ```
> source ~/.bash_aliases
> ```

Now that the *.bashrc* bash script has demonstrated the power of automating bash commands, the following section will show how to write your own bash scripts.

Scripting with Bash

Repeating processes on the command line is made easy with files (like *.bashrc*) that can store many commands and be executed at will. Any such series of commands can be placed into a file called a *script*. This type of file, like a program, can be written once and executed many times.

Bash scripts typically end in the *.sh* extension. So, the first step for creating a bash script is to create such a file. As we did earlier in this chapter, you can do this by opening a text editor like *nano* and supplying the filename as an argument:

```
~ $ nano explore.sh
```

Any commands that are valid in the terminal are valid in a bash script. Some text in a bash script might be for reference, however. This text, called a *comment*, must be denoted with a #.

If Lise would like to automate the process of exploring her directory tree, she might write a bash script for the task. A very simple bash script that enters three levels of parent directories and prints the contents as well as the directory names is only a few simple lines:

```
# explore.sh ❶
# explore the three directories above this one
# print a status message
echo "Initial Directory:"
# print the working directory
pwd
# list the contents of this directory
ls
echo "Parent Directory:"
# ascend to the parent directory
cd ..
pwd
ls
echo "Grandparent Directory:"
cd ..
pwd
ls
echo "Great-Grandparent Directory:"
cd ..
pwd
ls
```

❶ Comments are preceded by a # symbol. They are reference text only and are not executed.

After you save this file, there is only one more step required to make it a bona fide program. To run this script on the command line, its permissions must be set to *executable*. To make this the case, Lise must execute (in the terminal) the command:

```
~ $ chmod a+x explore.sh
```

Now, the *explore.sh* script is runnable. To run the command, Lise must either call it with its full path or add the location to her PATH environment variable. When we use a relative path, the execution looks like:

```
~ $ ./explore.sh
```

> **Exercise: Write a Simple Bash Script**
>
> 1. Create a file called *explore.sh*.
> 2. Copy the example script into that file.
> 3. Change the permissions of the file so that it is executable.
> 4. Run it and watch the contents of your filesystem be printed to the terminal.

Much more sophistication is possible in a bash script, but that is somewhat beyond the scope of this chapter. To learn more about sophisticated bash scripting, check out some of the O'Reilly books on the topic or sign up for a workshop like those run by Software Carpentry (*http://software-carpentry.org*).

> **The history Command**
>
> At the end of some series of bash commands, an effective physicist may want to create a bash script to automate that set of commands in the future. The `history` command provides a list of all the most recent commands executed in the terminal session. It is very helpful for recalling recent work and enshrining it in a bash script.

Command Line Wrap-up

This chapter has only just scratched the surface of the power the command line holds. It has covered:

- Navigating the filesystem
- Creating, deleting, and moving files and directories
- Finding help
- Running commands
- Handling permissions
- Writing scripts

If you want to find out more about the command line, many books and Internet resources are available. Cameron Newham's *Learning the bash Shell* (O'Reilly) is a good place to start, and Software Carpentry's workshops and online resources provide an excellent introduction to the deeper secrets of bash.

CHAPTER 2
Programming Blastoff with Python

A lot of people talk about the greatness of Python as a programming language. They are right! But what *is* Python, and why is it so awesome? Python is a general-purpose, dynamic, high-level language that is easy to learn. Python is also known as a glue language because it plays nicely with other languages, including C, C++, and Fortran. For these reasons it has established a strong foothold as a data analysis language. This makes it popular in science and engineering, and in physics-related fields in particular.

The main criticism of Python is its speed. Python is an interpreted language, which makes it more similar to R, Ruby, and MATLAB than it is to compiled languages like C, C++, or Fortran. Pythonistas everywhere live by the mantra that "premature optimization is bad." Concerns that it is too slow are often refuted with these arguments:

- Developer time—the time the programmer spends programming—is more valuable than execution time.[1]
- Most speed issues can be overcome by using the appropriate data structures and algorithms.
- If you really have a need for speed you can always write the performance-critical parts in a compiled language and then expose the functionality to Python.

The most important aspect of Python is it is *fun* to use! The more you learn, the more you want to learn, and the more you find there is to learn. The Python ecosystem is extraordinarily rich and the community members are, by and large, friendly. Unfortunately, there is no way that this book can fully cover all of the excellent

---

[1] For a detailed language comparison, please see Lutz Prechelt's article "An Empirical Comparison of Seven Programming Languages."

aspects of Python. This chapter is a meant as a first introduction to the basics of Python syntax. Many more detailed references and resources are available. For installation instructions, please refer back to the Preface.

Running Python

Python itself is a special type of program called an *interpreter*, because it translates Python *source code* into instructions that your computer's processor can understand. The Python interpreter can be fired up in a number of ways. The most basic (and least used) way is to type `python` at the command prompt in your terminal. This will normally display some information about Python itself and then return with a line that begins with >>>. This is the Python prompt, and from here you can start inputting Python code:

```
$ python
Python 2.7.5+ (default, Sep 19 2013, 13:48:49)
[GCC 4.8.1] on linux2
Type "help", "copyright", "credits" or "license" for more information.
>>>
```

Hitting Enter will execute what you type in and return a >>> prompt:

```
>>> print("Hello Sir Newton.")
Hello Sir Newton.
>>>
```

To get help at any time, use the `help()` function. To exit back to the command line, use the `exit()` function. If this looks a lot like bash, it is because this method of interacting with Python is the same as the one we used to interact with bash: a read-eval-print loop, or REPL.

However, for Python, the stock REPL is not the only one available. IPython (which stands for Interactive Python) provides a REPL that is in many ways superior to the default one. You can get IPython in one of the following ways:

1. Visit *ipython.org* and download the latest stable release.
2. If you are using the Conda package manager, as described in the Preface, and followed the instructions in "Installation and Setup" on page xxiii, you should already have IPython. If you like you can run the command `conda update ipython` to be sure you have the most recent version.
3. If you have Python installed, run the command `pip install ipython`.
4. If you are using Ubuntu, run the command `sudo apt-get install ipython`.

Starting up and executing code in IPython looks like this:

```
$ ipython
Python 2.7.5+ (default, Sep 19 2013, 13:48:49)
Type "copyright", "credits" or "license" for more information.

IPython 1.1.0 -- An enhanced Interactive Python.
?         -> Introduction and overview of IPython's features.
%quickref -> Quick reference.
help      -> Python's own help system.
object?   -> Details about 'object', use 'object??' for extra details.

In [1]: print("Good day, Madam Curie.")
Good day, Madam Curie.

In [2]:
```

In addition to the text-based REPL, IPython also comes with a web-browser-based notebook (*http://ipython.org/notebook.html*) that is similar in look and feel to the notebooks you find in Mathematica or MATLAB. These notebooks are an excellent platform for data analysis and are fast becoming a standard for creating and sharing information. It is *highly* encouraged that you check them out.

While REPLs are often useful, they have a couple of drawbacks. The first is that it is difficult, annoying, and error-prone to write multiline statements in them. The second is that it is hard to save and load work from them to a normal file. This makes it difficult to share what you have done in a REPL environment.

Most people write the majority of their Python code in text files. If you run the interpreter on a file whose name ends in *.py*, then Python will execute all of the code in the file exactly as if each line had been typed into the REPL one after another.

For example, say we have a file called *greetings.py* with the following contents:

```
print("Hey Isaac, what's Newton?!")
print("How is it going, Gottfried?")
```

This may be executed from bash with:

```
$ python greetings.py
Hey Isaac, what's Newton?!
How is it going, Gottfried?
```

Now that we can run Python code, it is time to jump in and learn how the language works!

Comments

All modern programming languages have comment characters. These indicate part of the code that should be skipped by the interpreter, allowing the programmer to write meaningful notes about the code right at the relevant locations. Python uses the #

character to denote comments. Any characters after a # on a line are skipped; there are no multiline comments in Python:

```
# this whole line is a comment
this_part = "is not a comment"   # this part is a comment
```

Variables

Variables consist of two parts: the name and the value. To assign a variable to a name, use a single equals sign (=). Put the variable name on the left of the = and the value on the right. Variable names may be made up of upper- and lowercase letters, digits (0–9), and underscores (_). Here, we give the reduced Planck constant as the variable h_bar:

```
h_bar = 1.05457e-34
```

Variable names cannot start with a digit, to prevent the clever user from redefining what literal numbers mean; they must begin with a letter or underscore.

Variable names that start with numbers are not allowed!

```
2plus_forty = 42   # bad
two_plus40 = 42    # good
```

Once a variable has been defined, you can use or manipulate it however you wish. Say we wanted to print Planck's constant. We could first define π and then multiply h_bar by 2π:

```
pi = 3.14159
h = 2 * pi * h_bar
print(h)
```

All variables in Python are typed. This means that the values have certain well-defined properties that dictate how they are used. Different types have different properties that satisfy different needs. Integers and floating-point numbers (int and float) are meant for mathematical operations. Strings (str) are helpful for textual manipulation. These are all *literal types* because Python provides a special syntax for creating them directly:

```
dims = 3                      # int, only digits
ndim = 3.0                    # float, because of the '.'
h_bar = 1.05457e-34           # float, because of the '.' or 'e'
label = "Energy (in MeV)"     # str, quotes surround the text
```

Integers and strings are sometimes known as *precise* types, because all variables of a precise type will exactly represent the underlying idea. The integer 1 is the only one, and there can be only one. Floats, on the other hand, are sometimes called *imprecise*.

42 | Chapter 2: Programming Blastoff with Python

In general, they are 64-bit *approximations* to real numbers.[2] Some floats, like `1.0`, may be exactly represented with a finite amount of data. Unfortunately you cannot count on this exact behavior. This leads to many gotchas in scientific computing. To learn more, please read What Every Computer Scientist Should Know About Floating-Point Arithmetic, by David Goldberg.

If you are ever unsure, you can always determine the type of a variable or a literal value by using the built-in `type()` function. To use this function, put the variable you want to know the type of in between the parentheses:

```
In [1]: type(h_bar)
Out[1]: float

In [2]: type(42)
Out[2]: int
```

You can use the type names to convert between types, in a similar fashion. First write the name of the type, then surround the variable you want to convert with parentheses:

```
In [1]: float(42)
Out[1]: 42.0

In [2]: int("28")
Out[2]: 28
```

In the expression `int("28")`, the string `"28"` is being converted to an integer. This is possible because the string only contains characters that happen to be digits. If the string has a value that makes no sense as an integer, then the conversion fails! For example:

```
In [1]: int("quark")
ValueError                                Traceback (most recent call last) ❶
<ipython-input-5-df7f23f9b45e> in <module>()  ❷
----> 1 int("quark")  ❸

ValueError: invalid literal for int() with base 10: 'quark'  ❹
```

❶ The type of error we have (here, `ValueError`).

❷ The location of the error—either the filename or (here) the interactive interpreter.

❸ The line number where the error occurred and a printout of the offending line.

---

2 As a mathematical aside, the set of all floats is not a subfield of the real numbers, or even the extended reals. In fact, floats are not a field at all! This is because floats contain a single element—NaN, or "Not a Number"—that does not admit an inverse. This element spoils it for the rest of the floats.

❹ The all-important error message. Read this to understand what the problem was. If the error message is not clear or you do not understand what is going on, search the Internet with the text of the error message.

This is a standard pattern in Python, which promotes exploration and creativity. If the action is not allowed, then the code should fail as early as possible and return a helpful error message. This "fail early and often" credo is central to the interactive development process. The programmer is encouraged to experiment, adjust the code in response to an error, try new code, and repeat until the code has converged on a working version. In the previous example, "quark" will never be a base-10 number. It's probably best to change the value to be a string composed of only digits.

Python is *dynamically typed*. This means that:

1. Types are set on the variable values and not on the variable names.
2. Variable types do not need to be known before the variables are used.
3. Variable names can change types when their values are changed.

The following is completely valid Python:

```
x = 3
x = 1.05457e-34
x = "Energy (in MeV)"
```

Here, the type of x changes every time it is assigned to a new value. The new value replaces the previous value, but the variable retains the same name. Such behavior differs significantly from statically typed languages, such as C, C++, Fortran, and Java, where:

1. Types are set on the variable names and not on the variable values.
2. Variable types must be specified (declared or inferred) before they are used.
3. Variable types can never change, even if the value changes.

We will not be discussing static languages much in this book, but it is important to note that many of the language features of Python evolved in order to mitigate some of the difficulty of working with lower-level languages. Variable typing is a great example of Python abstracting away strict requirements in lower-level languages. This flexibility comes with trade-offs, though, which will be presented as they come up.

Special Variables

Python has a few special variables that are so important that their values are built into the language: namely, True, False, None, and NotImplemented. Each of these variables exists only once whenever you start up a Python interpreter. For this reason, they are known as singletons. Let's dig into these special variables and their meanings now.

Boolean Values

The variables True and False make up the entirety of the Boolean type bool. Boolean variables are used to represent the truth value of other Python expressions and may be used directly by the programmer as flags for turning behavior on or off. Other data types can be converted into Booleans. In general, if the value is zero or the container is empty, then it is converted to False. If the value is nonzero or nonempty in any way, then it is converted to True. Luckily, these are the only two options!

```
In [1]: bool(0)
Out[1]: False

In [2]: bool("Do we need Oxygen?")
Out[2]: True
```

None Is Not Zero!

None is a special variable in Python that is used to denote that no value was given or that no behavior was defined. This is different than just using zero, an empty string, or some other nil value. Zero is a valid number, while None is not. If None happens to make it to a point in a program that expects an integer or float, then the program with rightfully break. With a zero, the program would have continued on. This fills the same role as NULL in C/C++ and null in JavaScript. Additionally, None has a special place in Python as the default return value of functions, which we will discuss more in upcoming chapters.

NotImplemented Is Not None!

Unlike None, the variable NotImplemented is used to signal not only that behavior is not defined but also that the action is impossible, nonsensical, or nonexistent. For example, NotImplemented is used under the covers when you are trying to divide a string by a float. This results in a TypeError:

```
In [1]: "Gorgus" / 2.718
TypeError                                 Traceback (most recent call last)
<ipython-input-1-8cdca6dc67bb> in <module>()
----> 1 "Gorgus" / 2.718

TypeError: unsupported operand type(s) for /: 'str' and 'float'
```

NotImplemented is important to know about when you are defining custom types of your own, which we will cover in more depth in Chapter 6.

Now that we know about types, variables, and Python's special variables, we are ready to talk about what we can *do* with variables. The next section is about what actions and operations are available as part of the Python language.

Operators

Operators are the syntax that Python uses to express common ways to manipulate data and variables. Formally, Python has three kinds of operators: unary, binary, and ternary. This means that these operators take one, two, or three variables as arguments, respectively.

Table 2-1 shows the operators you should know about for computational physics. Of course, not all operators are made equal; we'll discuss some of the most important ones here, and others will be discussed as they come up. Note that not all operators are valid for all types or all variables!

Table 2-1. Python operators using the variables x, y, and z

Name	Usage	Returns
Unary operators		
Positive	+x	For numeric types, returns x.
Negative	-x	For numeric types, returns -x.
Negation	not x	Logical negation; True becomes False and vice versa.
Bitwise Invert	~x	Changes all zeros to ones and vice versa in x's binary representation.
Deletion	del x	Deletes the variable x.
Call	x()	The result of x when used as a function.
Assertion	assert x	Ensures that bool(x) is True.
Binary Operators		
Assignment	x = y	Set the name x to the value of y.
Attribute Access	x.y	Get the value of y which lives on the variable x.
Attribute Deletion	del x.y	Remove y from x.
Index	x[y]	The value of x at the location y.

Name	Usage	Returns
Index Deletion	del x[y]	Remove the value of x at the location y.
Logical And	x and y	True if bool(x) and bool(y) are True, False otherwise.
Logical Or	x or y	x if bool(x) is True, otherwise the value of y.

Arithmetic Binary Operators

Name	Usage	Returns
Addition	x + y	The sum.
Subtraction	x - y	The difference.
Multiplication	x * y	The product.
Division	x / y	The quotient in Python 2 and true division in Python 3.
Floor Division	x // y	The quotient.
Modulo	x % y	The remainder.
Exponential	x ** y	x to the power of y.
Bitwise And	x & y	Ones where both x and y are one in the binary representation, zeros otherwise.
Bitwise Or	x | y	Ones where either x or y are one in the binary representation, zeros otherwise.
Bitwise Exclusive Or	x ^ y	Ones where either x or y but not both are one in the binary representation, zeros otherwise.
Left Shift	x << y	Shifts the binary representation of x up by y bits. For integers this has the effect of multiplying x by 2^y.
Right Shift	x >> y	Shifts the binary representation of x down by y bits. For integers this has the effect of dividing x by 2^y.
In-Place	x op= y	For each of the above operations, op may be replaced to create a version which acts on the variable 'in place'. This means that the operation will be performed and the result will immediately be assigned to x. For example, x += 1 will add one to x.

Comparison Binary Operators

Name	Usage	Returns
Equality	x == y	True or False.
Not Equal	x != y	True or False.

Name	Usage	Returns
Less Than	x < y	True or False.
Less Than or Equal	x <= y	True or False.
Greater Than	x > y	True or False.
Greater Than or Equal	x >= y	True or False.
Containment	x in y	True if x is an element of y.
Non-Containment	x not in y	False if x is an element of y.
Identity Test	x is y	True if x and y point to the same underlying value *in memory*.
Not Identity Test	x is not y	False if x and y point to the same underlying value *in memory*.
Ternary Operators		
Ternary Assignment	x = y = z	Set x and y to the value of z.
Attribute Assignment	x.y = z	Set x.y to be the value of z.
Index Assignment	x[y] = z	Set the location y of x to be the value of z.
Ternary Compare	x < y < z	True or False, equivalent to (x < y) and (y < z). The < here may be replaced by >, <=, or >= in any permutation.
Ternary Or	x if y else z	x if bool(y) is True and z otherwise. It's equivalent to the C/C++ syntax y?x:z.

Most of the operators presented in Table 2-1 can be *composed* with one another. This means that you can chain them together, nest them within one another, and set their order of operation by putting parentheses around them. This is exactly the same as the composition of mathematical operators. For example:

```
(x < 1) or ((h + y - f) << (m // 8) if y and z**2 else 42)
```

However, for certain classes of operators—namely, the assignment (=) and deletion (del) operators—composition is not possible. These must come on their own lines. This is because they directly modify the variables they are working with, rather than simply using their values. For example:

```
x = 1  # Create x
del x  # Destroy x
```

If an operator is fully composable, then it can be part of a Python *expression*. An expression is a snippet of code that does not require its own line to be executed. Many expressions may be on the same line. On the other hand, if an operator is not fully composable and requires its own line to work, then it is a *statement*. In essence, all Python code is a series of statements, which are themselves composed of expressions. Try running the following example:

```
x = (42 * 65) - 1
```

This is composed of the x = <code> assignment statement, which has the expression (42 * 65) - 1 to the right of the equals sign. This expression is itself composed of the subexpressions (42 * 65) and <code> - 1. Any of the subexpressions may be executed on their own, too, which is also worth trying out:

```
In [1]: 42 + 65
Out[1]: 107

In [2]: (42 + 65) + 1
Out[2]: 108
```

Next up, we investigate a data type that is a little different from the numeric ones we have seen so far. *Strings* are meant to represent text of all kinds. They are a critical piece of the programming puzzle.

Strings

Strings are one of the fundamental data types in Python. The type name is str, and as we saw earlier, it can be used to convert other types into strings. For example, str(42) will return "42". The simplest ways to define string *literals* are by using matching single quotes (') or matching double quotes ("):

```
x = "Nature abhors a vacuum"
y = 'but loves a mop!'
```

It is tempting but incorrect to think of a string as a sequence of individual characters. Historically, strings have been represented by computers in this way, and for the most part this remains a valid mental model. So what has changed about text processing?

Python has no character type, known as char in other languages. The char type is made up of 8 bits (1 byte). All 256 (2^8) permutations of these bits correspond to specific meanings given by extended ASCII. A quick Internet search will bring up the full ASCII table. As an example, the numbers 65–90 represent the uppercase letters A–Z. Strings used to be just bunches of these bytes living next to each other to form human-readable phrases. This is fine so long as your human reads English.

In all of history, people have invented far more than 256 characters, in a huge variety of languages. From the programmer's perspective, it would be great not to have to change the data type that is being used just to represent strings in a different natural language. In the late 1980s, programmers began to experiment with the idea of having one number-to-character mapping to rule them all. This came to be known as *Unicode*.

Unicode currently supports upward of 110,000 characters. In order to represent all of these additional characters, it must use more space than just 8 bits per character. Different *encodings* in Unicode use anywhere from 1 to 4 bytes, and the meaning of the bytes changes based on which encoding is used.

Python 3, rather than continuing to use extended ASCII, adopted Unicode for its implementation of strings. Because of this, strings are now not merely arrays of characters, like they used to be in Python 2 and before. A string in Python 3 is an array of bytes and an associated encoding. Python's strings have become a little more complicated to accommodate a more connected world. Thankfully, Python still makes them easy and enjoyable to use.

For most tasks in scientific computing, the fundamental data is numeric. Unlike web developers, we fret about the arcana of floating-point numbers rather than the intricacies of Unicode. Serious issues with strings should not arise often. The default string behavior is typically good enough. When in doubt, use the UTF-8 (*http://en.wikipedia.org/wiki/UTF-8*) encoding.

> In Python 2, which remains popular in scientific computing, Unicode is a separate `unicode` type while `str` remains ASCII. However, this line is blurred, and this can lead to a lot of confusion. If you want to be sure that you are using Unicode, simply add a `u` to the front of the string: `u"this bytes"`. In Python 3.3+, the `u` is ignored, making this expression Python 3–compatible as well.

String Indexing

Indexing (or "indexing into") a string is the process of retrieving data from part or all of a string. Indexing actually applies to all sequences in Python and uses square brackets ([]) to operate on the variable.

> The indexing techniques described here will come up again and again in later sections and chapters.

The simplest way to index a string is to put a single integer inside of the brackets and place the brackets after the string. Python is *zero-indexed*. This means the element count starts at 0, then 1, 2, etc. Therefore, to get the second element of a string, you would use the index 1. To try this out, open up IPython from the command line with the ipython command, and type in the following In lines. After hitting Enter, you should see the results show up an Out line:

```
In [1]: p = "proton"

In [2]: p[1]
Out[2]: 'r'
```

> If zero indexing seems a bit odd at first, do not worry. It is easy to get used to, and many other languages, such as C/C++, are also zero-indexed. MATLAB and others constitute a suite of languages that are one-indexed (the count starts at 1). Still other languages, such as Fortran, are arbitrarily indexed. This means that the programmer can declare the integer assigned to the first element. All other indices are scaled linearly from this point.

String elements can also be extracted with *negative indices*. Rather than counting from the front, negative indices count from the back. The last element is −1, the second to last is −2, and so on. This is a shortcut for having to write that you want to compute the length of the string and then walk back a certain number of elements. You can compute the length of a string s by writing len(s). Here are a couple of examples of implicit and explicit negative indexing:

```
In [3]: p[-1]
Out[3]: 'n'

In [4]: p[len(p)-2]   # also works, but why write len(p) all the time?
Out[4]: 'o'
```

Now suppose you want to pull out more than just a single element at a time. To extract a substring, you index with a *slice*. A slice is a sequence-independent way of defining a range of indices. In their simplest, literal form slices are spelled out as two integer indices separated by a colon: s[start:stop]. Continue to try this in IPython with the following:

```
In [5]: p[2:5]
Out[5]: 'oto'
```

Notice that the n at the end (p[5]) did not make it into the substring! This is because slices are defined to be inclusive on the lower end and exclusive on the upper end. In more mathematical terms, a slice is defined by [start,stop).

Slicing gets to the heart of *why* Python is zero-indexed. The difference between the stop and start values will always be the length of the subsequence. Or, in code, the following expression will always be true for any sequence s:

```
(stop - start) == len(s[start:stop])
```

It can be easier to think of the indices as living on either side of the element rather than on the element itself. The slice then traverses from `start` to `stop`, picking up elements as it goes. This can be seen in Figure 2-1.

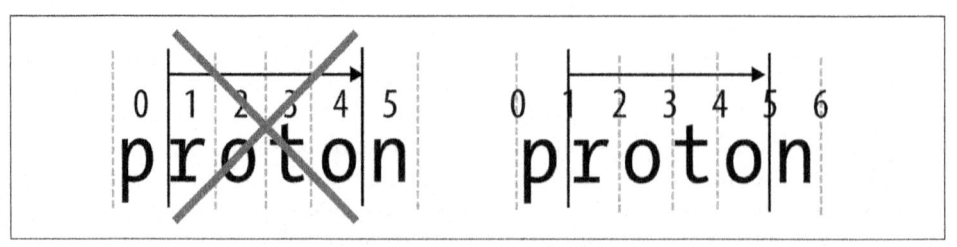

Figure 2-1. Indices are not on elements, but in between them

Feel free to mix positive and negative indices when slicing. However, the slice will not wrap around the left or right edges of the sequence. If you do try to wrap around, you get an empty sequence because there are no elements between the two indices:

```
In [6]: p[1:-1]
Out[6]: 'roto'

In [7]: p[-1:2]
Out[7]: ''
```

One of the greatest aspects of slicing is that the start and stop values are optional. If either or both of these values are left out of the slice, then sensible defaults are used. Namely, start becomes zero and stop becomes the length of the list. The colon (`:`) still has to be present to delimit start and stop and to differentiate this as a slice rather than an integer index, though. Here are some examples:

```
s[:2]   # the first two elements
s[-5:]  # the last five elements
s[:]    # the whole string!
```

Slicing has one last parameter: the *step*. The step represents how many elements to go in the sequence before picking up the next element for the slice. This is also sometimes known as a *stride*. Stepping is useful if you want to only see every other element, every third element, and so on. The step defaults to one, meaning to grab every element as we go and not skip any values. The syntax for stepping is very similar to that of starting and stopping: simply add a colon and an integer after the stop value. Thus, the full notation for slicing is `s[start:stop:step]`. Like the start and stop values, the step can also be negative. This just means to go backward through the sequence. Here are some stepping examples:

```
In [1]: q = "AaBbCcDdEeFfGgHhIiJjKkLlMmNnOoPpQqRrSsTtUuVvWwXxYyZz"

In [2]: q[2:-2:2]
Out[2]: 'BCDEFGHIJKLMNOPQRSTUVWXY'
```

```
In [3]: q[1::2]
Out[3]: 'abcdefghijklmnopqrstuvwxyz'

In [4]: q[::-3]
Out[4]: 'zYwVtSqPnMkJhGeDbA'
```

Because slicing is so easy, it comes up a lot in Python code. The most concise way to reverse a sequence is simply by slicing with a step size of: -1: s[::-1]. This allows us to write a very simple palindrome test:

```
In [1]: x = "neveroddoreven"

In [2]: x == x[::-1]
Out[2]: True
```

Slices are their own type and can be created independently of an indexing operation. They can be stored and used multiple times. To create a raw slice, use the expression slice(start, stop, step). If any of these need to have their default values, pass in None rather than an integer index:

```
In [3]: my_slice = slice(3, 1415, 9)  # my slice of the pi

In [4]: x[my_slice]
Out[4]: 'ee'
```

The indexing and slicing rules that we have just seen are very important. This is because they generally apply to all Python sequences. Strings are the most basic sequence, but we will be seeing more kinds of sequences in Chapter 3 and Chapter 9.

String Concatenation

Strings can be manipulated through a variety of operators. To start, consider the addition (+) operator, which in the context of strings is known as concatenation. This glues two strings together to make a bigger string:

```
In [1]: "kilo" + "meter"
Out[1]: 'kilometer'
```

Other data types will need to be converted to strings before they can be concatenated with a string. Take a numeric example:

```
In [1]: "x^" + str(2)
Out[1]: 'x^2'
```

Given that addition is defined and multiplication (*) is many additions, multiplying a string by an integer should yield that many copies of the string all concatenated together:

```
In [1]: "newto" * 10
Out[1]: 'newtonewtonewtonewtonewtonewtonewtonewtonewtonewto'
```

These tricks only work for addition and multiplication. Strings cannot be subtracted, divided, or exponentiated. The modulo (%) operator does apply to strings as a formatting mechanism. However, modulo does not follow the concatenation logic here, and its use is not recommended.

String Literals

What we have seen so far has been the most basic way to create strings. There are a few other mechanisms that are also useful. First, any two string literals that are next to each other are stuck together automatically:

```
In [1]: "H + H"    " -> H2"
Out[1]: 'H + H -> H2'
```

Newlines are ignored between parentheses. Long strings can be built up over multiple lines:

```
quote = ("Science is what we understand well enough to explain to a computer. "
         "Art is everything else we do. "
         "-Donald Knuth")
```

If a single- or double-quote character itself needs to be in the string, use the other kind of quote to define the string at the outermost level:

```
x = "It's easy!"
y = 'The computer said, "Does not compute."'
```

This works as long as both types of quote characters are not needed inside of the string. If they are, use the backslash character (\) to escape each quote character inside of the string:

```
"Bones said, \"He\'s dead, Jim.\""
```

There are a number of special escape characters that can be used. All escape characters start with a backslash and are interpreted as a single character even though they take two characters to write. The most important of these are seen in Table 2-2.

Table 2-2. String escape characters

Character	Interpretation
\\	Backslash
\n	Newline—start a new line
\r	Carriage return—go to the start of *this* line
\t	Tab
\'	Single quote

Character	Interpretation
\"	Double quote

String literals can also be prefixed with certain single characters that change how the string is interpreted. These prefixes are shown in Table 2-3.

Table 2-3. String prefixes

Prefix	Example	Meaning
r	r"escape!\n"	Raw string: all backslashes are escaped automatically. In the example, the \n is a \ and an n, not a newline.
b	b"this bytes"	Byte array: rather than becoming a string type, the value in quotes is interpreted as a raw series of bytes.
u	u"René Descartes"	Unicode string: the string is explicitly interpreted as a Unicode type. Useful in Python 2 and ignored in Python 3.

Finally, Python has support for multiline strings, which preserve the newlines that are inside of them. To create these, surround the text with either triple single quotes (''') or triple double quotes ("""). Triple double quotes are much preferred. Multiline string literals are essential for documentation, as we will see in future chapters. An example of such a string is shown here—note that the """ appear only at the beginning and end of the string, even though there are newlines present:

```
"""Humpty, he sat on a wall,
Then Humpty, he had a great fall.
But all the king's horses
And men with their forces
Couldn't render his entropy small.
"""
```

Now that we can make strings, we can learn about how to manipulate them. The next section describes string-based operations that are specific to the string type.

String Methods

Variables in Python have other variables that may "live on" them. These are known as attributes. Attributes, or *attrs* for short, are accessed using the dot operator (.). Suppose that x has a y; then the expression x.y means "Go into x and get me the y that lives there." Strings are no exception to this.

Additionally, some attributes are *function* types, which makes them *methods*. The details of what this means will be discussed in Chapter 5 and Chapter 6. For now, know that methods define special operations that you can perform on strings. To use methods, you *call* them with the parentheses (()) operator. In some cases, extra parameters will need to go inside of the parentheses.

> The following is not a comprehensive discussion of all string methods. It includes only the ones that are most useful to computational science.

The `strip()` method is incredibly useful for normalizing text-based data. It removes all leading and trailing whitespace while preserving internal whitespace. Whitespace is defined as spaces, tabs, newlines, and other blank characters. Suppose you had a flat data file, but the header had some very strange spacing. To trim the leading and trailing whitespace, you could fire up IPython and input the `header` string, then call `strip()` on it:

```
In [1]: header = "   temperature   pressure\t value \n"

In [2]: header.strip()
Out[2]: 'temperature   pressure\t value'
```

Here, we first define `header` to be the original string. Then we go into `header` and ask for `strip` by writing `header.strip`. Lastly, we compute the stripped string by calling the method immediately after accessing it, using the parentheses operator.

The `upper()` and `lower()` methods will return a version of the string with all alphabetical letters in uppercase or lowercase, respectively:

```
In [3]: header.upper()
Out[3]: '   TEMPERATURE   PRESSURE\t VALUE \n'
```

The `swapcase()` method will switch the existing case.

The `isdigit()` method returns `True` or `False` depending on whether or not the string contains only integer numbers:

```
In [1]: "10".isdigit()
Out[1]: True

In [2]: "10.10".isdigit()
Out[2]: False
```

Lastly, the `format()` method creates new strings from templates with the template values filled in. String formatting has its own mini-language, which will not be discussed in detail here but may be found in the Python string documentation (*https://docs.python.org/3.4/library/string.html#format-string-syntax*). The basic template form uses integers inside of curly braces ({}). The integers index into the values in the parentheses. For example:

```
In [1]: "{0} gets into work & then his {1} begins!".format("Hilbert", "commute")
Out[1]: 'Hilbert gets into work & then his commute begins!'
```

This helps convert data to strings without excess type conversion and concatenation. The following two expressions are equivalent, but the first one, using `format()`, is a lot shorter and easier to type than the second one:

```
In [1]: x = 42

In [2]: y = 65.0

In [3]: "x={0} y={1}".format(x, y)
Out[3]: 'x=42 y=65.0'

In [4]: "x=" + str(x) + " y=" + str(y)
Out[4]: 'x=42 y=65.0'
```

This covers the vast majority of string operations that you will need to perform for now. This is particularly true for physics software, which tends to be light on string manipulation. Probably the heaviest use of strings that you will do as a computational physicist is to generate input for other physics code and to read and parse their output. In these cases, most of the strings end up representing numbers anyway. Next, we will discuss how to access code from outside of the Python file or interpreter that we are currently writing or running.

Modules

Python code is typically written in files whose names end in the *.py* extension. When such a file is brought into a running Python interpreter, it is called a *module*. This is the in-memory representation of all of the Python code in the file. A collection of modules in a directory is called a *package*. It is worth noting that Python allows modules to be written in languages other than Python. These are called *extension modules* and are typically implemented in C.

Modules allow for a suite of related code files to all exist next to each other and to be accessed in a common way. They also provide a mechanism for saving and sharing code for use elsewhere and by other people. The Python standard library is itself an extensive collection of modules for a huge variety of common and not-so-common tasks. The batteries-included standard library is one of the things that makes Python so versatile. Using modules is how you get your hands on anything more than pure built-in Python.

Code from modules may be garnered in a number of different ways. All of these use the `import` keyword to pull in the module itself and allow you to access all of the variables that exist in that module. Modules may themselves use other modules.

Importing Modules

The `import` statement has four different forms. The first is just the `import` keyword followed by the module name without the trailing *.py*:

```
import <module>
```

Once a module has been imported, you can obtain variables in that module using the attribute access operator (.). This is exactly the same syntax that is used to get methods on an object. For example, say that there was one file, *constants.py*, which stored Planck's constant and pi. Another module could `import constants` and use it to compute h_bar:

constants.py

```
pi = 3.14159
h = 6.62606957e-34
```

physics.py

```
import constants

two_pi = 2 * constants.pi
h_bar = constants.h / two_pi
```

Importing Variables from a Module

Writing `constants.<var>` can be tedious if the `<var>` is going to be used many times. To alleviate this, Python has the `from-import` syntax that imports specific variables from a module. Either a single variable may be imported, or multiple comma-separated variable names may be imported simultaneously:

```
from <module> import <var>
from <module> import <var1>, <var2>, ...
```

This is equivalent to importing the module, setting a local variable to the name found in the module, and then deleting the module name, leaving only the local variable:

```
import <module>
<var> = <module>.<var>
del <module>
```

You can therefore think of `from-import` statements as renaming variable names for convenience. The `constants` import could be written as follows:

constants.py

```
pi = 3.14159
h = 6.62606957e-34
```

physics.py:

```
from constants import pi, h

two_pi = 2 * pi
h_bar = h / two_pi
```

Aliasing Imports

The next form of importing changes the name of the module on import. This is helpful if there is a local variable whose name would otherwise clash with the name of the module. (Generally, you control the names of the variables you use but do not have much say in the names of modules other people write.) This form uses the as keyword and has the following syntax:

```
import <module> as <name>
```

This is equivalent to importing the module, giving it a new name, and deleting the name it had when it was imported:

```
import <module>
<name> = <module>
del <module>
```

In the constants example, if there was a local variable that was also named constants, pi and h would only be accessible if the module was renamed. Here's how we would import the module using this syntax:

constants.py

```
pi = 3.14159
h = 6.62606957e-34
```

evenmorephysics.py

```
import constants as c

constants = 2.71828

two_pi = 2 * c.pi
h_bar = c.h / 2 / c.pi
```

In *evenmorephysics.py*, constants is Euler's number while the *constants.py* module is renamed to the variable c.

Aliasing Variables on Import

The final form of import combines elements of the form-import syntax and import aliasing to import only specific variables from a module, and rename them in the process. You can import and rename a single variable or a comma-separated list of variables, using the following syntax:

```
from <module> import <var> as <name>
from <module> import <var1> as <name1>, <var2> as <name2>, ...
```

This form of importing is equivalent to importing a variable from the module, renaming the variable locally, and deleting the original name:

```
from <module> import <var>
<name> = <var>
del <var>
```

Here's how we would import and rename the pi and h variables from the *constant.py* module:

constants.py

```
pi = 3.14159
h = 6.62606957e-34
```

yetmorephysics.py

```
from constants import pi as PI, h as H

two_pi = 2 * PI
h_bar = H / two_pi
```

Packages

As mentioned previously, a collection of modules in the same directory is called a *package*. For the package to be visible to Python, the directory must contain a special file named __*init*__.*py*. The main purpose of this file is to signal to Python that the directory is a package, and that other files in this directory whose names end in .*py* are importable. This file does not need to have any code in it. If it does, this code will be executed before any other modules in the package are imported.

The package takes on the name of the directory and may have subdirectories that are *subpackages*. For example, the filesystem for a compphys package may be laid out as follows:

```
compphys/
|-- __init__.py
|-- constants.py
|-- physics.py
|-- more/
|   |-- __init__.py
|   |-- morephysics.py
|   |-- evenmorephysics.py
|   |-- yetmorephysics.py
|-- raw/
|   |-- data.txt
|   |-- matrix.txt
|   |-- orphan.py
```

Here, `compphys` is the package name. This package has three modules (*__init__.py*, *constants.py*, and *physics.py*) and one subpackage (`more`). The *raw* directory does not count as a subpackage because it lacks an *__init__.py* file. This is true even though it contains other Python files, such as *orphan.py*, which are unreachable.

To import modules from a package, you use the attribute access operator (.). This is the same syntax used for importing variables from a module. Packages may be chained together with subpackage and module names, according to the filesystem hierarchy that the Python files live in. If you import a subpackage or module from a package, all of the packages above it in the hierarchy are automatically imported. However, you do not have access to the automatically imported packages unless you explicitly import them elsewhere. Once a module is imported, you can access all of the variables that are defined inside of it with the dot operator. For example:

```
import compphys.constants  ❶
import compphys.more.evenmorephysics  ❷

two_pi = 2 * compphys.constants.pi  ❸
```

❶ Import the `constants` module that lives in the `compphys` package.

❷ Import the `evenmorephysics` module that lives in the `more` subpackage of the `compphys` package.

❸ Access the `pi` variable of the `constants` module that lives in the `compphys` package by using the dot attribute access operator.

These are called *absolute imports* because the full paths to the modules are given.

> Absolute imports are recommended over all other styles of importing. This is because they provide the most clarity for the path to a module.

Inside of a package, you may import modules at the same level without giving the package name. This is called *implicit relative importing*. For example, *evenmorephysics.py* could import *morephysics.py* without your having to give the `compphys.more` prefix. The import would be:

```
import morephysics
```

Or, from *physics.py*, you could import modules from the subpackage using only the subpackage name:

```
import more.yetmorephysics
```

However, in modern Python, implicit relative imports are looked down upon.

> Implicit relative imports have been removed from Python 3. They are only available in Python 2. You probably shouldn't use them.

Explicit relative imports replace the need for implicit ones. Here, the `from` keyword must be used, and the module name is prefixed by either a single dot (`.`) or a double dot (`..`). The single dot refers to the current package level. The double dot refers to the package level one higher in the filesystem hierarchy. These have the same meaning that they do in bash.

For example, from *physics.py*, the following are valid imports:

```
from . import constants
from .constants import pi, h
from .more import morephysics
```

From *evenmorephysics.py*, the following imports would succeed:

```
from . import morephysics
from .. import constants
from ..constants import pi, h
```

Having more than two dots prefix the module name is not allowed. There is no way to go up more than one subpackage at a time. Oftentimes, it is best to rethink the filesystem layout if this much nesting is required.

Python enables you to write the modules and packages that you need to get your work done. However, you don't need to write everything from scratch yourself. The language itself comes prepackaged with a wide variety of tools for many situations, as we'll see next.

The Standard Library and the Python Ecosystem

One aspect that makes Python invaluable as a tool is its comprehensive standard library, which comes by default with the language. The standard library is a collection of packages and modules that combine to make performing most everyday tasks easy and Pythonic. It includes support for platform-independent operating system tasks, mathematical functions, compression algorithms, databases, and basic web servers. Wherever you have Python, you know you also have these standard tools. Table 2-4 describes some of the most useful Python modules that the standard library provides. This is by no means a complete listing.

Table 2-4. Important and useful modules in the Python standard library

Module	Description
os	Operating system abstractions: file path operations, file removal, etc.
sys	System-specific, gets into the guts of the Python interpreter
math	Everyday mathematical functions and constants
re	Regular expression library; see Chapter 8
subprocess	Spawns child processes and shells, good for running other command-line tools
argparse	Creates easy and beautiful command-line utilities
itertools	Helpful tools for looping
collections	Advanced collection types and tools for making custom collections
decimal	Arbitrary-precision integers and floats
random	Pseudo-random number generators
csv	Tools for reading and writing comma-separated value files
pdb	The Python debugger (similar to gdb for C/C++/Fortran)
logging	Utilities for logging the progress of a program while it is running

Another excellent aspect of Python is the fabulous ecosystem of third-party modules that has built up in support of the language. While these exist outside of the standard library, they are even more rich and diverse, and many of them are ideally suited for the needs of scientific computing and physics. Many of the other chapters in this book focus on the excellent external packages that are available.

Python Wrap-up

At this point, you should be familiar with the following:

- How to start up a Python interpreter
- Dynamically typed variables
- Basic data types such as `int`, `float`, and `str`
- How to manipulate variables with built-in operators

- String indexing and slicing
- How to import and use modules

From this foundation, you can start to build the more complicated representations of data and logic that are needed in scientific software. Up next, in Chapter 3, we will start tackling how to collect data together using mechanisms native to Python.

CHAPTER 3
Essential Containers

Let's now delve further into the tools of the Python language. Python comes with a suite of built-in data containers. These are data types that are used to hold many other variables. Much like you might place books on a bookshelf, you can stick integers or floats or strings into these containers. Each container is represented by its own type and has its own unique properties that define it. Major containers that Python supports are `list`, `tuple`, `set`, `frozenset`, and `dict`. All but `frozenset` come with their own literal syntax so that creating them is effortless. `dict` is by far the most important of these, for reasons that we'll see in "Dictionaries" on page 73 and in Chapter 6.

Before we dive in, there are two important Python concepts to understand:

- Mutability
- Duck typing

A data type is *mutable* if its value—also known as its *state*—is allowed to change after it has been created. On the other hand, a data type is *immutable* if its values are static and unchangeable once it is created. With immutable data you can create new variables based on existing values, but you cannot actually alter the original values. All of the data types we have dealt with so far—`int`, `float`, `bool`, and `str`—are immutable. It does not make sense to change the value of 1. It just *is* 1, and so integers are immutable. Containers are partially defined by whether they are mutable or not, and this determines where and how they are used.

Duck typing, on the other hand, is one of the core principles of Python and part of what makes it easy to use. This means that the type of a variable is less important than the interface it exposes. If two variables expose the same interface, then they should be able to be used in the same way. The argument goes, "If it looks like a duck and quacks like a duck, it is a duck!" Python believes that what a variable acts like at the moment it is used is more important than the actual underlying type. This is in stark contrast to lower-level languages, where it is more important what a variable "is" than what it does.

> **Interfaces**
>
> An *interface*, in programing terminology, is a set of rules, expectations, and protocols for how different pieces of software may interact with one another. Though these rules change from language to language, for a program to be able to run it must hook up all the interfaces in a valid way. Consider children's block playsets as a simple interface. Square pegs do not go into round holes; they only go into square holes. There are formal mathematical definitions of interfaces, but these rely on the notion of functions, which we'll meet in Chapter 5. Sometimes you will see the term *application programming interface* or *API* used. For our purposes here, "API" is synonymous with the word "interface."

We have already seen some examples of duck typing with indexing. The concept of indexing applies to any sequence, but "sequence" is not a fully defined type on its own. Instead, indexing can be applied to any variable that is sufficiently sequence-like. For example, we learned how to index strings in "String Indexing" on page 50. As will be seen shortly, the same indexing syntax may be used with lists and tuples. The idea that you can learn something once (string indexing) and use it again later for different types (list indexing and tuple indexing) is what makes duck typing so useful. If duck typing sounds generic, that is because it is. The whole point of duck typing is that the syntax of an operator should not change just because the type of the underlying variable changes. This notion highlights one of the things that makes Python easier to learn than other languages.

Lists

Lists in Python are one-dimensional, ordered containers whose elements may be any Python objects. Lists are mutable and have methods for adding and removing elements to and from themselves. The literal syntax for lists is to surround comma-separated values with square brackets ([]). The square brackets are a syntactic hint that lists are indexable. Here are some examples:

```
[6, 28]
[1e3, -2, "I am in a list."]
[[1.0, 0.0], [0.0, 1.0]]
```

In Python, unlike in other languages, the elements of a list do not have to match other in type. Anything can go into a list, including other lists! You can concatenate two lists together using the addition operator (+) to form a longer list:

```
In [1]: [1, 1] + [2, 3, 5] + [8]
Out[1]: [1, 1, 2, 3, 5, 8]
```

You can also append to lists in-place using the append() method, which adds a single element to the end:

```
In [2]: fib = [1, 1, 2, 3, 5, 8]

In [3]: fib.append(13)

In [4]: fib
Out[4]: [1, 1, 2, 3, 5, 8, 13]
```

Since building up a list element by element using append() can be tedious, whole sequences may be added to the end of a list in-place via the extend() method or the (+=) operator:

```
In [5]: fib.extend([21, 34, 55])

In [6]: fib
Out[6]: [1, 1, 2, 3, 5, 8, 13, 21, 34, 55]

In [7]: fib += [89, 144]

In [8]: fib
Out[8]: [1, 1, 2, 3, 5, 8, 13, 21, 34, 55, 89, 144]
```

List indexing is exactly the same as string indexing, but instead of returning strings it returns new lists. See "String Indexing" on page 50 for a refresher on how indexing works. Here is how to pull every other element out of a list:

```
In [9]: fib[::2]
Out[9]: [1, 2, 5, 13, 34, 89]
```

In addition to element access, indexes can also be used to set or delete elements in a list. This is because lists are mutable, whereas strings are not. Multiple list values may be set simultaneously as long as the new values are stored in a sequence of the same length as their destination. This can all be managed with the assignment (=) and del operators:

```
In [10]: fib[3] = "whoops"   ❶

In [11]: fib   ❷
Out[11]: [1, 1, 2, 'whoops', 5, 8, 13, 21, 34, 55, 89, 144]

In [12]: del fib[:5]   ❸

In [13]: fib   ❹
```

Lists | 67

```
Out[13]: [8, 13, 21, 34, 55, 89, 144]

In [14]: fib[1::2] = [-1, -1, -1]  ❺

In [15]: fib  ❻
Out[15]: [8, -1, 21, -1, 55, -1, 144]
```

❶ Set the fourth element of the `fib` list to *whoops*.

❷ See that the list was changed in-place.

❸ Remove the first five elements of `fib`.

❹ See that only the end of the original list remains.

❺ Assign -1 to each odd element.

❻ See how the odd elements have changed.

The same multiplication-by-an-integer trick for strings also applies to lists:

```
In [1]: [1, 2, 3] * 6
Out[1]: [1, 2, 3, 1, 2, 3, 1, 2, 3, 1, 2, 3, 1, 2, 3, 1, 2, 3]
```

You can also create lists of characters directly from strings by using the `list()` conversion function:

```
In [1]: list("F = dp/dt")
Out[1]: ['F', ' ', '=', ' ', 'd', 'p', '/', 'd', 't']
```

Another fascinating property is that a list will infinitely recurse if you add it to itself!

```
In [1]: x = []

In [2]: x.append(x)

In [3]: x
Out[3]: [[...]]

In [4]: x[0]
Out[4]: [[...]]

In [5]: x[0][0]
Out[5]: [[...]]
```

To explain how this is possible, we'll need to explore of how Python manages memory. Python is *reference counted*, which means that variable names are actually references to the underlying values. The language then keeps an internal count of how many times a reference has been used and what its names are. Think of this as there being data on the one hand, and names that are attached to data—like sticky notes—

on the other hand. Names cannot refer to other names, but only to the underlying data. Consider the following simple example:

```
x = 42
y = x
del x
```

In the code here, Python starts by first creating the number 42 in memory. It then sets the name x to refer to the point in memory where 42 lives. On the next line, Python then sees that y should point to the same place that x is pointing to. Now, 42 has two names: x and y. Then x is deleted, but Python sees that 42 still has at least one name (y), so it keeps both y and 42 around for later use. This can be seen in Figure 3-1.

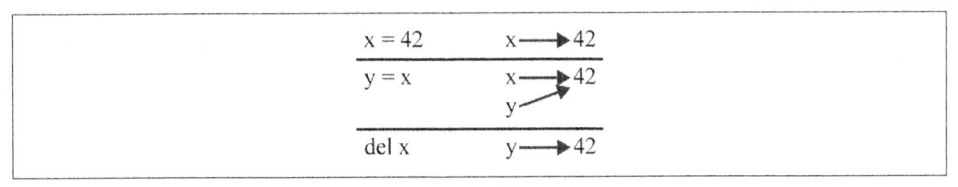

x = 42	x ⟶ 42
y = x	x ⟶ 42 y ↗
del x	y ⟶ 42

Figure 3-1. Reference counting of simple variables

So how does this apply to lists? Lists are collections of names, not values! The name a list gives to each of its elements is the integer index of that element. Of course, the list itself also has a name. This means that when a list itself has two or more variable names and any of them has an element changed, then all of the other variables also see the alteration. Consider this example:

Code	Output

`x = [3, 2, 1, "blast off!"]`
`y = x`
`y[1] = "TWO"`
`print(x)`
	`[3, "TWO", 1, "blast off!"]`

Here, when y's second element is changed to the string 'TWO', this change is reflected back onto x. This is because there is only one list in memory, even though there are two names for it (x and y). Figure 3-2 shows this graphically.

x = [3, 2, 1 "blast off!"]	x ⟶ [3, 2, 1 "blast off!"]
y = x	x ⟶ [3, 2, 1 "blast off!"] y ↗
y[1] = "TWO"	x ⟶ [3, "TWO", 1 "blast off!"] y ↗

Figure 3-2. Reference counting with lists

This is the spooky action at a distance of programming. But it is also how Python containers work. Python is not alone here; this is how all reference-counted languages act. In compiled languages, this is what makes smart pointers smart. The reason this technique is used is that memory volume is handled much more efficiently, though this often comes at the cost of increased CPU usage.

Now that you understand how Python is handling memory, it is not hard to use it correctly. Though this is often a "gotcha" for people new to dynamic languages, it becomes second nature very quickly.

> The Python statement x = y = [] means that there is one new empty list with two names (x and y). If you come from a C/C++ background, it is tempting to read this as meaning to create two new empty lists with two names. However, this is incorrect because of how Python's memory management works.

Returning to the example of the infinitely recurring list, this is a list that holds a reference to itself. This means that names in the series x, x[0], x[0][0], ... point to exactly the same place. This is partially drawn out in Figure 3-3.

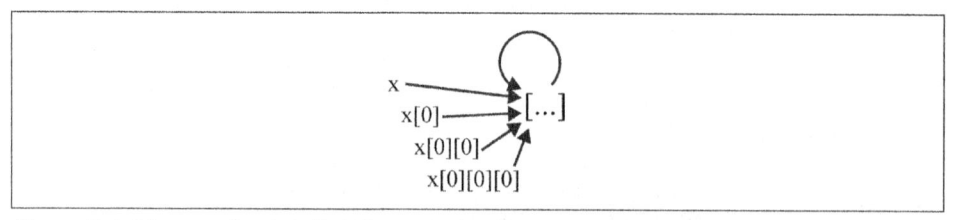

Figure 3-3. Memory in an infinitely recurring list

Now that we've covered the basics of lists, we can move on to a very similar container that differs from lists in a single, but significant, way.

Tuples

Tuples are the immutable form of lists. They behave almost exactly the same as lists in every way, except that you cannot change any of their values. There are no append() or extend() methods, and there are no in-place operators.

They also differ from lists in their syntax. They are so central to how Python works that *tuples are defined by commas* (,). Oftentimes, tuples will be seen surrounded by parentheses. These parentheses serve only to group actions or make the code more readable, not to actually define the tuples. Some examples include:

```
a = 1, 2, 5, 3    # length-4 tuple
b = (42,)         # length-1 tuple, defined by comma
```

```
c = (42)       # not a tuple, just the number 42
d = ()         # length-0 tuple- no commas means no elements
```

You can concatenate tuples together in the same way as lists, but be careful about the order of operations. This is where the parentheses come in handy:

```
In [1]: (1, 2) + (3, 4)
Out[1]: (1, 2, 3, 4)

In [2]: 1, 2 + 3, 4
Out[2]: (1, 5, 4)
```

The tuple converter is just called `tuple()`. If you have a list that you wish to make immutable, use this function:

```
In [1]: tuple(["e", 2.718])
Out[1]: ('e', 2.718)
```

Note that even though tuples are immutable, they may have mutable elements. Suppose that we have a list embedded in a tuple. This list may be modified in-place even though the list may not be removed or replaced wholesale:

```
In [1]: x = 1.0, [2, 4], 16

In [2]: x[1].append(8)

In [3]: x
Out[3]: (1.0, [2, 4, 8], 16)
```

Other than immutability, what are the differences between lists and tuples? In principle, there are not any. In practice, they tend to be used in different ways. However, there are no strict rules and there are no predominant conventions. There is a loose guideline that lists are for *homogeneous* data (all integers, all strings, etc.) while tuples are for *heterogeneous* data with semantic meaning in each element (e.g., ("C14", 6, 14, 14.00324198843)). Other, more sophisticated data structures that we'll see in future chapters add semantic meaning to their elements. For tuples and lists, though, this rule of thumb is only a suggestion.

Tuples are integral to functions and will be seen much more in Chapter 5. Both tuples and lists may have duplicated elements. Sometimes this is exactly what you want. Up next, though, is a container that ensures that each of its elements is unique.

Sets

Instances of the `set` type are equivalent to mathematical sets. Like their math counterparts, literal sets in Python are defined by comma-separated values between curly braces ({}). Sets are unordered containers of unique values. Duplicated elements are ignored. Because they are unordered, sets are not sequences and cannot be indexed.

Containment—is x in y?—is much more important for sets than how the elements are stored. Here are some examples:

```
# a literal set formed with elements of various types
{1.0, 10, "one hundred", (1, 0, 0,0)}

# a literal set of special values
{True, False, None, "", 0.0, 0}

# conversion from a list to a set
set([2.0, 4, "eight", (16,)])
```

One common misconception of new Python programmers deals with the set of a single string. The set of a string is actually the set of its characters. This is because strings are sequences. To have a set that actually contains a single string, first put the string inside of another sequence:

```
In [1]: set("Marie Curie")
Out[1]: {' ', 'C', 'M', 'a', 'e', 'i', 'r', 'u'}

In [2]: set(["Marie Curie"])
Out[2]: {'Marie Curie'}
```

Sets may be used to compute other sets or be compared against other sets. These operations follow mathematical analogies that can be seen in Table 3-1 (take s = {1, 2, 3} and t = {3, 4, 5} for the examples).

Table 3-1. Set operations

Operation	Meaning	Example
s | t	Union	{1, 2, 3, 4, 5}
s & t	Intersection	{3}
s - t	Difference—elements in s but not in t	{1, 2}
s ^ t	Symmetric difference—elements not in s or t	{1, 2, 4, 5}
s < t	Strict subset	False
s <= t	Subset	False
s > t	Strict superset	False
s >= t	Superset	False

The uniqueness of set elements is key. This places an important restriction on what can go in a set in the first place. Namely, the elements of a set must be *hashable*. The core idea behind hashing is simple. Suppose there is a function that takes any value and maps it to an integer. If two variables have the same type and map to the same integer, then the variables have the same value. This assumes that you have enough integers and a reasonable mapping function. Luckily, Python takes care of those details for us. Whether or not something is allowed to go into a set depends only on if it can be unambiguously converted to an integer.

In Python, the hash function is just called `hash()`, and you can try using it on any variable. If this function fails for some reason, that value cannot be placed in a set. If two variables are hashable, though, the following logic statement is roughly true:

```
hash(x) == hash(y) implies that x == y
```

This assumption breaks down across type boundaries. Python handles differently typed variables separately because it knows them to be different. For example, an empty string and the float `0.0` both hash to `0` (as an `int`, because hashes are integers). However, an empty string and the float `0.0` clearly are not the same value, because they have different types:

```
hash("") == hash(0.0) == 0 does not imply that "" == 0.0
```

What makes a type hashable? Immutability. Without immutability there is no way to reliably recompute the hash value. As a counterexample, say you could compute the hash of a list. If you were then to add or delete elements to or from the list, its hash would change! If this list were already in a set, list mutability would break the guarantee that each element of the set is unique. This is why lists are not allowed in sets, though tuples are allowed if all of their elements are hashable.

Lastly, sets themselves are mutable. You can `add()`, `discard()`, or otherwise modify sets in-place. This means that you cannot have a set inside of another set. To get around this, there is an immutable version of the `set` type called `frozenset`. The relationship between sets and frozensets is the same as that between lists and tuples: frozensets are exactly the same as sets, except that they cannot be modified once they are created.

Highly related to sets and based on the same notion of hashability are the ever-present dictionaries. We'll look at these next.

Dictionaries

Dictionaries are hands down *the most important* data structure in Python. Everything in Python is a dictionary. A dictionary, or `dict`, is a mutable, unordered collection of unique key/value pairs—this is Python's native implementation of a *hash table*. Dictionaries are similar in use to C++ maps, but more closely related to Perl's hash type, JavaScript objects, and C++'s `unordered_map` type. We will cover hash tables and how they are implemented in much greater detail in "Hash Tables" on page 258. Right

now, we will see just enough to understand how to use dictionaries, so we can move
forward with learning Python.

In a dictionary, keys are *associated* with values. This means that you can look up a
value knowing only its key(s). Like their name implies, the keys in a dictionary must
be unique. However, many different keys with the same value are allowed. They are
incredibly fast and efficient at looking up values, which means that using them incurs
almost no overhead.

Both the keys and the values are Python objects. So, as with lists, you can store any‐
thing you need to as values. Keys, however, must be hashable (hence the name "hash
table"). This is the same restriction as with sets. In fact, in earlier versions of Python
that did not have sets, sets were faked with dictionaries where all of the values were
None. The syntax for dictionaries is also related to that for sets. They are defined by
outer curly brackets ({}) surrounding key/value pairs that are separated by commas
(,). Each key/value pair is known as an *item*, and the key is separated from the value
by a colon (:). Curly braces are treated much like parentheses, allowing dictionaries
to be split up over multiple lines. They can also be defined with a list of 2-tuples. Here
are some examples:

```
# A dictionary on one line that stores info about Einstein
al = {"first": "Albert", "last": "Einstein", "birthday": [1879, 3, 14]}

# You can split up dicts onto many lines
constants = {
    'pi': 3.14159,
    "e": 2.718,
    "h": 6.62606957e-34,
    True: 1.0,
    }

# A dict being formed from a list of (key, value) tuples
axes = dict([(1, "x"), (2, "y"), (3, "z")])
```

You pull a value out of a dictionary by indexing with the associated key. If we had
typed each of these dicts into IPython, we could then access their values as follows:

```
In [1]: constants['e']
Out[1]: 2.718

In [2]: axes[3]
Out[2]: 'z'

In [3]: al['birthday']
Out[3]: [1879, 3, 14]
```

Since dictionaries are unordered, slicing does not make any sense for them. However, items may be added and deleted through indexing. Existing keys will have their values replaced:

```
constants[False] = 0.0
del axes[3]
al['first'] = "You can call me Al"
```

Because dictionaries are mutable, they are not hashable themselves, and you cannot use a dictionary as a key in another dictionary. You may nest dictionaries as values, however. This allows for the infinitely recurring trick, like with lists:

```
In [4]: d = {}
```

```
In [5]: d['d'] = d
```

```
In [6]: d
Out[6]: {'d': {...}}
```

Note that `dicts` predate `sets` by over a decade in the language. Since the syntax collides, empty `dicts` are defined by just curly braces while an empty `set` requires the type name and parentheses:

```
{}     # empty dict
set()  # empty set
```

Tests for containment with the `in` operator function only on dictionary keys, not values:

```
In [7]: "N_A" in constants
Out[7]: False
```

Dictionaries have a lot of useful methods on them as well. For now, content yourself with the `update()` method. This incorporates another dictionary or list of tuples in-place into the current `dict`. The update process overwrites any overlapping keys:

```
In [8]: axes.update({1: 'r', 2: 'phi', 3: 'theta'})
```

```
In [9]: axes
Out[9]: {1: 'r', 2: 'phi', 3: 'theta'}
```

This is only enough to get started. Dictionaries are more important than any other data type and will come up over and over again. Their special place in the Python language will be seen in Chapter 5 and Chapter 6.

Containers Wrap-up

Having reached the end of this chapter, you should now be familiar with the following concepts:

- Mutability and immutability
- Duck typing
- Lists and tuples
- Hash functions
- Sets and dictionaries

These data containers and their underlying concepts are the building blocks for higher-level, more complex structures. They let you represent your data in the way that makes the most sense for the problem at hand. They also enable a wide variety of expressive Python syntax, which we will start to explore in Chapter 4.

CHAPTER 4
Flow Control and Logic

Flow control is a high-level way of programming a computer to make decisions. These decisions can be simple or complicated, executed once or multiple times. The syntax for the different flow control mechanisms varies, but what they all share is that they determine an *execution pathway* for the program. Python has relatively few forms of flow control. They are *conditionals*, *exceptions*, and *loops*.

As someone primarily interested in physical reality, you might wonder why you should care about flow control and logic. In some ways, this is like asking why arithmetic is important. Logic presents rules that allow you to build up and represent more complex ideas. This enables the physics modeling you want to do by giving you a means to express the choices and behavior of your model to the computer. With basic flow control syntax, your models can make simple decisions. With more advanced flow control, your models can make more sophisticated choices more easily. In other situations, flow control allows you to reuse the same code many times. This makes the software model faster to write and easier to understand, because it has fewer total lines of code. Logic and flow control are indispensible to doing any significant amount of work with computers. So, without further delay, let's jump into conditionals, our first bit of flow control.

Conditionals

Conditionals are the simplest form of flow control. In English, they follow the syntax "if x is true, then do something; otherwise, do something else." The shortest conditional is when there is only an `if` statement on its own. The format for such a statement is as follows:

```
if <condition>:
    <if-block>
```

Here, the Python keyword `if` is followed by an expression, <condition>, which is itself followed by a colon (:). When the Boolean representation of the condition, `bool(condition)`, is `True`, the code that is in the <if-block> is executed. If `bool(condition)` is `False`, then the code in the block is skipped. The condition may be composed of any of the comparison operators (or a combination of these operators) that were listed in Table 2-1. For convenience, just the comparison operators are shown again here in Table 4-1.

Table 4-1. Python logical operators useful for comparing the variables x, y, and z

Name	Usage	Returns
Unary operators		
Negation	`not x`	Logical negation—`True` becomes `False`, and vice versa.
Bitwise invert	`~x`	Changes all zeros to ones and vice versa in x's binary representation.
Binary operators		
Logical and	`x and y`	`True` if `bool(x)` and `bool(y)` are `True`; `False` otherwise.
Logical or	`x or y`	x if `bool(x)` is `True`; otherwise the value of y.
Comparison binary operators		
Equality	`x == y`	`True` or `False`.
Not equal	`x != y`	`True` or `False`.
Less than	`x < y`	`True` or `False`.
Less than or equal	`x <= y`	`True` or `False`.
Greater than	`x > y`	`True` or `False`.
Greater than or equal	`x >= y`	`True` or `False`.
Containment	`x in y`	`True` if x is an element of y.
Non-containment	`x not in y`	`False` if x is an element of y.
Identity test	`x is y`	`True` if x and y point to the same underlying value in memory.

Name	Usage	Returns
Not identity test	`x is not y`	`False` if x and y point to the same underlying value in memory.
Ternary operators		
Ternary compare	`x < y < z`	`True` or `False`, equivalent to `(x < y)` and `(y < z)`. The `<` here may be replaced by `>`, `<=`, or `>=` in any permutation.

For example, if we wanted to test if Planck's constant is equal to one and then change its value if it is, we could write the following:

```
h_bar = 1.0
if h_bar == 1.0:
    print("h-bar isn't really unity! Resetting...")
    h_bar = 1.05457173e-34
```

Here, since `h_bar` is `1.0` it is reset to its actual physical value (`1.05457173e-34`). If `h_bar` had been its original physical value, it would not have been reset.

A key Pythonism that is part of the `if` statement is that Python is *whitespace separated*. Unlike other languages, which use curly braces and semicolons, in Python the contents of the `if` block are determined by their indentation level. New statements *must* appear on their own lines. To exit the `if` block, the indentation level is returned back to its original column:

```
h_bar = 1
if h_bar == 1:
    print("h-bar isn't really unity! Resetting...")
    h_bar = 1.05457173e-34
h = h_bar * 2 * 3.14159
```

The last line here (the one that defines h) indicates that the `if` block has ended because its indentation level matches that of the `if` on the second line. The last line will always be executed, no matter what the conditional decides should be done for the `if` block.

While we are on the subject, it is important to bring up the distinction between the equality operator (`==`) and the identity operator (`is`). The equality operator tests if two values are equivalent. For example, `1 == 1.0` is `True` even though `1` is an integer and `1.0` is a float. On the other hand, the identity operator tests if two variable names are references to the same underlying value in memory. For example, `1 is 1.0` is `False` because the types are different, and therefore they cannot actually be references to the same value. `is` is much faster than `==`, but also much more strict. In general, you want to use `is` for singletons like `None` and use the safer `==` in most other situations. The following examples show typical use cases and gotchas:

Code	Output
`1 == 1`	True
`1 == 1.0`	True
`1 is 1.0`	False
`1 is 1` ❶	True
`10**10 == 10**10`	True
`10**10 is 10**10` ❷	False
`None is None`	True
`0 is None` ❸	False
`0 == None`	False

❶ To help with performance, Python only stores a single copy of small integers. So for small `int`s, every usage will be the same value in memory.

❷ However, for big integers a new copy is computed each time.

❸ Only None is None.

Before we move on, it is important to note that, by tradition, Python uses four spaces per level to indent all code blocks. Two spaces, eight spaces, or any other spacing is looked down upon. Tabs cause many more problems than they are worth. Most text editors have an option to automatically convert tabs to spaces, and enabling this can help prevent common errors. Some people find the whitespace syntax a little awkward to begin with, but it becomes easy and natural very quickly. The whitespace-aware aspect of Python is a codification of what is a best-practice coding style in other languages. It forces programmers to write more legible code.

if-else Statements

Every `if` statement may be followed by an optional `else` statement. This is the keyword `else` followed by a colon (`:`) at the same indentation level as the original `if`. The `<else-block>` lines following this are indented just like the `if` block. The code in the `else` block is executed when the condition is `False`:

```
if <condition>:
    <if-block>
else:
    <else-block>
```

For example, consider the expression sin(1/x). This function is computable everywhere except a x = 0. At this point, L'Hôpital's rule shows that the result is also zero. This could be expressed with an `if-else` statement as follows:

```
if x == 0:
    y = 0
else:
    y = sin(1/x)
```

This is equivalent to negating the conditional and switching the `if` and `else` blocks:

```
if x != 0:
    y = sin(1/x)
else:
    y = 0
```

However, it is generally considered a good practice to use positive conditionals (==) rather than negative ones (!=). This is because humans tend to think about an expression being true rather than it being false. This is not a hard and fast rule, but it does help eliminate easy-to-miss logic bugs.

if-elif-else Statements

Python also allows multiple optional `elif` statements. The `elif` keyword is an abbreviation for "else if," and such statements come after the `if` statement and before the `else` statement. The `elif` statements have much the same form as the `if` statement, and there may be as many of them as desired. The first conditional that evaluates to `True` determines the block that is entered, and no further conditionals or blocks are executed. The syntax is as follows:

```
if <condition0>:
    <if-block>
elif <condition1>:
    <elif-block1>
elif <condition2>:
    <elif-block2>
...
else:
    <else-block>
```

Suppose that you wanted to design a simple mid-band filter whose signal is 1 if the frequency is between 1 and 10 Hertz and 0 otherwise. This could be done with an `if-elif-else` statement:

```
if omega < 1.0:
   signal = 0.0
elif omega > 10.0:
   signal = 0.0
```

```
else:
    signal = 1.0
```

A more realistic example might include ramping on either side of the band:

```
if omega < 0.9:
    signal = 0.0
elif omega > 0.9 and omega < 1.0:
    signal = (omega - 0.9) / 0.1
elif omega > 10.0 and omega < 10.1:
    signal = (10.1 - omega) / 0.1
elif omega > 10.1:
    signal = 0.0
else:
    signal = 1.0
```

if-else Expression

The final syntax covered here is the ternary conditional operator. It allows simple `if-else` conditionals to be evaluated in a single expression. This has the following syntax:

```
x if <condition> else y
```

If the condition evaluates to `True`, then x is returned. Otherwise, y is returned. This turns out to be extraordinarily handy for variable assignment. Using this kind of expression, we can write the h_bar conditional example in one line:

```
h_bar = 1.05457173e-34 if h_bar == 1.0 else h_bar
```

Note that when using this format you must always include the `else` clause. This fills the same role as the `condition?x:y` operator that is available in other languages. Writing out `if` and `else` arguably makes the Python way much more readable, though also more verbose.

Exceptions

Python, like most modern programming languages, has a mechanism for *exception handling*. This is a language feature that allows the programmer to work around situations where the unexpected and catastrophic happen. Exception handling is for the truly exceptional cases: a user manually types in an impossible value, a file is deleted while it is being written, coffee spills on the laptop and fries the motherboard.

> Exceptions are not meant for normal flow control and dealing with expected behavior! Use conditionals in cases where behavior is anticipated.

The syntax for handling exceptions is known as a `try-except` block. Both `try` and `except` are Python keywords. `try-excepts` look very similar to `if-else` statements, but without the condition:

```
try:
    <try-block>
except:
    <except-block>
```

The `try` block will attempt to execute its code. If there are no errors, then the program skips the `except` block and proceeds normally. If any error at all happens, then the `except` block is immediately entered, no matter how far into the `try` block Python has gone. For this reason, it is generally a good idea to keep the `try` block as small as possible. Single-line `try` blocks are strongly preferred.

As an example, say that a user manually inputs a value and then the program takes the inverse of this value. Normally this computes just fine, with the exception of when the user enters 0:

```
In [1]: val = 0.0

In [2]: 1.0 / val

ZeroDivisionError                         Traceback (most recent call last)
<ipython-input-2-3ac1864780ca> in <module>()
----> 1 1.0 / val

ZeroDivisionError: float division by zero
```

This error could be handled with a `try-except`, which would prevent the program from crashing:

```
try:
    inv = 1.0 / val
except:
    print("A bad value was submitted {0}, please try again".format(val))
```

The `except` statement also allows for the precise error that is anticipated to be caught. This allows for more specific behavior than the generic catch-all exception. The error name is placed right after the `except` keyword but before the colon. In the preceding example, we would catch a `ZeroDivisionError` by writing:

```
try:
    inv = 1.0 / val
```

```
    except ZeroDivisionError:
        print("A zero value was submitted, please try again")
```

Multiple except blocks may be chained together, much like elif statements. The first exception that matches determines the except block that is executed. The previous two examples could therefore be combined as follows:

```
try:
    inv = 1.0 / val
except ZeroDivisionError:
    print("A zero value was submitted, please try again")
except:
    print("A bad value was submitted {0}, please try again".format(val))
```

Raising Exceptions

The other half of exception handling is raising them yourself. The raise keyword will *throw* an exception or error, which may then be *caught* by a try-except block elsewhere. This syntax provides a standard way for signaling that the program has run into an unallowed situation and can no longer continue executing.

raise statements may appear anywhere, but it is common to put them inside of conditionals so that they are not executed unless they need to be. Continuing with the inverse example, instead of letting Python raise a ZeroDivisionError we could check for a zero value and raise it ourselves:

```
if val == 0.0:
    raise ZeroDivisionError
inv = 1.0 / val
```

If val happens to be zero, then the inv = 1.0 / val line will never be run. If val is nonzero, then the error is never raised.

All errors can be called with a custom string message. The helps locate, identify, and squash bugs. Error messages should be as detailed as necessary while remaining concise and readable. A message that states "An error occurred here" does not help anyone! A better version of the preceding code is:

```
if val == 0.0:
    raise ZeroDivisionError("taking the inverse of zero is forbidden!")
inv = 1.0 / val
```

Python comes with 150+ error and exception types. (This is not as many as it seems at first glance—these exceptions are sufficient to cover the more than one million lines of code in Python itself!) Table 4-2 lists some of the most common ones you will see in computational physics.

Table 4-2. Common Python errors and exceptions

Exception	Description
`AssertionError`	Used when the `assert` operator sees `False`.
`AttributeError`	Occurs when Python cannot find a variable that lives on another variable. Usually this results from a typo.
`ImportError`	Occurs when a package or module cannot be found. This is typically the result of either a typo or a dependency that hasn't been installed.
`IOError`	Happens when Python cannot read or write to an external file.
`KeyboardInterrupt`	Automatically raised when an external user kills the running Python process with Ctrl-c.
`KeyError`	Raised when a key cannot be found in a dictionary.
`MemoryError`	Raised when your computer runs out of RAM.
`NameError`	Occurs when a local or global variable name cannot be found. Usually the result of a typo.
`RuntimeError`	Generic exception for when something, somewhere has gone wrong. The error message itself normally has more information.
`SyntaxError`	Raised when the program tries to run non-Python code. This is typically the result of a typo, such as a missing colon or closing bracket.
`ZeroDivisionError`	Occurs when Python has tried to divide by zero, and is not happy about it.

It is often tempting to create custom exceptions for specific cases. You'll find more information on how to do this in Chapter 6. However, custom exception types are rarely necessary—99% of the time there is already a built-in error that covers the exceptional situation at hand. It is generally better to use message strings to customize existing error types rather than creating brand new ones.

Loops

While computers are not superb at synthesizing new tasks, they are *very* good at performing the same tasks over and over. So far in this chapter, we've been discussing the single execution of indented code blocks. *Loops* are how to execute the same block multiple times. Python has a few looping formats that are essential to know: `while` loops, `for` loops, and *comprehensions*.

while Loops

`while` loops are related to `if` statements because they continue to execute "while a condition is true." They have nearly the same syntax, except the `if` is replaced with the `while` keyword. Thus, the syntax has the following format:

```
while <condition>:
    <while-block>
```

The condition here is evaluated right before every loop *iteration*. If the condition is or remains `True`, then the block is executed. If the condition is `False`, then the `while` block is skipped and the program continues. Here is a simple countdown timer:

Code	Output

`t = 3`
`while 0 < t:`
    `print("t-minus " + str(t))`
    `t = t - 1`
`print("blastoff!")`
	`t-minus 3`
	`t-minus 2`
	`t-minus 1`
	`blastoff!`

If the condition evaluates to `False`, then the `while` block will never be entered. For example:

Code	Output

`while False:`
    `print("I am sorry, Dave.")`
`print("I can't print that for you.")`
	`I can't print that for you.`

On the other hand, if the condition always evaluates to `True`, the `while` block will continue to be executed no matter what. This is known as an *infinite* or *nonterminating* loop. Normally this is not the intended behavior. A slight modification to the countdown timer means it will never finish on its own:

Code	Output
| ```
t = 3
while True:
    print("t-minus " + str(t))
    t = t - 1
print("blastoff!")
``` | ```
t-minus 3
t-minus 2
t-minus 1
t-minus 0
t-minus -1
t-minus -2
t-minus -3
t-minus -4
t-minus -5
...
# blastoff is never reached
``` |

Integers counting down to negative infinity is not correct behavior in most situations.

> Interestingly, it is impossible to predict whether a loop (or any program) will terminate without actually running it. This is known as the *halting problem* and was originally shown by Alan Turing. If you do happen to accidentally start an infinite loop, you can always hit Ctrl-c to exit the Python program.

The `break` statement is Python's way of leaving a loop early. The keyword `break` simply appears on its own line, and the loop is immediately exited. Consider the following `while` loop, which computes successive elements of the Fibonacci series and adds them to the `fib` list. This loop will continue forever unless it finds an entry that is divisible by 12, at which point it will immediately leave the loop and not add the entry to the list:

Code	Output
| ```
fib = [1, 1]
while True:
    x = fib[-2] + fib[-1]
    if x%12 == 0:
        break
    fib.append(x)
``` | `[1, 1, 2, 3, 5, 8, 13, 21, 34, 55, 89]` |

This loop does terminate, because 55 + 89 == 144 and 144 == 12**2. Also note that the `if` statement is part of the `while` block. This means that the `break` statement needs to be additionally indented. Additional levels of indentation allow for code blocks to be *nested* within one another. Nesting can be arbitrarily deep as long as the correct flow control is used.
```

for Loops

Though while loops are helpful for repeating statements, it is typically more useful to iterate over a container or other "iterable," grabbing a single element each time through and exiting the loop when there are no more elements in the container. In Python, for loops fill this role. They use the for and in keywords and have the following syntax:

```
for <loop-var> in <iterable>:
    <for-block>
```

The <loop-var> is a variable name that is assigned to a new element of the iterable on each pass through the loop. The <iterable> is any Python object that can return elements. All containers (lists, tuples, sets, dictionaries) and strings are iterable. The for block is a series of statements whose execution is repeated. This is the same as what was seen for while blocks. Using a for loop, we could rewrite our countdown timer to loop over the list of integers [3, 2, 1] as follows:

```
for t in [3, 2, 1]:
    print("t-minus " + str(t))
print("blastoff!")
```

Again, the value of t changes on each iteration. Here, though, the t = t - 1 line is not needed because t is automatically reassigned to the next value in the list. Additionally, the 0 < t condition is not needed to stop the list; when there are no more elements in the list, the loop ends.

The break statement can be used with for loops just like with while loops. Additionally, the continue statement can be used with both for and while loops. This exits out of the current iteration of the loop only and *continues on* with the next iteration. It does not break out of the whole loop. Consider the case where we want to count down every t but want to skip reporting the even times:

Code	Output

`for t in [7, 6, 5, 4, 3, 2, 1]:`
`    if t%2 == 0:`
`        continue`
`    print("t-minus " + str(t))`
`print("blastoff!")`
	t-minus 7
	t-minus 5
	t-minus 3
	t-minus 1
	blastoff!

Note that containers choose how they are iterated over. For sequences (strings, lists, tuples), there is a natural iteration order. String iteration produces each letter in turn:

Code	Output
`for letter in "Gorgus":` `    print(letter)`	G o r g u s

However, unordered data structures (sets, dictionaries) have an unpredictable iteration ordering. All elements are guaranteed to be iterated over, but when each element comes out is not predictable. The iteration order is not the order that the object was created with. The following is an example of set iteration:

Code	Output
`for x in {"Gorgus", 0, True}:` `    print(x)`	0 True Gorgus

Dictionaries have further ambiguity in addition to being unordered. The loop variable could be the keys, the values, or both (the items). Python chooses to return the keys when looping over a dictionary. It is assumed that the values can be looked up normally. It is very common to use **key** or **k** as the loop variable name. For example:

Code	Output

`d = {"first": "Albert",`
`     "last": "Einstein",`
`     "birthday": [1879, 3, 14]}`
`for key in d:`
`    print(key)`
`    print(d[key])`
`    print("======")`
	birthday
	[1879, 3, 14]
	======
	last
	Einstein
	======
	first
	Albert
	======

Dictionaries may also be explicitly looped through their keys, values, or items using the `keys()`, `values()`, or `items()` methods:

Code	Output
| ```
d = {"first": "Albert",
     "last": "Einstein",
     "birthday": [1879, 3, 14]}
print("Keys:")
for key in d.keys():
    print(key)

print("\n======\n")

print("Values:")
for value in d.values():
    print(value)

print("\n======\n")

print("Items:")
for key, value in d.items():
    print(key, value)
``` | ```
Keys:
birthday
last
first

======

Values:
[1879, 3, 14]
Einstein
Albert

======

Items:
('birthday', [1879, 3, 14])
('last', 'Einstein')
('first', 'Albert')
``` |

When iterating over items, the elements come back as key/value tuples. These can be *unpacked* into their own loop variables (called `key` and `value` here for consistency, though this is not mandatory). Alternatively, the items could remain packed, in which case the loop variable would still be a tuple:

Code	Output
| ```
for item in d.items():
    print(item)
``` | ```
('birthday', [1879, 3, 14])
('last', 'Einstein')
('first', 'Albert')
``` |

It is a very strong idiom in Python that the loop variable name is a singular noun and the iterable is the corresponding plural noun. This makes the loop more natural to read. This pattern expressed in code is shown here:

```
for single in plural:
    ...
```

For example, looping through the set of quark names would be done as follows:

```
quarks = {'up', 'down', 'top', 'bottom', 'charm', 'strange'}
for quark in quarks:
    print(quark)
```

Comprehensions

`for` and `while` loops are fantastic, but they always take up at least two lines: one for the loop itself and another for the block. And often when you're looping through a container the result of each loop iteration needs to be placed in a new corresponding list, set, dictionary, etc. This takes at least three lines. For example, converting the quarks set to a list of uppercase strings requires first setting up an empty list:

Code	Output

`upper_quarks = []`
`for quark in quarks:`
`    upper_quarks.append(quark.upper())`
	`upper_quarks = ['BOTTOM', 'TOP',`
	`                'UP', 'DOWN',`
	`                'STRANGE', 'CHARM']`

However, it seems as though this whole loop could be done in one line. This is because there is only one meaningful expression where work is performed: namely `upper_quarks.append(quark.upper())`. Enter *comprehensions*.

Comprehensions are a syntax for spelling out simple for loops in a single expression. List, set, and dictionary comprehensions exist, depending on the type of container that the expression should return. Since they are simple, the main limitation is that the for block may only be a single expression itself. The syntax for these is as follows:

```
# List comprehension
[<expr> for <loop-var> in <iterable>]

# Set comprehension
{<expr> for <loop-var> in <iterable>}

# Dictionary comprehension
{<key-expr>: <value-expr> for <loop-var> in <iterable>}
```

Note that these comprehensions retain as much of the original container syntax as possible. The list uses square brackets ([]), the set uses curly braces ({}), and the dictionary uses curly braces {} with keys and values separated by a colon (:). The upper_quarks loop in the previous example can be thus transformed into the following single line:

```
upper_quarks = [quark.upper() for quark in quarks]
```

Sometimes you might want to use a set comprehension instead of a list comprehension. This situation arises when the result should have unique entries but the expression may return duplicated values. For example, if users are allowed to enter data that you know ahead of time is categorical, then you can *normalize* the data inside of a set comprehension to find all unique entries. Consider that users might be asked to enter quark names, and lowercasing the entries will produce a common spelling. The following set comprehension will produce a set of just {'top', 'charm', 'strange'}, even though there are multiple spellings of the same quarks:

```
entries = ['top', 'CHARm', 'Top', 'sTraNGe', 'strangE', 'top']
quarks = {quark.lower() for quark in entries}
```

It is also sometimes useful to write dictionary comprehensions. This often comes up when you want to execute an expression over some data but also need to retain a mapping from the input to the result. For instance, suppose that we want to create a dictionary that maps numbers in an entries list to the results of x**2 + 42. This can be done with:

```
entries = [1, 10, 12.5, 65, 88]
results = {x: x**2 + 42 for x in entries}
```

Comprehensions may optionally include a *filter*. This is a conditional that comes after the iterable. If the condition evaluates to True, then the loop expression is evaluated and added to the list, set, or dictionary normally. If the condition is False, then the iteration is skipped. The syntax uses the if keyword, as follows:

```
# List comprehension with filter
[<expr> for <loop-var> in <iterable> if <condition>]

# Set comprehension with filter
{<expr> for <loop-var> in <iterable> if <condition>}

# Dictionary comprehension with filter
{<key-expr>: <value-expr> for <loop-var> in <iterable> if <condition>}
```

Thus, list comprehensions with a filter are effectively shorthand for the following code pattern:

```
new_list = []
for <loop-var> in <iterable>:
    if <condition>:
        new_list.append(<expr>)
```

Suppose you had a list of words, pm, that represented the entire text of *Principia Mathematica* by Isaac Newton and you wanted to find all of the words, in order, that started with the letter t. This operation could be performed in one line with the following list comprehension with a filter:

```
t_words = [word for word in pm if word.startswith('t')]
```

Alternatively, take the case where you want to compute the set of squares of Fibonacci numbers, but only where the Fibonacci number is divisible by five. Given a list of Fibonacci numbers fib, the desired set is computable via this set comprehension:

```
{x**2 for x in fib if x%5 == 0}
```

Lastly, dictionary comprehensions with filters are most often used to retain or remove items from another dictionary. This is often used when there also exists a set of "good" or "bad" keys. Suppose you have a dictionary that maps coordinate axes to indexes. From this dictionary, you only want to retain the polar coordinates. The corresponding dictionary comprehension would be implemented as follows:

```
coords = {'x': 1, 'y': 2, 'z': 3, 'r': 1, 'theta': 2, 'phi': 3}
polar_keys = {'r', 'theta', 'phi'}
polar = {key: value for key, value in coords.items() if key in polar_keys}
```

Comprehensions are incredibly powerful and expressive. The reasoning goes that if the operation cannot fit into a comprehension, then it should probably be split up into multiple lines in a normal `for` loop anyway. It is possible to nest comprehensions inside of one another, just like loops may be nested. However, this can become pretty convoluted to think about since two or more loops are on the same line. Python allows for simple looping situations to be dealt with simply, and encourages complex loops to be made readable.

Flow Control and Logic Wrap-up

Having reached the end of this chapter, you should be familiar with the following big ideas:

- How to make decisions with `if-else` statements
- Handling the worst situations with exceptions
- Reusing code with loops
- The `for single in plural` loop pattern
- Using comprehensions to write concise loops

And now that you have seen the basics of decision making and code reuse, it is time to step those ideas up to the next level with functions in Chapter 5.

CHAPTER 5
Operating with Functions

Successful software development is all about code reuse. This is neatly summarized by the "don't repeat yourself" (DRY) principle. Code reuse is great not just because it involves less typing, but also because it reduces the opportunity for errors. This in turn makes debugging easier. When you're performing the same sequence of operations multiple times, it is best to encapsulate this code into a *function*.

In the majority of modern high-level languages, the function is the most basic form of code reuse. Functions are ubiquitous throughout programming languages, though some languages have more primitive forms of code reuse, too, such as *macros*, *jumps*, and *blocks*. Functions in code are based on the mathematical notion of functions. A function (e.g., f(x)) is a name or reference to a sequence of possibly parameterized operations. Just like mathematical functions play an essential role in pen-and-paper physics, functions as a programming construct are crucial to computational physics.

Once a function is defined, it may be *called* as many times as desired. Calling a function will execute all of the code inside of that function. What actions are performed by the function depends on its definition and on what values are *passed* into the function as its arguments. Functions may or may not *return* values as their last operation. The logic that makes up a function may be simple or complex and is entirely at the discretion of the programmer. As a concrete trigonometric example, take the sine function: `sin(0)` returns 0 and `sin(pi/2)` returns 1. Using arguments with software functions is much the same. Understanding and implementing any given function is the hard part!

Let's start with how to define functions in Python.

Functions in Python

Throughout history, there have evolved many ways to write down functions in mathematics. Programming languages have created a host of new ways to define functions in a much shorter amount of time. Each language seems to have its own unique take on the subject. In Python, the first line of a function is defined with the `def` keyword followed by the function name, the argument listing, and a colon (`:`). On the following lines, the operations that make up the function live in an indented block of code. This indentation is the same as with "Conditionals" on page 77 and the function definition ends when the block ends. A simple function that takes no arguments has the following form:

```
def <name>():
    <body>
```

Here, the empty parentheses indicate that the function is not parameterized in any way; that is, it takes no arguments when called. For example, the following is a function that creates a string and assigns it to a variable:

```
def sagan():
    quote = "With insufficient data it is easy to go wrong."
```

Since every function must have a `<body>`, the `pass` keyword is available for when a do-nothing function is truly desired. This is the Python no-op statement. Using the `pass` statement, we can create a `null()` function as follows:

```
def null():
    pass
```

Pretty simple, just like a no-op should be.

> Functions such as `null()` are very minimalistic. In general, you don't want your code to be this sparse. This is because for more complex situations it is hard for users or your future self to figure out what the function is meant to do. To mitigate issues of understanding functions, there exist a variety of conventions and best practices that you should adopt when writing software. We will introduce these as they come up over the rest of the chapter and the book. The minimal examples that you see here are so that you can learn how functions work and be able to understand the upcoming conventions.

Functions can also return values by using the `return` keyword followed by an expression to be returned. A function may have more than one `return` statement. However, the function will not execute any further operations after the first `return` it encounters. Because of this, it is generally advisable to only have one `return` statement per function. As a simple example, consider a function that simply returns 42 each time it is called:

Code	Output

`# define the function`
`def forty_two():`
`    return 42`
	`# Defining a function does not`
	`# return any output.`

`# call the function` `forty_two()`	
`# call the function, and print the result` `print(forty_two())`	`42`

`# call the function, assign the result`
`# to x, and print x`
`x = forty_two()`
`print(x)`
	`42`

As seen here, the same function may be called multiple times after it is defined. The return value may be ignored, used immediately, or assigned to another variable for later use. The `return` statement itself is always part of the function body. You call a function with the unary operator ( ), as seen in Table 2-1.

Functions, like their math counterparts, may take arguments. These are comma-separated variable names that may be used in the function body and *parameterize* the function. You call a function with arguments by separating the input values by commas as well. Functions may have as many arguments as required. The format for functions with arguments is as follows:

```
def <name>(<arg0>, <arg1>, ...):
    <body>
```

As an example, here is a function that either prints out 42 or 0, based on the value of a single argument x:

Functions in Python | 97

Code	Output
| ```
def forty_two_or_bust(x):
    if x:
        print(42)
    else:
        print(0)

# call the function
forty_two_or_bust(True)

bust = False
forty_two_or_bust(bust)
``` | ```
# Parameterizing functions is
# important to code reuse.



42


0
``` |

As a demonstration of using multiple arguments, here is a reimplementation of the power() function that takes both a base and an exponent. Note that the order in which the arguments are defined determines the order in which they need to be provided. For this reason, Python classifies these as *positional arguments*:

```
def power(base, x):
    return base**x
```

Functions may also call other functions from within their bodies. They would not be very useful if they could not! Consider again the mathematical expression sin(1/x). This is well defined everywhere except at x = 0. However, it is easy to show that at this point the expression converges to zero, even though the computer will fail to evaluate it. A common strategy in cases like these is to special-case the troublesome value. This may all be wrapped up into a single function:

```
from math import sin

def sin_inv_x(x):
    if x == 0.0:
        result = 0.0
    else:
        result = sin(1.0/x)
    return result
```

The last bit of basic function syntax is that all functions may have optional documentation embedded within them. Such documentation is written as the first unassigned string literal that occurs before any other operations in the function body. For this reason these literals are known as *docstrings*. It is standard practice for docstrings to be triple-quoted so that they may span multiple lines. Functions with docstrings have the following format:

```
def <name>(<args>):
    """<docstring>"""
    <body>
```

The docstring should be descriptive and concise. This is an incredibly handy way to convey the intended use of the function to users. The docstring itself is available at runtime via Python's built-in `help()` function and is displayed via IPython's ? magic command. A docstring could be added to the `power()` function as follows:

```
def power(base, x):
    """Computes base^x. Both base and x should be integers,
    floats, or another numeric type.
    """
    return base**x
```

Note that because Python is duck typed, as we saw in Chapter 3, the exact type of each argument is neither specified nor verified. This means that a user could pass a string and a list to the `power()` function, although it would not work. The flip side to this is that if someone were to come up with a new numeric type this function *would* work with it automatically, thereby making your code partially future-proof. With duck typing being as prevalent as it is in Python, it is always a good idea to give users a sense of what kinds of types the function arguments are intended for, even if it may work in other contexts; the docstring is useful for this purpose.

Functions provide a lot of flexibility and, when used correctly, a lot of efficiency. When properly leveraged, they form the basis of any good physics solver. Logic and data generation are implemented in the function bodies, while functions themselves are strung together to implement an algorithm. They are so powerful that the entirety of computational physics could be implemented just with the basic function syntax. However, doing so would not be pretty, which is why there are more complex programming structures (such as classes, as we will see in Chapter 6). Functions also have more sophisticated features that aid in their development and use, as we will see in the following sections of this chapter.

Keyword Arguments

Default values for arguments are a feature for when arguments should have a standard behavior. In Python these are known as *keyword arguments*, and they have three main advantages:

1. Keyword arguments are optionally supplied when the function is called, reducing what must be explicitly passed to the function.
2. When used by name they may be called in any order and are not restricted to the order in which they were defined.
3. They help define the kinds of values that may be passed into the function.

Keyword arguments are defined by the argument name (the key), an equals sign (=), and the default value that is used if the argument is not supplied when the function is called. All keyword arguments must come after all of the regular positional arguments. The format of a function with default arguments is as follows:

```
def <name>(<arg0>, <arg1>, ..., <kwarg0>=<val0>, <kwarg1>=<val1>, ...):
    """<docstring>"""
    <body>
```

As an example, consider the first-order polynomial ax + b. This could be implemented as a function with a and b having default values 1 and 0:

```
def line(x, a=1.0, b=0.0):
    return a*x + b
```

The line() function could then be called with neither a nor b, either a or b, or both. Since x is a positional argument, a value for it must always be given when the function is called. If a keyword argument is used out of the order of the original function, then the name of the argument *must* be given in the function call. Here are a few variations on how line() may be called:

```
line(42)                 # no keyword args, returns 1*42 + 0
line(42, 2)              # a=2, returns 84
line(42, b=10)           # b=10, returns 52
line(42, b=10, a=2)      # returns 94
line(42, a=2, b=10)      # also returns 94
```

Note that mutable data types such as lists, sets, and dictionaries should *never* be used as default values. This is because they retain their state from one call to the next. This is usually not intended and can lead to hard-to-find bugs. Consider the case of a custom append function that optionally takes a list as well as the value to append. By default, an empty list is provided:

```
# Do not do this!
def myappend(x, lyst=[]):
    lyst.append(x)
    print(lyst)
    return lyst
```

Since lyst is not copied in the function body, the list that is defined in the default argument is reused on each function call, even though it is also returned. This function will display the following behavior:

Code	Output
myappend(6) # seems right	[6]
myappend(42) # hmmm...	[6, 42]
myappend(12, [1, 16])	[1, 16, 12]
myappend(65) # nope, not right!	[6, 42, 65]

An ever-growing list is clearly not the right thing to have when the intention was to have a new empty list each time. A common pattern for having mutable containers as

100 | Chapter 5: Operating with Functions

default values is to have the default value be None in the function definition. Then this keyword argument is set to be the actual default value in the function body if and only if it is None when the function is called. Here's how we would rewrite myappend() with this pattern:

```
def myappend(x, lyst=None):
    if lyst is None:
        lyst = []
    lyst.append(x)
    print(lyst)
    return lyst
```

This will have the intended behavior of creating a new empty list each time the function is called if a list is not otherwise provided:

Code	Output
myappend(6)	[6]
myappend(42)	[42]
myappend(12, [1, 16])	[1, 16, 12]
myappend(65)	[65]

Variable Number of Arguments

Some functions may take a variable number of arguments. To see why this is useful, consider Python's built-in max() function. This function will take any number of arguments that are thrown at it and always return the highest value:

Code	Returns
max(6, 2)	6
max(6, 42)	42
max(1, 16, 12)	16
max(65, 42, 2, 8)	65

As you can see, this is useful for preventing excessive nesting of data structures just to make a function call. We did not need to put all of the values into a list first and then call max() on that list.

To write a function that takes a variable number of arguments, you must define the function with a single special argument that may have any name but is prefixed by an asterisk (*). This special argument must come after all other arguments, including keyword arguments. The format for such a function is thus:

```
def <name>(<arg0>, ..., <kwarg0>=<val0>, ..., *<args>):
    """<docstring>"""
    <body>
```

When the function is called, the <args> variable is a tuple into which all of the extra arguments are packed. For example, we could write our own version of a minimum() function:

```
def minimum(*args):
    """Takes any number of arguments!"""
    m = args[0]
    for x in args[1:]:
        if x < m:
            m = x
    return m
```

This could be called in the same way that the max() function was earlier. However, since args is a tuple we can also unpack an existing sequence into it when we call the function. This uses the same single asterisk (*) prefix notation, but during the function call rather than the definition. Here are a couple ways that minimum() could be called:

Code	Returns
minimum(6, 42)	6
data = [65, 42, 2, 8] minimum(*data)	2

This is a great feature to have because it allows users to build up their data before calling the function. Data preparation can be a clean, separate step from data analysis.

A variable number of unknown keyword arguments may also be supplied. This works similarly to supplying a variable number of positional arguments, but with two key differences. The first is that a double asterisk (**) is used to prefix the variable name. The second is that the keyword arguments are packed into a dictionary with string keys. The definition of this dictionary in the function signature must follow the *<args> definition and all other arguments. Therefore, the form of such a function is:

```
def <name>(<arg0>, ..., <kwarg0>=<val0>, ..., *<args>, **<kwargs>):
    """<docstring>"""
    <body>
```

Much like with tuples and args, you can pack existing dictionaries (or other mappings) into the kwargs variable using a double asterisk (**) prefix when the function is called. Hence, the most general function signature—one that takes any positional and any keyword arguments—is simply:

```
def blender(*args, **kwargs):
    """Will it?"""
    print(args, kwargs)
```

All keyword arguments must still come after all of the positional arguments in the function call. Here are some examples of calling `blender()` to show that `args` and `kwargs` are truly tuples and dictionaries:

Code	Returns
blender("yes", 42)	('yes', 42) {}
blender(z=6, x=42)	() {'x': 42, 'z': 6}
blender("no", [1], "yes", z=6, x=42)	('no', [1], 'yes') {'x': 42, 'z': 6}
t = ("no",) d = {"mom": "ionic"} blender("yes", kid="covalent", *t, **d)	('yes', 'no') {'mom': 'ionic', 'kid': 'covalent'}

Notice that the tuples and dictionaries that are unpacked in the function call are merged with the other arguments present in the call itself.

Multiple Return Values

In Python, as with many languages, only one object may be returned from a function. However, the packing and unpacking semantics of tuples allow you to mimic the behavior of multiple return values. That is, while the statement `return x, y, z` appears to return three variables, in truth a 3-tuple is created and that tuple is returned. Upon return, these tuples may either be unpacked inline or remain as tuples. Consider a function that takes a mass and a velocity and returns the momentum and energy. Such a function may be defined via:

```
def momentum_energy(m, v):
    p = m * v
    e = 0.5 * m * v**2
    return p, e
```

This may be called in either of the following two ways:

Code	Output

# returns a tuple 	
p_e = momentum_energy(42.0, 65.0)
print(p_e)
	(2730.0, 88725.0)

# unpacks the tuple mom, eng = momentum_energy(42.0, 65.0) print(mom)	2730.0

Having multiple return values is a useful feature for when a function computes two or more things simultaneously. A good example of this situation is a Newton's method root finder. A function implementing this method should return not only the solution but also the numerical error associated with that solution. Multiple return values let you obtain both the solution and the error with one function call and without jumping through hoops.

Scope

Function scope is key to understanding how functions work and how they enable code reuse. While the exact implementation of functions is language dependent, all functions share the notion that variables defined inside of a function have lifetimes that end when the function returns. This is known as *local scope*. When the function returns, all local variables "go out of scope," and their resources may be safely recovered. Both function arguments and variables created in the function body have local scope.

Variables defined outside the function have *global scope* with respect to the function at hand. The function may access and modify these variables, so long as their names are not overridden by local variables with the same names. Global scope is also sometimes called *module scope* because variables at this level are global only to the module (the *.py* file) where they live; they are not global to the entire Python process.

The following example illustrates function scoping graphically. In this example, the variables a, b, and c all share global scope. The x, y, and z variables are all local to func():

```
# global scope
a = 6
b = 42

def func(x, y):
    # local scope
    z = 16
    return a*x + b*y + z

# global scope
c = func(1, 5)
```

Functions are *first-class objects* in Python. This means that they have two important features:

1. They may be dynamically renamed, like any other object.
2. Function definitions may be nested inside of other function bodies.

Point 1 will be discussed in "Decorators" on page 112. Point 2 has important implications for scope. The rule is that inner functions share scope with the outer function, but the outer function does not share scope with the inner functions. Both the inner and outer functions have access to global variables. For example, consider the following:

```
# global scope
a = 6
b = 42

def outer(m, n):
    # outer's scope
    p = 10

    def inner(x, y):
        # inner's scope
        return a*p*x + b*n*y

    # outer's scope
    return inner(m+1, n+1)

# global scope
c = outer(1, 5)
```

Functions may be nested indefinitely, building up a hierarchy of scope. With each nesting, the inner functions maintain the scope of the outer functions, while the outer functions cannot peek into the scopes of the inner functions. Again, we'll see decorators later in this chapter that provide an important use case for nested functions.

Now suppose that a function assigns a variable to a name that already exists in the global scope. The global value is overridden for the remainder of the function call. This means that the global variable remains unchanged, though locally the new value is used when the variable name is accessed. For example, say that there is a variable a at both global and local scope:

Code	Output
| ```
a = 6

def a_global():
    print(a)

def a_local():
    a = 42
    print(a)

a_global()
a_local()
print(a)
``` | ```
# set the global value of 'a'




6
42
6
``` |

This shows that when a_local() assigns a value to a, it creates its own local version of a and the global variable named a remains unperturbed. In fact, whenever a function defines a variable name, that variable becomes local. This is true even if the variable has yet to be assigned and there exists a global variable by the same name. The following code will result in an unbound local error when func() is called. This error means that a was used before it existed because a is reserved as local:

Code	Output
| ```
a = "A"

def func():
    # you cannot use the global 'a' because...
    print("Big " + a)
    # a local 'a' is eventually defined!
    a = "a"
    print("small " + a)

func()
``` | ```
Traceback (most recent call last):
  File "<file>", line 8, in <module>
    func()
  File "<file>", line 4, in func
    print("Big " + a)
UnboundLocalError: local variable 'a' referenced before assignment
``` |

Python provides a global keyword that lets functions promote local variables to the global scope. This can help fix issues like the unbound local error just shown. However, it also means that the variable is truly at the global scope. The global keyword is typically used at the top of the function body and is followed by a comma-separated listing of global variable names. There may be as many global statements in a function as desired. However, the global statement for a variable must precede any other uses of that variable name. This is why putting these statements at the top of the function is the best practice. It is also generally considered a best practice to not use the global keyword at all and instead pick different variable names. Still, there are a few rare situations where it is unavoidable.

Modifying our func() example to use the global keyword allows the function to be called but also changes the value of the global a:

Code	Output
| ```
a = "A"

def func():
    global a
    print("Big " + a)
    a = "a"
    print("small " + a)

func()
print("global " + a)
``` | ```
Big A
small a
global a
``` |

Scoping rules can sometimes be a little tricky to understand when you're approaching a new language. However, after a bit of use they become simple and painless. Scope is also needed for the great and wonderful feature of recursion, which we'll look at next.

Recursion

A function name is part of the surrounding scope. Therefore, a function has access to its own name from within its own function body. This means that a function may call itself. This is known as *recursion*. The simplest example of a recursive function is:

```
# DO NOT RUN THIS
def func():
    func()
```

The classic recursive function is one that produces the nth value in the Fibonacci sequence:

```
def fib(n):
    if n == 0 or n == 1:
        return n
    else:
        return fib(n - 1) + fib(n - 2)
```

Here, for all cases where n > 1, the fib() function is called for n - 1 and n - 2. However, zero and one are fiducial cases for which further calls to fib() do not occur. This recursion terminating property makes zero and one *fixed points* of the Fibonacci function. More mathematically, fixed points are defined such that x is a fixed point of f if and only if x == f(x).

Fixed points are an important part of recursive functions because without them these functions will recurse and execute forever. It is very easy to get the fixed points of a function wrong, which leads to fairly painful (but obvious) bugs. In practice, Python has a maximum recursion depth (this defaults to 1,000) such that if a function calls itself this many times Python will raise an exception. This is a helpful feature of Python that not all languages share. To get and set the recursion limit, use the appropriate functions from the standard library sys module:

```
import sys
sys.getrecursionlimit()         # return the current limit
sys.setrecursionlimit(8128)     # change the limit to 8128
```

Recursion can be used to implement very complex algorithms with very little effort. As was seen with fib(), any time a mathematical recursion relation is available, recursive functions are a natural fit. However, recursion also has a special place in computing itself. Most languages make heavy use of recursion in their implementations. This has to do with how various language features, such as type systems, work. Unfortunately, the details of language design and implementation are outside the scope of this book. To read more about recursion, please see "Recursion" on page 107.[1]

---

[1] Get it?!

Now that you have seen functions in their full generality, we can discuss a few specific types of functions: lambdas, generators, and decorators. Each of these can make your life easier in its own way by manipulating the properties of being a function.

Lambdas

Lambdas are a special way of creating small, single-line functions. They are sometimes called *anonymous functions* because they are defined in such a way as to not have explicit names. Unlike normal functions, lambdas are expressions rather than statements. This allows them to be defined on the righthand side of an equals sign, inside of a literal list or dictionary, in a function call or definition, or in any other place that a Python expression may exist.

Lambdas have a couple of important restrictions that go along with their flexibility. The first is that lambdas must compute only a single expression. Because statements are not allowed, they cannot assign local variables. The second restriction is that the evaluation of this expression is *always* returned.

You define a lambda by using the *lambda* keyword followed by the function arguments, a colon (:), and then the expression that makes up the entirety of the function body. The format of the lambda is thus:

```
lambda <args>: <expr>
```

The argument syntax for lambdas follows exactly the same rules as for normal functions. The expression may be as simple or as complex as desired, as long as it is a single expression. The following demonstrates common examples of lambdas in action. Try running these yourself in IPython to see what they return:

```
# a simple lambda
lambda x: x**2

# a lambda that is called after it is defined
(lambda x, y=10: 2*x + y)(42)

# just because it is anonymous doesn't mean we can't give it a name!
f = lambda: [x**2 for x in range(10)]
f()

# a lambda as a dict value
d = {'null': lambda *args, **kwargs: None}

# a lambda as a keyword argument f in another function
def func(vals, f=lambda x: sum(x)/len(x)):
    f(vals)

# a lambda as a keyword argument in a function call
func([6, 28, 496, 8128], lambda data: sum([x**2 for x in data]))
```

One of the most common use cases for lambdas is when sorting a list (or another container). The Python built-in `sorted()` function will sort a list based on the values of elements of the list. However, you can optionally pass in a *key function* that is applied to each element of the list. The sorting then occurs on the return value of the key function. For example, if we wanted to sort integers based on modulo-13, we could write the anonymous function `lambda x: x%13`. The following code sorts a list of perfect numbers with and without this key function:

Code	Output
`nums = [8128, 6, 496, 28]`	`[8128, 6, 496, 28]`
`sorted(nums)`	`[6, 28, 496, 8128]`
`sorted(nums, key=lambda x: x%13)`	`[496, 28, 8128, 6]`

Historically, lambdas come from the *lambda calculus*, which helps form the mathematical basis for computation. Their importance cannot be overstated. This topic has spawned its own language paradigm called *functional programming*. Unlike in object-oriented languages, where everything is an object, in a functional language everything is a function. Functional languages have been gaining in popularity recently, and well-established examples include Lisp, Haskell, and OCaml. For more information about the lambda calculus, please see Henk P. Barendregt and Erik Barendsen's article "Introduction to Lambda Calculus."

Lambdas may seem like a simple bit of unnecessary syntax, but as is true of many language constructs, they have a subtle beauty when you start using them. This is also the case with generators, discussed next.

Generators

When a function returns, all execution of further code in the function body ceases. *Generators* answer the question, "What if functions paused, to be unpaused later, rather than stopping completely?" A generator is a special type of function that uses the `yield` keyword in the function body to return a value and defer execution until further notice.

When a function that has a `yield` statement is called, rather than returning its return value—it does not have one—it returns a special generator object that is bound to the original function. You can obtain the values of successive `yield` statements by calling Python's built-in `next()` function on the generator. As you would expect, using `yield` statements is mutually exclusive with using `return` statements inside of a function body.

Generators are very important for representing custom, complex data. In particular, they are needed for efficient custom container types that are variants of lists, dictionaries, and sets. We'll touch on this use case more in Chapter 6. For now, though, consider a simple countdown generator function:

```
def countdown():
    yield 3
    yield 2
    yield 1
    yield 'Blast off!'
```

Calling this will return a generator object, and calling next() on this object will tease out the yielded values:

Code	Output

# generator 	
g = countdown()
 	
next(g)
x = next(g)
print(x)
y, z = next(g), next(g)
print(z)
next(g)
	# yielded values
	
	2
	
	"Blast off!"
	<StopIteration Error>

As you can see, generators are only valid for as many yield statements as they execute. When there are no more yield statements, the generator raises a StopIteration error. This belies that generators are iterable and may be the antecedents of for loops. In fact, using generators in for loops is much more common than using the next() function. The countdown() generator would more commonly be used as follows:

Code	Output

for t in countdown():
    if isinstance(t, int):
        message = "T-" + str(t)
    else:
        message = t
    print(message)
	T-3
	T-2
	T-1
	Blast off!

This demonstrates that any amount of work may take place between successive yield calls. This is true both where the generator is defined and where it is called. As a more complex example, take the case where you wish to return the square plus one of all numbers from zero to n. The generator function for this could be written as:

```
def square_plus_one(n):
    for x in range(n):
```

```
x2 = x * x
yield x2 + 1
```

Using such a generator is just as simple as placing the generator in a `for` loop:

Code	Output
`for sp1 in square_plus_one(3):` `    print(sp1)`	1 2 5

Note that in Python v3.3 and later, generators were extended with the `yield from` semantics. This allows a generator to delegate to subgenerators. This makes `yield from` statements shorthand for using multiple generators in a row. As a relatively simple example of `yield from` usage, we can create a palindrome by yielding each element of a sequence in its forward direction and then yielding each element in the backward direction. A use case for this kind of functionality would be a symmetric matrix where only half of the elements are stored, but you want to iterate through all elements as if they actually existed. The palindrome generator may be written as follows:

```
# define a subgenerator
def yield_all(x):  ❶
    for i in x:
        yield i

# palindrome using yield froms
def palindromize(x):  ❷
    yield from yield_all(x)  ❸
    yield from yield_all(x[::-1])  ❹

# the above is equivalent to this full expansion:
def palindromize_explicit(x):  ❺
    for i in x:  ❻
        yield i
    for i in x[::-1]:  ❼
        yield i
```

❶ This subgenerator yields all of the values, in order, from a list or other iterable.

❷ The generator will yield every element from a list and then reverse the list and yield each reversed element, thereby generating a palindrome.

❸ The `yield from` to generate the forward direction.

❹ The `yield from` to generate the backward direction.

❺ An alternative palindrome generator that does not use `yield from`s must explicitly yield each element itself.

❻ Loop and yield in the forward direction.

❼ Loop and yield in the backward direction.

The `yield from` expression also enables communication between generators, but such communication is too advanced to cover here. For more information, please refer to the Python documentation.

Generators and lambdas both introduce new pieces of Python syntax that enable functions to be more expressive in more situations. Up next are decorators, which again add a small piece of syntax to great effect.

Decorators

A *decorator* is a special flavor of function that takes only one argument, which is itself another function. Decorators may return any value but are most useful when they return a function. Defining a decorator uses no special syntax other than the single-argument restriction. Decorators are useful for modifying the behavior of other functions without actually changing the source code of the other functions. This means that they provide a safe way of changing other people's software. This makes decorators especially useful in analysis libraries and toolkits. For instance, NumPy (see Chapter 9) has a decorator called `vectorize()` that you may occasionally find useful when the time comes. Here are a few primitive decorators that are of questionable usefulness but are good for demonstration purposes:

```
def null(f):
    """Always return None."""
    return

def identity(f):
    """Return the function."""
    return f

def self_referential(f):
    """Return the decorator."""
    return self_referential
```

Python uses the at sign (@) as a special syntax for applying a decorator to a function definition. On the line above the function definition, you place an @ followed by the decorator name. This is equivalent to:

1. Defining the function normally with the `def` keyword
2. Calling the decorator on the function

3. Assigning the original function's name to the return value of the decorator

For example, here we define a function `nargs()` that counts the number of arguments. In addition to its definition, it is decorated by our `null()` decorator:

```
@null   ❶
def nargs(*args, **kwargs):   ❷
    return len(args) + len(kwargs)
```

❶ Decorate the `nargs()` function with `null()`.

❷ Regular definition of `nargs()` function.

This performs the same operations as the following snippet, but with less repetition of the function name:

```
def nargs(*args, **kwargs):   ❶
    return len(args) + len(kwargs)
nargs = null(nargs)   ❷
```

❶ Regular definition of `nargs()` function.

❷ Manual decoration of `nargs()` with `null()` by passing `nargs` into `null()` and overwriting the variable name `nargs` with the return value of `null()`.

Decoration is only possible because functions are first-class objects in Python. This is what lets us pass functions as arguments to other functions (as in the preceding example, where `nargs()` is passed into `null()`). Functions being first-class objects is also what allows the original function names to be overwritten and reused. All of these pieces make functions very dynamic: you can modify, rename, and delete them well after their creation.

In "Scope" on page 104, we discussed how function definitions can be nested inside of other functions. This is important to decorators that wish to modify the arguments or return values of the functions they are decorating. To do so, the decorator must create its own new wrapper function and then return the wrapper. The wrapper typically—though not necessarily—calls the original function. Even though the original function is replaced where it is defined, this works because the scoping rules make the original function a local variable to the decorator. Consider a decorator that adds one to the return value of a function:

```
def plus1(f):   ❶
    def wrapper(*args, **kwargs):   ❷
        return f(*args, **kwargs) + 1   ❸
    return wrapper   ❹
```

❶ The decorator takes one argument: another function, `f()`.

❷ Nested inside of the decorator, we create a wrapper() function that accepts any and all arguments and keyword arguments.

❸ The wrapper() function calls the function f() that was passed into the decorator with all arguments and keyword arguments. It then adds one to the result, and returns it.

❹ The decorator itself returns the wrapper() function.

Typically, the signature (args and kwargs, in this example) of the original function is not known. In order to ensure that a decorator is useful in as many places as possible it is a good practice to always use (*args, **kwargs) as the arguments to the wrapper function, because it is the one-size-fits-all signature. Suppose we wanted to write a power() function and add one to the result. We could apply the plus1() decorator we just created to the power() definition:

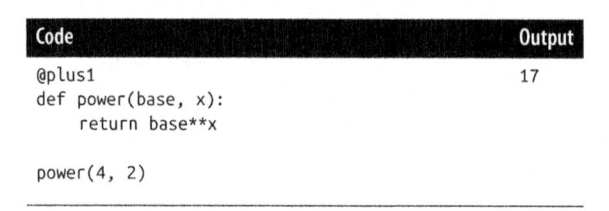

Code	Output
| ```
@plus1
def power(base, x):
    return base**x

power(4, 2)
``` | 17 |

You can chain decorators together by stacking them on top of each other. For chaining to really work at runtime, it assumes that each decorator returns a wrapper function of its own. Here's an example using decorators that have been defined here on a newly defined square root() function:

Code	Output
| ```
@plus1
@identity
@plus1
@plus1
def root(x):
    return x**0.5

root(4)
``` | 5.0 |

Decorators only being able to accept one argument can feel fairly restrictive. With the plus1() decorator, the behavior of adding one to the returned value was hardcoded. Adding two to the value instead of one would require a separate decorator, as there is no mechanism to parameterize the decorator itself to add n instead of one at decoration time. However, you can accomplish this flexibility by nesting the decorator definition *inside of another function*. When the outermost function is called, it should return the decorator. The decorator in turn returns the wrapper function. The outermost function is sometimes called a *decorator factory*, *decorator generator* (no relation to "Generators" on page 109), or some other term that indicates that it creates a decorator. The decorator factory is not itself a decorator, even though it will be used in much the same way. Decorator factories may accept as many arguments and keyword arguments as you wish. The only real restriction on decorator factories is that they actually return a decorator. A plus_n() decoration function, parameterized by the argument n, may be defined as follows:

```
def plus_n(n):    ❶
    def dec(f):    ❷
        def wrapper(*args, **kwargs):    ❸
            return f(*args, **kwargs) + n    ❹
        return wrapper    ❺
    return dec    ❻
```

❶ The decorator factory that takes the number n to add.

❷ The decorator dec() must adhere to the same rules as other decorators and only accepts a single function argument, f().

❸ The decorator dec() creates a wrapper function.

❹ The wrapper function calls f(), adds n, and returns the value.

❺ The decorator dec() still returns the wrapper function.

❻ The decorator factory returns the decorator dec().

Again, this process works because of function scoping rules and because functions are first-class objects and may be manipulated as you wish. For example, the following defines the root() function, here decorated by @plus_n(6). We could not decorate by just @plus_n because plus_n is a not a decorator, it is a decorator factory. We must call plus_n() with a valid argument (6) in order to obtain the actual decorator:

Code	Output

`@plus_n(6)`
`def root(x):`
`    return x**0.5`
 	
`root(4)`
	8.0

This may be used seamlessly with other decorators. Further nesting for decoration is redundant and is not required. Three levels—decorator factory, decorator, and wrapper—are the most you will ever need.

If you provide a decorator as part of one of your modules or packages, other people can use this decorator to modify the behavior of their functions. Your users do not have to know how the decorator is implemented, and you do not have to know how the decorator will be applied. This strategy is used frequently in some packages. Some major web frameworks, such as Flask, make extensive use of decorators to indicate that a function actually returns the body of a web page. In scientific computing, decorators can be used to automatically validate the input of functions and ensure that the arguments are on a physically valid range.

You can also use decorators to modify other people's functions without them even knowing. In this case, you cannot use the @ symbol syntax because the function has already been defined; instead, you need to call the decorator like it was a normal function. For example, if we wanted to always add one to the return value of Python's built-in max() function, we could use our plus1() decorator manually as follows:

```
max = plus1(max)
```

Here, max() is the argument to plus1(), and the returned wrapper function overwrites the name max locally. We highly recommend that you look into the Python standard library functools module for a few invaluable decorators.

Function Wrap-up

This chapter has covered the basic, intermediate, and even some advanced features of functions in Python. You should now understand:

- How functions enable code reuse
- The difference between positional and keyword arguments
- How to have a variable number of arguments and multiple return values
- That scoping rules determine what variable names are visible in a function body
- That recursion is possible because a function body can see the name of the function itself
- That lambdas provide anonymous function expressions
- That generators make looping more customizable
- That decorators allow you to modify other people's functions without changing their source code

Functions form the basis of computation, and their usefulness to physics is apparent through comparison to mathematical functions. Python's take on functions, with all of the bells and whistles, makes programming even easier. Still, functions are not the only big idea in how to represent and manipulate data. At last, in Chapter 6, we will be introduced to classes and object-oriented programming.

CHAPTER 6
Classes and Objects

When physicists create a mathematical model of a physical process, they rely on the mathematical framework that can represent that process as closely as possible. When Newton developed a model of forces and motion, the appropriate mathematical framework was calculus. When Einstein developed a model of wave-particle motion, he relied on the mathematics of wave equations and eigenvalues. For many models in scientific computation, the computational framework that best aligns with our need is object orientation.

The universe presents itself as a collection of objects that humans (in particular, scientists) tend to classify based on their attributes and behaviors. Similarities, relationships, and hierarchies among these objects further help to structure our perception of the world. In alignment with that conceptual model, object orientation facilitates representation of the world as classes of objects that possess attributes, behaviors, and hierarchical relationships. Classes in object orientation organize data, methods, and functions. Those classes manifest themselves as specific objects. We will discuss these two concepts in great detail here.

This chapter will describe how object orientation allows the scientist to cleanly organize behaviors and data. It will also mention the many objects this book has used already and how they operate as particular instances of distinct classes. To demonstrate the use of classes for modeling physics, this chapter will implement some of the classes needed to simulate the taxonomy of particles in the Standard Model. Along the way, we will address many notions underlying object orientation, such as attributes, methods, and interfaces. But first, let's delve slightly deeper into what object orientation is.

Object Orientation

Object orientation, emphasizing descriptive classification of data and behaviors, will feel familiar to the scientist, as it captures the basic concepts behind reductionism, a fundamental driving philosophy underpinning science in general. In this sense, object orientation is a computational version of the reductionist frameworks that scientists have relied on for centuries to formulate conceptual models of physical systems.

Object orientation models systems in the same way that scientists have always approached complex systems: by breaking them into their fundamental parts. In this way, object orientation reduces the scientist's cognitive load. As scientists increasingly write large, complex models and simulations, the need for object orientation increases. In the same way that the human brain is not effective at comprehending more than approximately a paragraph of text at a time, it also is not effective at comprehending enormous code blocks or endless variable lists. To solve this, from a simulation perspective, object orientation provides a framework for classifying distinct concepts into comprehensible sizes. These smaller conceptual units facilitate cleaner, more scalable modeling.

> **Main Ideas in Object Orientation**
>
> Very helpfully, the website Software Carpentry (*http://software-carpentry.org*) breaks down object orientation into five main ideas:
>
> - Classes and objects combine functions with data to make both easier to manage.
> - A class defines the behaviors of a new kind of thing, while an object is a particular thing.
> - Classes have constructors that describe how to create a new object of a particular kind.
> - An interface describes what an object can do; an implementation defines how.
> - One class can inherit from another and override just those things that it wants to change.

Classically, object orientation is described by the following three features:

- *Encapsulation* is the property of owning data, which we will discuss in "Classes" on page 123.
- *Inheritance* establishes a relationship hierarchy between models, which we will discuss in "Superclasses" on page 137.

- *Polymorphism* allows for models to customize their own behavior even when they are based on other models, which we will discuss in "Polymorphism" on page 135.

That said, the fundamental notion of object orientation is that data, methods, and functions are best organized into classes. Furthermore, classes in a simulation should be able to manifest themselves as specific objects. Let's learn more about those.

Objects

At this point in the book, you have encountered many objects already, because *everything in Python is an object*. That includes all variables, even simple integer variables. After all, integers have attributes and methods—and anything with both of those is certainly an object. All objects in Python have both attributes and methods.

To see this in action, open up a Python interpreter (by typing python or ipython on the command line), create some simple variables, and request their docstrings with the help() function. In the following example, we investigate the number 1. We already know 1 is an integer. With the help() function, we can learn more about integers:

Code	Output
a = 1 help(a)	Help on int object: ❶ class int(object) ❷ | int(x=0) -> int or long | int(x, base=10) -> int or long | | Convert a number or string to an integer, | or return 0 if no arguments | ...

❶ The help() function clearly states that this integer is an object.

❷ Aha. Integers are *objects* of the int *class*.

The results of help(a) in this example are a pretty clear indication that the integer is an object. According to the rules mentioned earlier in the chapter, that must mean it has data and behaviors associated with it. In Python, the dir() function lists all of the attributes and methods associated with the argument that is passed into it. We can therefore use the dir() function to see a list of the data and behaviors associated with the int class:

Code	Output
a = 1 dir(a)	['__abs__', '__add__', '__and__', . . . 'bit_length', 'conjugate', 'denominator', 'imag', 'numerator', 'real']

Indeed, the `dir()` function lists attributes possessed by 1, an instance of the integer (`int`) class. These attributes can be requested from any integer; any integer has an absolute value (`__abs__`), can be added to another (`__add__`), can have its real (`real`) and imaginary (`imag`) components teased out, etc.

> ### What Are the Underscores For?
>
> The first entries that appear when `dir()` is called are usually attributes named with two leading and two trailing underscores. This is a meaningful naming convention in Python. According to the PEP8 Style Guide (*http://bit.ly/pep8-style/*), this naming convention is used for "magic objects or attributes that live in user-controlled namespaces. E.g. __init__, __import__ or __file__. Never invent such names; only use them as documented." In the Python parlance, these are called "dunder," which stands for the mouthful that is "double underscore."

Here, it is important to note that the data attributes of *this* integer (a=1) take specific values according to the value of a. Its specific absolute value is 1. A different integer object instance, however, may have a different value for __abs__:

Code	Output
a = 1 a.__abs__() ❶ b = -2 b.__abs__() ❷	1 2

❶ The absolute value method is called on the integer a.

❷ The absolute value method is called on the integer b.

So, the data types we have seen in previous chapters are really objects under the hood. The same is true for all data types and data structures in Python.

However, only in *exceedingly rare* circumstances should you ever call a dunder method directly, like in the preceding examples. Instead, you should always call the built-in function abs(). This implicitly goes to the underlying __abs__() method and performs some additional safety and sanity checks. Calling a.__abs__() was done solely for demonstration purposes to show how the Python magic works. The correct version of the previous two examples is:

Code	Output
a = 1 abs(a)	1
b = -2 abs(b)	2

You should always try to rely on the built-in Python functions or operators rather than the dunder methods. For a detailed description of which Python dunder methods map to which built-in functions and operators, please see the Python data model (*http://bit.ly/py-dm*) documentation.

> **Exercise: Explore the Native Objects of Python**
>
> 1. Open IPython.
> 2. Create some variables of the types introduced in Chapter 2.
> 3. Use the dir() and help() functions to learn more about the attributes associated with them. Do you understand why some data structures have certain methods that others are missing? Are there any surprises?

Given that all the data structures that we have seen are objects in Python, what about functions? Are they objects too? If we execute the dir() function on a function instead, the response clearly shows that functions possess attributes as well. In Python, *even functions are objects*.

Code	Output
import math ❶ dir(math.sin) ❷	['__call__', '__class__', '__cmp__', '__delattr__', '__doc__', . . . '__self__', '__setattr__', '__sizeof__', '__str__', '__subclasshook__']

❶ Import the math module.

❷ Request the attribute listing for the sin function.

Looking over this list, you may notice that the sin() function has a __doc__ (or docstring) attribute. This is how Python stores the docstrings we have been using to learn about functions, as you can see in the following example:

Code	Output
import math math.sin.__doc__	'sin(x)\n\nReturn the sine of x (measured in radians).'

Docstrings, then, are not magic. They are simply attributes of the built-in function class, stored by and explicitly associated with the function objects that they describe!

The results of dir() confirm a few things about the sin() function:

- Many attributes can be accessed with the dot (.) syntax.
- Some, like the __doc__ attribute, are very specific to the sin() function.
- A __self__ attribute is available for implicit passing into methods.

These things together indicate that the sin() function is explicitly associated with attributes and methods. *It must be an object.*

Exercise: Turtles All the Way Down

1. Open IPython.
2. Import the math module.
3. Use the dir() function to determine whether the docstring of the sin() function is an object. (Hint: use dir(math.sin.__doc__).)

Everything in Python truly is an object, functions included. Despite being simple, built-in Python objects such as integers, lists, dictionaries, functions, and modules are fully fledged, first-class objects. In particular, they encapsulate data and behaviors within their attributes and methods. This section has shown examples of objects and demonstrated how all entities in Python are objects. Earlier in the chapter, however, we defined objects as particular instances of classes. So, what is a class, exactly?

Classes

Classes define logical collections of attributes describing a kind of object. They also define how to create a particular object of that kind. Additionally, to capture the hierarchical nature of types, subtypes, and supertypes of objects in a system, classes can inherit from one another. This section will describe all of these features of classes by exploring the way a physicist might use classes to abstract away implementation details of objects in a particle physics simulation.

First, the physicist must decide what classes to create. Classes should be chosen to ensure that the internal data and functions related to different types of objects are separated (*encapsulated*) from one another.

In particle physics, for example, particles are an obvious choice for our objects. The first class we will need to create is, therefore, the Particle class. We can begin to describe the notion of particles in a simulation by defining the Particle class. A class definition begins with the class keyword.

> The way a class definition begins with the class keyword is analogous to the way function definitions use the def keyword.

Our Particle class definition will take this form:

```
class Particle(object):  ❶
    """A particle is a constituent unit of the universe."""  ❷
    # class body definition here  ❸
```

❶ Begin the class definition and give it a name, Particle.

❷ A docstring documents the class, just like the function docstrings we met in Chapter 5.

❸ The remainder of the class definition lives in a whitespace-indented block, just like a function body.

We will explain in more detail later why the word `object` appears. For now, just trust that this helps distinguish this as a class that defines a certain type of object.

A well-formed class can include many types of attributes:

Class variables
    Data associated with the class itself.

Constructors
    Special methods that initialize an object that is an instance of the class. Inside of the constructor, instance variables and data that is associated with a specific object may be assigned.

Methods
    Special functions bound to a specific object that is an instance of the class.

The following sections will address these features, beginning with class variables.

Class Variables

Class variables should be thought of as *data universally applicable to all objects of the class*. These class-level attributes are declared in the class definition. They are held in the scope of the class. This means that they can be accessed even without reference to a specific instance (object) of that class. However, it is also true that each object has access to every class-level variable.

Particles have a lot of interesting features. Those that are true for every particle in the universe should be included as class-level variables. The particles in Figure 6-1 do not have a lot in common. However, every particle in the universe should be able to say, "I am a particle!"

To begin the definition of the `Particle` class, then, we create a class-level variable:

```
# particle.py
class Particle(object):
    """A particle is a constituent unit of the universe."""
    roar = "I am a particle!"  ❶
```

❶ A class-level attribute, `roar`, is set equal to a string.

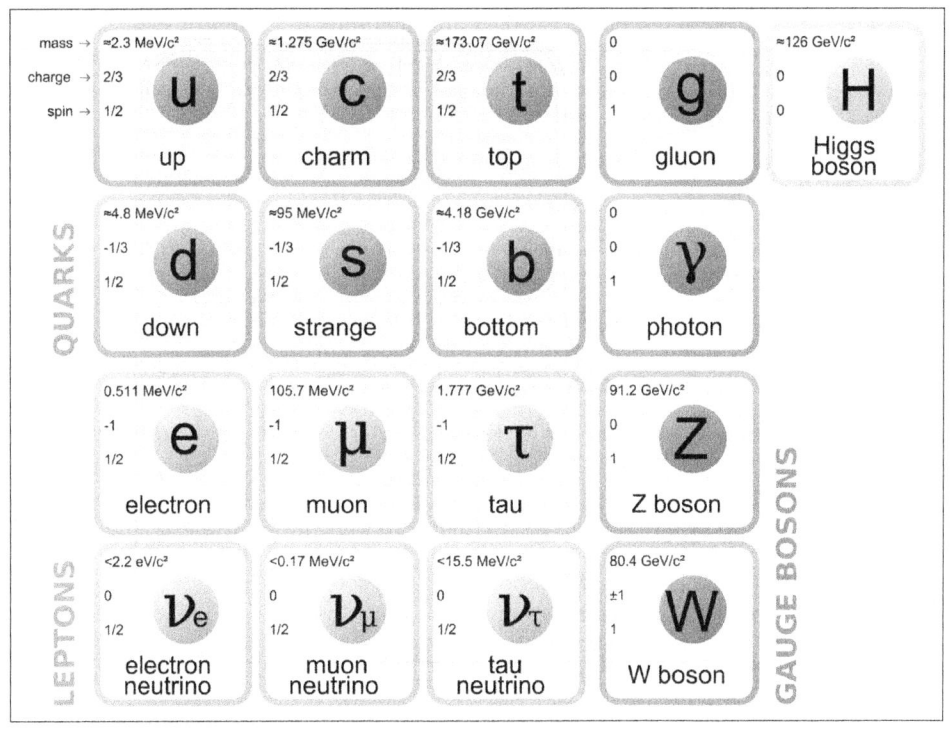

Figure 6-1. The Standard Model of Elementary Particles (source: Wikipedia (http://bit.ly/1xPc9w3))

This example makes the `roar` string an attribute that is accessible across all `Particle` objects. To access this variable, it is not necessary to create a concrete instance of the class. Rather, you are able to obtain `roar` directly from the class definition:

Code	Output

`# import the particle module`
`import particle as p`
`print(p.Particle.roar)`
	`'I am a particle!'`

This class variable, `p.roar`, can also be accessed by any object instance of the class, as seen in the following example. For now, to create a `Particle` instance, call the class definition like you would call a function with no arguments (i.e., `Particle()`):

Code	Output

`# import the particle module`
`import particle as p`
`higgs = p.Particle()`
`print(higgs.roar)`
	`'I am a particle!'`

Class-level attributes are excellent for data and methods that are universal across all instances of a class. However, some attributes are unique to each object and should not be changed by other objects, even those of the same class. Such attributes are called *instance variables*.

Instance Variables

Every particle in the universe has a physical position, r, in the coordinate system. Thus, position should certainly be an attribute of the Particle class. However, each particle must have a *different* physical position at any particular time (see also the "identity of indiscernibles" principle). So, an attribute storing the position data should be bound specifically to each individual particle.

havior is not unlike the example earlier in the chapter concerning the different absolute values of different integers. That is, if the class is defined properly, it should be possible to set the position variable uniquely for each particle. Using the class to create a list of observed Particle objects might, in that case, be achieved as shown here:

```
# import the Particle class from the particle module
from particle import Particle as p

# create an empty list to hold observed particle data
obs = []

# append the first particle
obs.append(p.Particle())

# assign its position
obs[0].r = {'x': 100.0, 'y': 38.0, 'z': -42.0}

# append the second particle
obs.append(p.Particle())

# assign the position of the second particle
obs[1].r = {'x': 0.01, 'y': 99.0, 'z': 32.0}

# print the positions of each particle
print(obs[0].r)
print(obs[1].r)
```

This code outputs:

```
{'y': 38.0, 'x': 100.0, 'z': -42.0}
{'y': 99.0, 'x': 0.01, 'z': 32.0}
```

This behavior is exactly what can be accomplished with instance variables. While the task of describing the data and behaviors of particles may seem daunting at first, if we start with the basics, it will soon become clear how object orientation simplifies the cognitive task. For starters, all particles have position, mass, charge, and spin—much of the rest can be derived.

Thus, given these instance variables, not much more complexity is necessary to store all the data associated with a particle observation. This example shows how, in an object-oriented model, the individual data associated with multiple observations can be kept impeccably organized in the instance variables of many objects.

The value of this reduced complexity using instance variables should be obvious, but how is it accomplished in the class definition? To associate data attributes with a specific instance of the class in Python, we use the special __init__() function, the *constructor*. Implementation of constructors is addressed in the following section.

Constructors

A constructor is a function that is executed upon instantiation of an object. That is, when you set higgs = p.Particle(), an object of the Particle type is created and the __init__() method is called to initialize that object.

> In Python, the constructor is always named __init__(), because it sets the *initial* state of the object.

The constructor is one of the methods defined inside of the class definition. A user-written constructor is not required to exist for a class definition to be complete. This is because every class automatically has a *default constructor*. Furthermore, if the __init__() method does exist, it needs only to perform constructor actions specific to defining objects of this class. However, because it is *always* run when an object is created, best practice is to make this function responsible for initializing all of the instance variables of the object. That way, every time an object is created, it is guaranteed to be fully initialized.

> Introducing instance variables outside of an __init__() function is somewhat risky, because there is no guarantee that they'll be initialized by the time you need to use them. Try to avoid it.

As an example, a Particle() constructor can be introduced that defines and initializes a few instance variables that should be specific to particular particles:

```
# particle.py
class Particle(object):
    """A particle is a constituent unit of the universe.

    Attributes
    ---------
```

```
        c : charge in units of [e]
        m : mass in units of [kg]
        r : position in units of [meters]
        """

        roar = "I am a particle!"

        def __init__(self):  ❶
            """Initializes the particle with default values for
            charge c, mass m, and position r.
            """
            self.c = 0  ❷
            self.m = 0
            self.r = {'x': 0, 'y': 0, 'z': 0}
```

❶ The self argument is required since this function is a method. More details will be discussed in "Methods" on page 129. This parameter is used so the method is bound to a specific instance of the class.

❷ The instance attribute c is introduced (and assigned to self) with an initial value of 0.

The instance variables c, m, and r introduced in the __init__() method are assigned to the current object, called self, using the syntax self.<var> = <val>.

> Note how the self parameter is passed to the __init__() method. This argument represents the instance of the class. The function becomes a method by being part of the class definition. All methods are required to accept at least one argument, and the first argument is the instance of the class. By a *very strong* convention, this first argument is named self. However, since this is only a convention, nothing prevents you from using me, this, x, or any other variable name other than social pressure.

In the previous example, to set actual values for the instance variables we would have to assign them outside of the constructor, just as we did with the positions in "Instance Variables" on page 126. That's a bit inefficient, though. This constructor would be more powerful if it were capable of specifying specific data values upon initialization. Then, it would take only one line of code to fully specify all of the data attributes of the particle. To do just that, the __init__() method can instead be written to accept arguments that can be used directly to initialize the object. To achieve this, we must replace the previous example with the following:

```
# particle.py
class Particle(object):
    """A particle is a constituent unit of the universe.
```

```
    Attributes
    ----------
    c : charge in units of [e]
    m : mass in units of [kg]
    r : position in units of [meters]
    """

    roar = "I am a particle!"

    def __init__(self, charge, mass, position):    ❶
        """Initializes the particle with supplied values for
        charge c, mass m, and position r.
        """
        self.c = charge    ❷
        self.m = mass
        self.r = position
```

❶ The `self` parameter remains the first argument. However, it is followed by the three positional arguments with which the `__init__()` method will initialize the object.

❷ The instance attribute `c` is introduced (and assigned to `self`) with an initial value, `charge`, provided in the method call.

We've mentioned that `__init__()` is a "method," and arguably the most important one since it is the constructor in Python. The next section will explain further what methods are and how they are different from functions, and will give some examples of other kinds of methods that can be defined in the class definition.

Methods

The constructor, as mentioned previously, is a special *method* in Python, but many other methods can exist in a class definition. Methods are functions, like those covered in Chapter 5. However, not all functions are methods. Methods are distinguished from functions purely by the fact that they are *tied to a class definition*. Specifically, when a method is called, the object that the method is found on is implicitly passed into the method as the first positional argument. For this reason, methods may operate on data contained by the object. Let's add another method, `hear_me()`, to our Particle class definition:

```
# particle.py
class Particle(object):
    """A particle is a constituent unit of the universe.

    Attributes
    ----------
    c : charge in units of [e]
    m : mass in units of [kg]
```

Classes | 129

```
    r : position in units of [meters]
    """

    roar = "I am a particle!"

    def __init__(self, charge, mass, position):
        """Initializes the particle with supplied values for
        charge c, mass m, and position r.
        """
        self.c = charge
        self.m = mass
        self.r = position

    def hear_me(self):    ❶
        myroar = self.roar + (
            "  My charge is:     " + str(self.c) +   ❷
            "  My mass is:       " + str(self.m) +
            "  My x position is: " + str(self.r['x']) +
            "  My y position is: " + str(self.r['y']) +
            "  My z position is: " + str(self.r['z']))
        print(myroar)    ❸
```

❶ The object is passed to the `hear_me()` method as `self`.

❷ The `self` argument is used to access the instance variable c.

❸ The `myroar` string is printed from within the method. The roar is heard.

This example uses the global roar string. The `self` argument (representing the concrete object) allows the attribute to be accessed from the `hear_me()` method. The instance variables—roar, c, m, and r[*]—are used to construct a string that is specific to this particle. All of this is done in the `hear_me()` method, which then prints the string:

Code	Output

`from scipy import constants`
 	
`import particle as p`
 	
`m_p = constants.m_p`
`r_p = {'x': 1, 'y': 1, 'z': 53}`
`a_p = p.Particle(1, m_p, r_p)`
`a_p.hear_me()`
	I am a particle!
	My mass is: 1.672621777e-27
	My charge is: 1
	My x position is: 1
	My y position is: 1
	My z position is: 53

In this example, a proton is described. Note that the mass of the proton was retrieved from the `scipy.constants` module.

Don't hardcode well-known scientific constants into your program. Use the constants provided by the `scipy.constants` module instead.

Methods can also alter instance variables. As an example, let us imagine a `Quark` class that has an instance variable called `flavor`. Quarks and leptons have flavors. The "weak interaction" can alter that flavor, but symmetry must be preserved. So, in some quantum superposition interactions, a flavor can flip, but only to its complementary flavor. A method on the `Quark` class could flip the flavor. That `flip()` method would be defined to reset the `flavor` variable from `up` to `down`, `top` to `bottom`, or `charm` to `strange`:

```python
def flip(self):
    if self.flavor == "up":
        self.flavor = "down"
    elif self.flavor == "down":
        self.flavor = "up"
    elif self.flavor == "top":
        self.flavor = "bottom"
    elif self.flavor == "bottom":
        self.flavor = "top"
    elif self.flavor == "strange":
        self.flavor = "charm"
    elif self.flavor == "charm":
        self.flavor = "strange"
    else :
        raise AttributeError("The quark cannot be flipped, because the "
                             "flavor is not valid.")
```

In this example, the `flip()` method is able to access *and* alter an attribute of a `Quark` object. To witness this in action, we can create a `Quark`, set it to one flavor, and flip it to the other:

Code	Output
``# import the class
from quark import Quark

# create a Quark object
t = Quark()

# set the flavor
t.flavor = "top"

# flip the flavor
t.flip()

print the flavor
print(t.flavor)``` | bottom |

Classes | 131

Because they can access attributes of an object, methods are very powerful functions. With this power, the scientist can begin to do impressive things with object orientation. For example, the `Particle` class should capture the relationship between uncertainty in momentum and uncertainty in position. The Heisenberg Uncertainty Principle states:

$$\Delta x \Delta p_x \geq \frac{\hbar}{2}$$

A method that returns the minimum possible value of Δx can be added to the class definition:

```
from scipy import constants

class Particle(object):
    """A particle is a constituent unit of the universe."""

    # ... other parts of the class definition ...

    def delta_x_min(self, delta_p_x):
        hbar = constants.hbar
        delx_min = hbar / (2.0 * delta_p_x)
        return delx_min
```

Now the physics can really take over. What other methods would you add to a `Particle` class? Do all of them need to be tied inextricably to the object instance? Sometimes methods have a place in a class but don't need to be associated with any other attributes.

Static Methods

As just stated, functions are different from methods by virtue of being unattached to a class. That is, a feature of the `Quark` class could be a function that lists all possible values of quark flavor. Irrespective of the flavor of a specific instance, the possible values are static. Such a function would be:

```
def possible_flavors():
    return ["up", "down", "top", "bottom", "strange", "charm"]
```

Now, suppose that you wanted to have a method that was associated with a class, but whose behavior did not change with the instance. The Python built-in decorator `@staticmethod` allows for there to be a method on the class that is never bound to any object. Because it is never bound to an object, a static method does not take an implicit `self` argument. However, since it lives on the class, you can still access it from all instances, like you would any other method or attribute. The following demonstrates how to bring `possible_flavors()` into the class definition as a static method:

```python
from scipy import constants

def possible_flavors():
  return["up","down","top","bottom","strange","charm"]

class Particle(object):
    """A particle is a constituent unit of the universe."""

    # ... other parts of the class definition ...

    def delta_x_min(self, delta_p_x):
        hbar = constants.hbar
        delx_min = hbar / (2.0 * delta_p_x)
        return delx_min

    @staticmethod
    def possible_flavors():
        return ["up", "down", "top", "bottom", "strange", "charm"]
```

All of the attributes described in the last few sections have defined the class interface. The methods and data that are associated with an object present an interface to the simulation. In Python, that interface is very important. Python interfaces rely heavily on the notion of duck typing, which we first encountered in Chapter 3 and will go into more detail on now.

Duck Typing

This peculiar name comes from the saying, "When I see a bird that walks like a duck and swims like a duck and quacks like a duck, I call that bird a duck." In the Python context, this refers to checking at runtime whether or not an object quacks when it is asked to quack. If instead the object is asked to swim, Python will check if it can swim. The full duck-ish nature of an object is never checked automatically.

That is, Python does not explicitly check for object types in the way that other programming languages do. Python neither requires that variable types be declared upon instantiation nor guarantees the types of variables passed into functions as parameters. Object behavior, but not object type, is checked when a method is called or an attribute is accessed, and not before. In this way, Python only performs duck-type checking. If two different object types (birds, say) implement identical interfaces ("quack like ducks"), then they can be treated identically within a Python program. In this paradigm, an object need not be of a certain type in order for its methods to be invoked. It must merely possess those methods.

All particles with a valid `charge()` method, for example, can be used identically. You can implement a function such as the following, calculating the total charge of a collection of particles, without knowing any information about the types of those particles:

```
def total_charge(particles):
    tot = 0
    for p in particles:
        tot += p.c
    return tot
```

If the function is parameterized with a collection of Quarks, Protons, and Electrons, it will sum the charges irrespective of the particle types. For two electrons and one proton, the total charge is −1e.

Code	Output

`p = Proton()`
`e1 = Electron()`
`e2 = Electron()`
`particles = [p, e1, e2]`
`total_charge(particles)`
	-1.602176565e-19

Note that the type of container holding the collection of particles is also irrelevant. Since the Python elemental iteration syntax for x in y is the same for many data structures, the exact same behavior would result whether the container were a list, a tuple, a set, or any other container with that iteration method. This, too, is an example of duck typing.

Explicit typing is sometimes helpful, though considered unpythonic. To trigger different methods for different types of object or to trigger a warning in the event of an unsupported type, explicit typing can be used. Note that, when needed, the built-in isinstance() function can be used to achieve explicit type checking. This function takes the object that you want to test and the type you want to test against.

For example, in this case the letter c is somewhat ambiguous. Perhaps some other object slips into the collection that possesses a method c with a *different* meaning (perhaps c is the speed of light). Rather than accidentally allowing the addition of the speed of light to one's calculation, the cautious developer could choose to ignore any objects that are not Particles:

```
def total_charge(collection):
    tot = 0
    for p in collection:
        if isinstance(p, Particle):
            tot += p.c
    return tot
```

In this way, duck typing can be overruled when it is inconvenient. However, usually duck typing adds flexibility and scalability to code. It is therefore highly desirable, and cousins of duck typing are accordingly part of the object-orientation paradigm. First up is polymorphism.

Polymorphism

In biology, polymorphism refers to the existence of more than one distinct phenotype within a single species. In object-oriented computation, polymorphism occurs when a class inherits the attributes of a parent class. As a general rule, what works for a parent class should also work for the subclass, but the subclass should be able to execute its own specialized behavior as well. This rule will be tempting to break, but should be respected.

A quark, for example, should behave like any other elementary particle in many ways. Like other elementary particles (e.g., an electron or a muon), a quark has no distinct constituent particles. Additionally, elementary particles have a type of intrinsic angular momentum called *spin*. Based on that spin, they are either fermions (obeying Fermi-Dirac statistics) or bosons (obeying Bose-Einstein statistics). Given all this, and making use of Python's modulo syntax, we might describe the `ElementaryParticle` class thus:

```
# elementary.py
class ElementaryParticle(Particle):

    def __init__(self, spin):
        self.s = spin
        self.is_fermion = bool(spin % 1.0)
        self.is_boson = not self.is_fermion
```

Note that the `ElementaryParticle` class seems to accept the `Particle` class instead of `object`. This is in order to denote that the `ElementaryParticle` class is a subclass of the `Particle` class. That relationship is called inheritance because the `ElementaryParticle` class *inherits* data and behaviors from the `Particle` class. The inheritance diagram for this relationship is shown in Figure 6-2.

Distinct from `ElementaryParticle`s, however, `CompositeParticle`s exist. These are particles such as protons and neutrons. They are composed of elementary particles, but do not share their attributes. The only attributes they share with `ElementaryParticle`s are captured in the parent (`Particle`) class. `CompositeParticle`s have all the qualities (charge, mass, position) of the `Particle` class and one extra, a list of constituent particles:

```
# composite.py
class CompositeParticle(Particle):

    def __init__(self, parts):
        self.constituents = parts
```

```
┌─────────────────────────────────┐
│           Particle              │
├─────────────────────────────────┤
│ c                               │
│ m                               │
│ roar : str                      │
│ constituents : list             │
│ r : dict                        │
├─────────────────────────────────┤
│ hear_me()                       │
│ charge()                        │
│ constituents()                  │
│ delta_x_min()                   │
│ roar()                          │
│ spin()                          │
└─────────────────────────────────┘
              ▲
              │
┌─────────────────────────────────┐
│      ElementaryParticle         │
├─────────────────────────────────┤
│ constituents : NoneType         │
│ is_boson                        │
│ is_fermion : bool               │
│ s                               │
└─────────────────────────────────┘
```

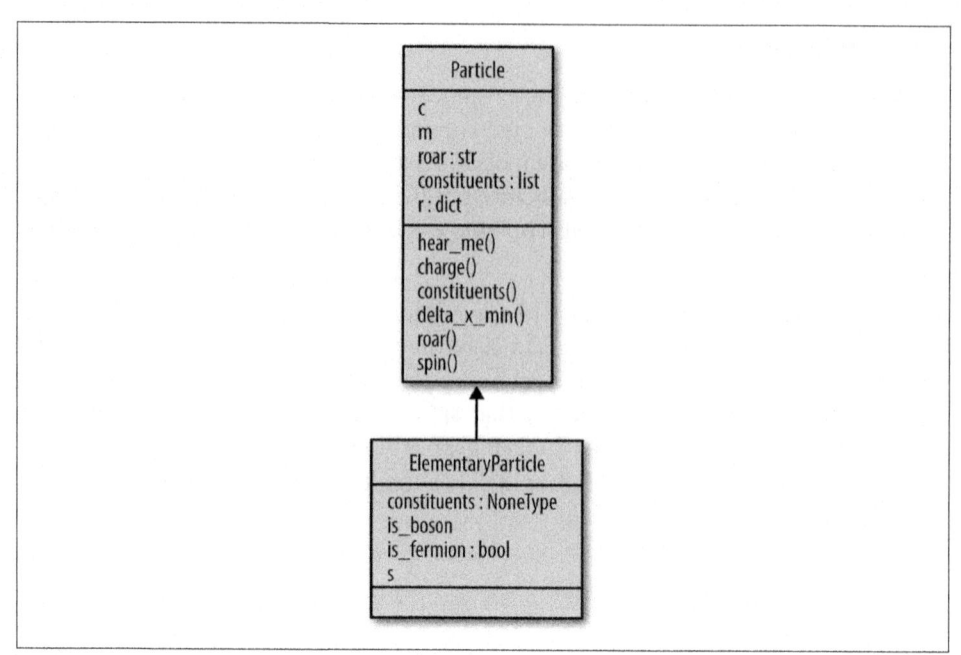

Figure 6-2. The ElementaryParticle class inherits from the Particle class

As a simulator or other physics software becomes more detailed, additional classes like `ElementaryParticle` and `CompositeParticle` can be created in order to capture more detailed resolution of `Particle` types. Additionally, since attributes vary depending on the type of particle (e.g., electrons and protons have charge, but neutrons do not), these classes may need to represent the various subtypes of particles as well.

Subclasses

Because they inherit from the `Particle` class, `ElementaryParticle` objects and `CompositeParticle` objects *are* `Particle` objects. Therefore, an `ElementaryParticle` has all of the functions and data that were previously assigned in the `Particle` class, but none of that code needs to be rewritten. In this way, the code defining the `Particle` class is reused.

Additionally, the `ElementaryParticle` class can override that data and those behaviors, if desired. For example, the `ElementaryParticle` class inherits the `hear_me()` function from the `Particle` class. However, it can override the `roar` string in order to change its behavior. If the `ElementaryParticle` class is defined thus:

```
# elementary.py
class ElementaryParticle(Particle):
```

```
        roar = "I am an Elementary Particle!"

        def __init__(self, spin):
            self.s = spin
            self.is_fermion = bool(spin % 1.0)
            self.is_boson = not self.is_fermion
```

Then the following is the resulting behavior:

Code	Output

`from elementary import ElementaryParticle`
 	
`spin = 1.5`
`p = ElementaryParticle(spin)`
`p.s`
`p.hear_me()`
	`1.5`
	`I am an Elementary Particle!`
	`My charge is:   -1`
	`My mass is:      0`
	`My x position is: 0`
	`My y position is: 0`
	`My z position is: 0`

The power here lies in the customization of behavior. `ElementaryParticle` objects have all of the attributes of `Particle` objects, so they can interact exactly as `Particles` do. However, they also have additional attributes only possessed by `Elementary Particles`, *and* they have specialized certain attributes of the `Particle` superclass.

Superclasses

Any class, including a subclass, can be a superclass or *parent* class. The subclass is said to *inherit from* its parent. In the preceding examples, the `Particle` class is a superclass and the `ElementaryParticle` class is a subclass. However, the `ElementaryParticle` class can also be a superclass. Since quarks are a type of elementary particle, the `Quark` class might inherit from the `ElementaryParticle` class.

The superclass of `Quark` is `ElementaryParticle`. But since the `ElementaryParticle` class still inherits from `Particle`, the `Particle` class is therefore a superclass of both the `ElementaryParticle` class and the `CompositeParticle` class.

Quarks have typical `ElementaryParticle` characteristics (`spin`, `is_fermion`, `is_boson`), as well as those that `ElementaryParticle` inherits from `Particle` (`charge`, `mass`, `position`). However, Quarks also have *flavor*. The `flavor` of the quark can take one of six values (`up`, `down`, `top`, `bottom`, `strange`, and `charm`).

```
        import randphys as rp

        class Quark(ElementaryParticle):

            def __init__(self):
                phys = rp.RandomPhysics()
                self.color = phys.color()
                self.charge = phys.charge()
```

```
self.color_charge = phys.color_charge()
self.spin = phys.spin()
self.flavor = phys.flavor()
```

Polymorphism, subclasses, and superclasses are all achieved with inheritance. The concept of inheritance is subtly distinct from polymorphism, however—a class is called polymorphic if it has more than one subclass. Both of these concepts are distinct from multiple inheritance, which we'll look at next.

Multiple inheritance

Multiple inheritance is when a subclass inherits from more than one superclass. For example, the quantum-mechanical phenomenon of wave-particle duality may need to be modeled in the ElementaryParticle class.

In their behavior as waves, ElementaryParticles should possess Wave-like attributes such as amplitude and frequency. These attributes rely on the energy of the ElementaryParticle object. Also, as a Wave, an ElementaryParticle should implement interference methods based on this class when interacting with other Wave functions.

All that said, ElementaryParticles should also continue to exhibit the attributes of a Particle (such as charge). To capture both its Particle nature and its Wave nature at the same time, the ElementaryParticle class can inherit from both the Wave and Particle classes.

```
# elementary.py
class ElementaryParticle(Wave, Particle):

    def __init__(self, spin):
        self.s = spin
        self.is_fermion = bool(spin % 1.0)
        self.is_boson = not self.is_fermion
```

This is multiple inheritance. The new, quantum-friendly model of the ElementaryParticle is shown here. With only one changed line, it now adopts the behavior of the Wave class as well as that of the Particle class.

If all of this is confusing, never fear. Figure 6-3 should clear things up somewhat. Looking at or drawing an inheritance diagram is always useful for understanding the relationships that exist in a class hierarchy. In practice, these hierarchies can become quite deep and convoluted as a code base grows. If this isn't enough, many resources are available that go into greater detail. We will mention a few at the end of the chapter.

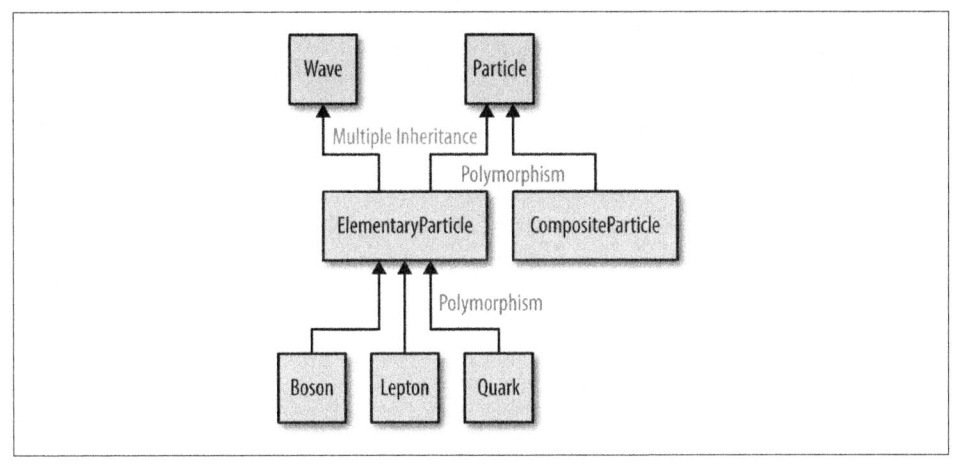

Figure 6-3. Inheritance, polymorphism, and multiple inheritance

Decorators and Metaclasses

Metaprogramming is when the definition of a class (or function) is specified, in part or in full, by code outside of the class definition itself. We saw a bit of this in "Decorators" on page 112, where decorators were used to alter the behavior of functions. In some programming languages, such as C++, metaprogramming is a central part of the language. (If you have worked in C++, then you know it is impossible to get by without templates for any length of time.) In Python, the need for metaprogramming is less pervasive since everything is an object. However, it is there when you need it. Admittedly this is rare in physics-based programming, but it does comes up when you're writing analysis frameworks.

The overwhelming majority of your metaprogramming needs can be handled by class decorators. These work in exactly the same way as function decorators: just place an `@<decorator>` above the class definition. This works because class definitions, like everything else in Python, are objects. Thus they can be used as arguments to a function and also returned from a function.

Inside of a class decorator we can add attributes or methods to an existing class. Or we could throw out the original class entirely and return a brand new class. The decorator itself is still a function. However, instead of returning a new function, a class decorator will typically return a class.

Suppose that we wanted to add an `is_particle` class attribute to our `Particle` class. We could do this through the following `add_is_particle()` decorator:

```
def add_is_particle(cls):  ❶
    cls.is_particle = True  ❷
    return cls  ❸
```

```
@add_is_particle ❹
class Particle(object):
    """A particle is a constituent unit of the universe."""

    # ... other parts of the class definition ...
```

❶ Defines the class decorator, which takes one argument that is the class itself.

❷ Modifies the class by adding the is_particle attribute.

❸ Returns the class.

❹ Applies the decorator to the class. This uses the same syntax as a function decorator.

Of course, nothing is stopping us from going all out and adding methods to the class in the decorator, or removing them. For example, we could add a distance() method that computes the distance between the current particle and another particle, as follows:

```
from math import sqrt

def add_distance(cls):
    def distance(self, other):  ❶
        d2 = 0.0
        for axis in ['x', 'y', 'z']:
            d2 += (self.r[axis] - other.r[axis])**2
        d = sqrt(d2)
        return d
    cls.distance = distance
    return cls  ❷


@add_distance
class Particle(object):
    """A particle is a constituent unit of the universe."""

    # ... other parts of the class definition ...
```

❶ It is probably still a good idea to call the first argument self, since this function will be a method, even though it is defined outside of the class.

❷ Since we are modifying the class in-place, we again want to return the class.

In the unlikely event where, for some reason, class decorators are not enough, there are always metaclasses. Class decorators are a relatively new addition to the Python language, so you are more likely to see metaclasses floating around in legacy code. In the way that classes create objects, metaclasses generate classes. All metaclasses must inherit from type, like all classes inherit from object. The type of type is, of course, type:

```
In [1]: type(type)
Out[1]: type
```

Thus, defining a new metaclass is as easy as inheriting from type. A common pattern in Python is for metaclasses to be prefixed with the phrase Is or Has. Let's create an IsParticle metaclass:

```
class IsParticle(type):
    pass
```

This can then be applied to our Particle class by passing it in as the metaclass keyword argument to the inheritance listing:

```
class Particle(metaclass=IsParticle):
    """A particle is a constituent unit of the universe."""

    # ... other parts of the class definition ...
```

Note that now the Particle class is an instance of the IsParticle metaclass. However, any instance p of Particle is not an instance of the metaclass:

```
In [1]: isinstance(Particle, IsParticle)
Out[1]: True

In [2]: p = Particle()

In [3]: isinstance(p, IsParticle)
Out[3]: False
```

Metaclasses are mainly used to override the special __new__() method, which is used to create new instances of a class prior to them being initialized via __init__(). The __new__() method prepares an object for initialization. By modifying __new__() you can dramatically change how a class creates instances of itself. Almost no one needs to do this, though. Where it does arise is often in the context of describing data in databases. But even then, there are more intuitive ways than metaclasses. Since the type of the class is being modified, metaclasses can be tricky to get right. For now, just know that they exist, they aren't scary, and you can look them up in more detail in the Python documentation.

Object Orientation Wrap-up

This chapter has covered object orientation from the perspective of simulating physics and understanding Python. In particular, it covered what objects and classes are, how they are defined, how classes create objects, and how classes relate to one another. With all of that, the reader should now:

- Understand how object orientation can be useful for the reductionist, encapsulated modeling necessary in physics simulation
- Know how to create a simple physical model using classes and objects in Python

Empowered with this knowledge, you can now apply object orientation to simulations and analyses in your subfield of physics. As you go forward, note that many additional resources on object orientation in Python and other languages are available online and in books; for example, Software Carpentry (*http://software-carpentry.org*) offers an excellent tutorial, both online and in person, concerning object orientation in Python.

Finally, any discussion of object orientation would be remiss if it failed to mention the existence and importance of design patterns. Many "patterns" have emerged over the years in object-oriented programming. Some patterns enable efficient or robust behavior. Others are common failure modes (these are often called antipatterns). The book responsible for laying the groundwork and establishing the early vocabulary around design patterns in object-oriented code is *Design Patterns*, by Erich Gamma, Richard Helm, Ralph Johnson, and John Vlissides (Pearson). These four authors are colloquially known as "The Gang of Four."

PART II
Getting It Done

CHAPTER 7
Analysis and Visualization

Churning out terabytes of data from simulations or experiments does not, on its own, constitute science. Only analysis and visualization can transform raw data into true scientific insight. Unanalyzed data is merely data—only interpretation and communication can sufficiently illuminate and clarify the scientific meaning of results. When analysis and visualization succeed, compelling data becomes a convincing result.

There was an era in the physical sciences when data was collected in laboratory notebooks, and when the time to publish plots of that data came about, it was done by hand. Legend has it that this was sometimes begun on enormous sheets of graph paper on the wall of a lab and scaled down to a reasonable publishable size by big, slow scanning machines. Many physicists and mathematicians, Roger Penrose not least among them, continue to make plots and diagrams with pen and paper. Nonetheless, it is an increasingly lost art.

While it is tempting to feel a nostalgia for freehand drawings of the complex plane, this chapter should inspire you to embrace the future instead. This chapter will provide an overview of principles and tools for data preparation, analysis, and visualization appropriate for publication-quality results in the physical sciences. Finally, a few examples of analysis and visualization using Python tools will be addressed. This chapter will provide a taste of the analysis tools that will then be discussed in detail in Chapters 9, 10, and 11.

Preparing Data

Researchers encounter data in many formats, from many sources. They accordingly can spend significant effort loading, cleaning, collapsing, expanding, categorizing, and generally "munging" data into consistent formats that can be used for analysis and plotting. Some basic methods for retrieving data from files will be covered in

Chapter 10, as will more advanced data manipulation. Depending on the source of the data, advanced data munging for myriad formats may be necessary. A few will be discussed here.

Faced with imperfect data in one or more raw formats, the researcher must perform several steps before even beginning to analyze or visualize it:

- Load the data into an analysis-ready format.
    - Possibly convert the data to an intermediate data storage format (CSV, HDF5, SQL, FITS, ENDF, ENSDF).
    - Convert the data to an easy-access data structure (NumPy arrays, Pandas data frames).
- Clean the data.
    - Handle missing values.
        - Drop them from analysis.
        - Replace them with defaults.
        - Count them for statistics.
    - Fix mangled or erroneous entries.
        - Detect errors.
        - Sort disordered data.
        - Standardize data formats.
    - Handle stylistic issues.
        - Rename long or irregular fields.
        - Reformat irregularly formatted dates, times, numbers, strings, etc.
- Combine the data with metadata.
    - Populate results tables with additional/external data.
    - Add identifying numbers and dates, for provenance.
    - Merge results from independent detectors, etc.

A visual representation of the aspects of this process appears in Figure 7-1.

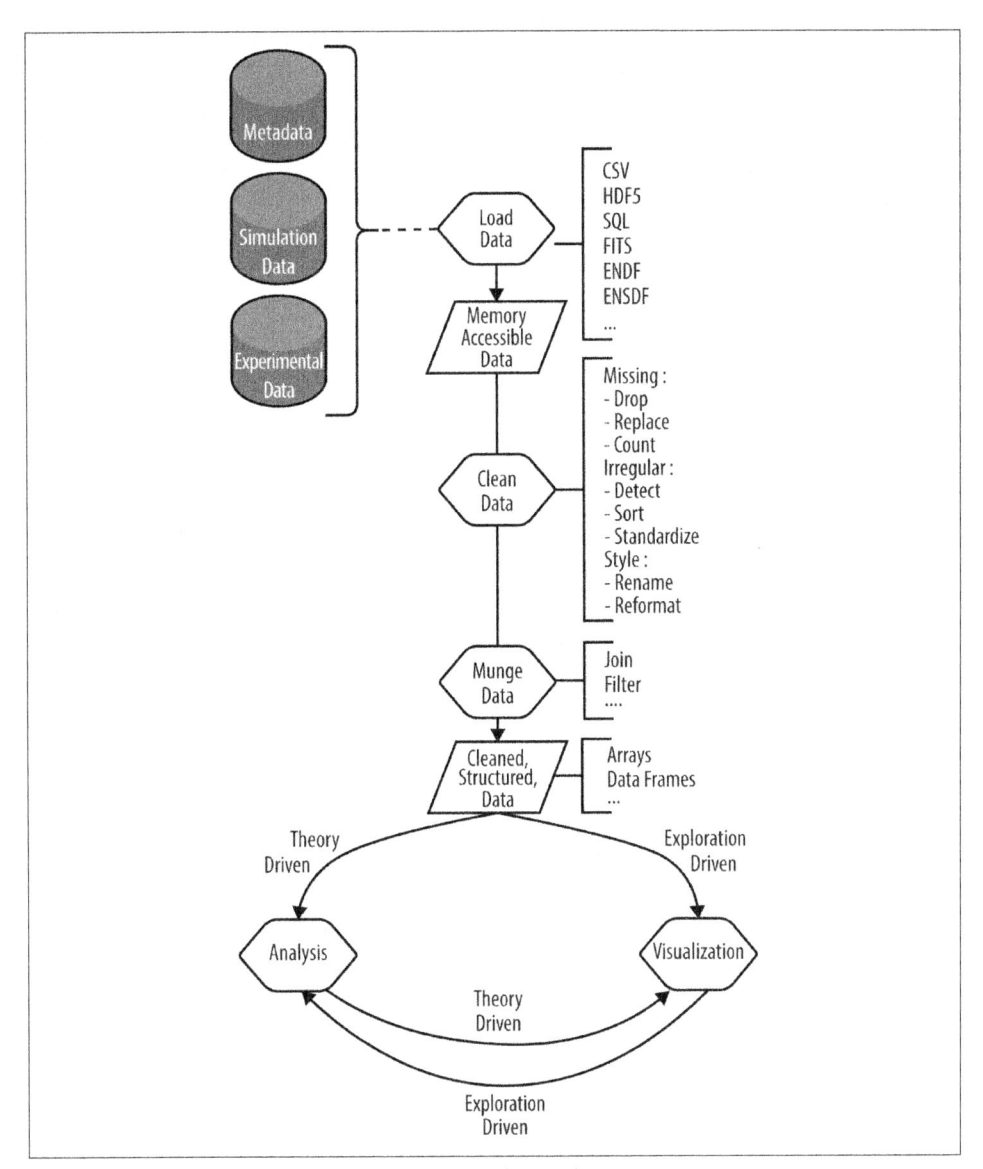

Figure 7-1. Data preparation for analysis and visualization

Due to this involved sequence of necessary tasks, many scientists spend vast portions of their research careers:

- Cleaning data by hand
- Executing analysis steps one at a time
- Using a mouse when creating plots

Preparing Data | 147

- Repeating the whole effort when new data appears or the process must be tweaked

However, more efficient scientists automate these processes and have more time for research. These efficient scientists spend their research careers doing the following:

- Scripting their data pipeline
- Gaining confidence in their results
- Doing additional research
- Publishing more papers

These scientists must invest extra effort up front, but they benefit from increased efficiency later. By taking the time to generate scripts that automate their pipelines (data cleaning, processing, analysis, plotting, etc.), efficient scientists can more rapidly incorporate new data into their analyses, test new analysis steps, and tweak plots. The pipeline can be simple or complex and may include a wide array of actions, such as:

- Data entry
- Data cleanup
- Building any necessary software
- Running simulation analyses
- Data post-processing
- Uncertainty estimation
- Generating tables and plots for publication
- Building papers around that work

An astute reader may note that the former kind of scientist may publish results faster if the datasets are pristine and reviewers are generous. Fools and optimists are invited to rely on these two miracles. Realists should automate. Furthermore, those who are sickened by the rampant lack of reproducibility in scientific computation should move to automate their pipelines, irrespective of the cost-benefit ratio. Fully scripted analysis and visualization is a necessary feature of reproducible science. Though incentive structures in the research sciences currently fall short of rewarding reproducibility, the tables are turning. Be on the righteous side of history—heed the words of Mario Savio, and automate your methods:

> There's a time when the operation of the machine becomes so odious, makes you so sick at heart, that you can't take part! You can't even passively take part! And you've got to put your bodies upon the gears and upon the wheels… upon the levers, upon all the apparatus, and you've got to make it stop! And you've got to indicate to the people who

run it, to the people who own it, that unless you're free, the machine will be prevented from working at all!

When you're automating your methods, the first thing to automate is the processing of raw data. How the data processing step is performed varies greatly depending on whether your work is based on experimentation or simulation. The next sections will cover each of those cases, as well as the metadata that is associated with them.

Experimental Data

Experimental data presents unique analysis challenges to the scientist. Such data usually takes the form of detector responses and physical measurements resulting from various experiments in the physical sciences.

Experiments that observe a variable changing over time will produce time series data, where the independent variable is time and the dependent variable is the observation. Time series data such as this is often stored in flat tabular plain-text or simple CSV files. An example might be an experiment that seeks to identify the decay constant of an isotope based on its dwindling radiation signature over time. Such data might be structured as in Table 7-1.

Table 7-1. Isotope decay data (decays.csv)

Time(s)	Decays (#)
0	10.0
1	1.353352832
2	0.183156389
3	0.024787522
4	0.003354626
5	0.000453999
6	6.1442e-05
7	8.315e-06
8	1.126e-06
9	1.52e-07
...	...

In its raw form, this data may be stored in a comma-separated or otherwise delimited plain-text format, as seen here:

```
# Time (s), Decays (#)
0,10.0
1,1.353352832
2,0.183156389
3,0.024787522
4,0.003354626
5,0.000453999
6,6.1442e-05
7,8.315e-06
8,1.126e-06
9,1.52e-07
```

Experiments that seek certain parametric relationships between variables, however, may produce multidimensional, structured, or tabular data. Many experiments have qualities of both and produce multidimensional, structured time series data. The possibilities are endless, but the structure of data often dictates the format it must be stored in. While time series can be stored easily in CSV format, very complex structured data typically calls for a standardized database format.

In special cases, scientific communities adopt their own very domain-specific file formats for storing experimental data. Astrophysicists, for example, store their enormous libraries of telescope images and telescope calibration metadata together in the specialized Flexible Image Transport System (FITS) file format. In nuclear physics, on the other hand, physicists do not deal with images. Rather, they deal primarily with particle interaction probabilities called *cross sections*. Many international standards exist for storing this type of data, but the most common nuclear data format is the Evaluated Nuclear Data File (ENDF) format.

Once formatted, evaluated data like the data that appears in ENDF or FITS formats is ready to be used in simulations.

Simulation Data

Simulations are just experiments *in silico*. A wise man, Professor Paul P.H. Wilson, used to tell his students and colleagues that scientific computation is just experimentation for scientists who like to control more variables. An experimentalist, he would explain, begins with the whole world of uncontrolled variables and designs an experiment that carefully controls those variables one at a time, step by step, until only the experimental variables are left. The computationalist, on the other hand, starts with a completely controlled simulation and carefully, one at a time, *releases* variables.

Because of the higher degree of control in simulation, simulation data output formats are often already quite clean and well controlled. Raw simulation data typically resides in databases.

For provenance, databases may need to be accompanied by data about the simulation, such as:

- The date the simulation was run
- The names and contents of the input files
- The version numbers of libraries used

This type of data, present in both experiments and simulations, is called *metadata*.

Metadata

Metadata is data about data. It is important to include metadata because the results produced by an experiment or a simulation sometimes fail to capture all of its features.

In an experiment, detector parameters, the date of the experiment, background radiation readings from another detector, and more. can all affect the interpretation of results, but these factors may not have been captured in the experimental data output stream. They can instead be captured in the metadata.

Metadata is not limited to experimentation, however. Metadata that may affect the interpretation of the results of a simulation include simulation ID numbers, library dependency version numbers, input file content, and more.

For reproducibility, as much of this data should be included in the workflow as possible. From a data processing perspective, this metadata also often needs to be joined with the experimental or simulation database. The steps necessary for preparing metadata for this process will vary from application to application. However, metadata should be held to the same standards of formatting and reproducibility as simulation or experimental data.

Of course, once all of the experimental, simulation, and/or metadata has been prepared, it must be loaded into a form that can be analyzed. In Figure 7-1, this next step is the "Load Data" step. It will be covered in the following section.

Loading Data

Many packages in Python enable data to be loaded quickly into a memory-accessible data structure ready for cleaning, munging, analysis, and visualization. More about data structures will be covered in Chapter 11. The choice of appropriate data structure depends profoundly on the size and type of the data as well as its analysis and visualization destiny. This section will merely cover how to load data into various analysis-ready forms using four tools in the Python ecosystem:

- NumPy

- PyTables
- Pandas
- Blaze

When you are choosing among tools like these, a number of factors come into play.

Size is the first parameter to consider. For big, dense arrays of numbers or for enormous suites of images, loading all of the data into memory at once is not recommended. Rather, loading *chunks* of the data into memory while cleaning, processing, or exploring it might be advised. For this and similar out-of-core computations on data exceeding the bounds of system memory, a database choice that emphasizes these features is warranted. On that topic, later sections will address loading data into PyTables and Blaze.

The type of the data also may determine the appropriate data structure. Structured data in a relational database format may be easiest to handle using extraction and exploration tools in the `sqlite3` or `pandas` packages.

All of that said, some data is small enough to fit into memory all at once, and much of the data that physicists encounter or produce is best represented as an array of numbers. For this application, the right tool is the numerical Python package, `numpy`.

NumPy

Due to their power, many solutions involve NumPy arrays. NumPy arrays will be covered in detail in Chapter 9. For now, simply know that NumPy arrays are data structures for holding numbers. One easy way to transform a file into a NumPy array is with the `loadtxt` function. Using this function, plain-text files holding columns of text delimited by commas, spaces, or tabs can be loaded directly into a NumPy array. The decay data in our earlier CSV-formatted example can be loaded directly into a NumPy array shown here:

```
import numpy as np ❶
decays_arr = np.loadtxt('decays.csv', delimiter=",", skiprows=1) ❷
```

❶ Import `numpy` with the alias `np`.

❷ Create a NumPy array object called `decays_arr` using the `loadtxt()` function.

In this example, the `numpy` package is loaded and given the short alias `np`. Next, a variable called `decays_arr` is declared and set equal to the output of the `loadtxt()` function. The variable `decays_arr` is a NumPy array. In this case, the `loadtxt()` function is parameterized with only one mandatory variable, the filename. The two optional variables are the delimiter (a comma), and the number of rows to skip (the header row, which is not numbers). Though not all were used in this example, many other options are available to customize the way a file is read with `loadtxt()`. To learn more about those, consult the `numpy.loadtxt()` documentation using the `help` command in IPython:

```
In [1]: import numpy as np  ❶

In [2]: help(np.loadtxt)  ❷
```

❶ Import `numpy` with the alias `np`.

❷ Learn more about the `loadtxt()` function.

Once data has been loaded into a NumPy array, one of the fastest ways to deal with that data for large-scale problems is to enlist the help of PyTables.

PyTables

As you will learn in Chapter 10, PyTables provides many tools for converting HDF5 data into analysis-ready NumPy arrays. Indeed, because PyTables can help load, modify, and manipulate HDF5 data in the form of NumPy arrays, it is a strong motivator for the use of HDF5 as a raw data format. Perhaps the decay data in the previous example could be more easily manipulated in the future if it were stored in an HDF5 database—PyTables can help with that. Any data loaded into a NumPy array can be quickly and easily saved as an HDF5 database. So, once data has been loaded as a NumPy array, it is ready for use with PyTables. This allows for faster filtering, joins, and analysis later. However, PyTables and HDF5 are most useful for storing and manipulating dense arrays of numbers, so if your data is heterogeneous or sparse, or contains structured, relational information, it may be best stored in another format. If that is the case, a multiformat Python package like `pandas` may be more appropriate than PyTables. For information on when and how to load data into Pandas, read on.

Pandas

Pandas is currently the easiest to use and most broadly applicable of all of the data analysis tools in the Python ecosystem. It is a good alternative to the previously discussed tools, especially if your data is not in a format that is supported by NumPy (CSV or plain text) or PyTables (HDF5). Also, Pandas may be the right choice even for those formats if your data is not arrays of numbers or is the kind of data that you would like to view, filter, browse, and interact with.

Pandas cleanly handles reading and writing of many of the data formats encountered by scientists in the wild:

- CSV
- Excel

- HDF
- SQL
- JSON
- HTML
- Stata
- Clipboard
- Pickle

Also, loading data into a Pandas format is very straightforward. Note that the capability of numpy.loadtxt() can be repeated in Pandas with very similar syntax:

```
import pandas as pd  ❶

decays_df = pd.read_csv('decays.csv')  ❷
```

❶ Import the pandas package and alias it as pd.

❷ Create a data frame object that holds the data loaded by the read_csv() function.

A lovely quality in Pandas is that once data has been loaded, it can be converted into any of the other supported formats. To write this data to an HDF5 file, we need to add just one line to the previous example:

```
import pandas as pd  ❶

decays_df = pd.read_csv('decays.csv')  ❷
decays_df.to_hdf('decays.h5', 'experimental')  ❸
```

❶ Import the pandas package and alias it as pd.

❷ Create a data frame object that holds the data loaded by read_csv().

❸ Convert it to HDF5, giving it the filename *decays.h5*, and create a group node called "experimental" where this data will be stored.

Pandas is a top-notch tool for data analysis with Python. To learn how to fully wield the power of Pandas, refer to *Python for Data Analysis* by the lead developer of Pandas, Wes McKinney (O'Reilly). The data analysis in that book goes way beyond the scope of this section. Here, we simply mean to introduce the existence of this tool, alongside a few other tools that might also be considered. The final such data analysis tool that we will introduce is Blaze. Like Pandas, it is intended for easily loading data into an analysis-ready format and emphasizes ease of conversion between formats.

Blaze

Blaze is another Python tool capable of converting data from format to format. This tool is still in active development, but possesses impressive capabilities already. In Blaze, the CSV data might be dealt with as a Blaze data descriptor or as a Blaze `Table` object. The following example shows the transformation from CSV to data descriptor, and an additional transformation from data descriptor to Blaze `Table` object:

```
import blaze as bz                        ❶
csv_data = bz.CSV('decays.csv')           ❷
decays_tb = bz.Table(csv_data)            ❸
```

❶ The `blaze` package is imported and given the alias `bz`.

❷ Next, the CSV data is transformed into Blaze data with the `CSV()` constructor.

❸ Finally, that data descriptor, `csv_data`, is transformed into a Blaze `Table`.

This example illustrates how one type of Blaze object can be quickly converted to another data structure within Blaze quite straightforwardly. Since the aim of Blaze is to support conversion between many data formats (or "backends," in Blaze-speak), it may be the right tool to use if your data files must be converted from one memory-accessible format to another.

> Blaze is still under active development. Unlike the other tools discussed here (NumPy and PyTables in particular), it is not yet fully stable. However, the features discussed here are quite mature, and it will be a tool to watch closely as it improves.

This flexibility is likely to make Blaze very handy for certain applications, as this tool not only provides an interface for converting between many data formats (CSV, HDF5, SQLite, etc.) but also provides an interface to support workflows using many computational engines. Blaze uses symbolic expression and typing systems to communicate with other tools including Pandas, NumPy, SQL, Mongo, Spark, and PyTables. Access to computational engines like those, which are capable of manipulating the data, is essential for the next step in the process of data analysis: *cleaning and munging*.

Cleaning and Munging Data

Data *munging* (or *wrangling*) is a term used to mean many different things within the broad scope of *dealing with data*. Typically, as in Figure 7-1, the term refers to the process of converting data from a raw form to a more well-structured form appropriate for plotting and mathematical transformation.

The scientist may *wrangle* the data by grouping, filtering, aggregating, collapsing, or expanding it. Depending on your particular data, this step may need to happen before the data is *cleaned*, or may not have to happen until after. Cleaning data can also take many forms. Typically, this task deals with imperfect, incomplete, and disorganized data.

Of course, ideally, experimentalists in the physical sciences use sophisticated, automated, comprehensive data acquisition systems that produce clean, flawless datasets in intuitive formats. However, even such systems can produce imperfect data in extreme conditions.

The decay data being used in the previous examples, for instance, might have errors if other radioactivity were present in the laboratory. Additionally, if the power to the detector were cut off in an electrical blackout, data would be unavailable for a period of time.

To explore this, let's consider a more realistic version of the data we dealt with before. It may have machine-generated timestamps instead of integer numbers of seconds, and it may have missing or imperfect data. Imagine, for example, that about 15 seconds into the experiment, a colleague walked through the room carrying a slightly more stable radioactive source, emitting two decays per second. Additionally, imagine that a few seconds later, the lights in the room flashed off for a few seconds—the storm outside must have interrupted power to the lab. The resulting data stream looks like this:

```
#Time,Decays
2014-11-08T05:19:31.561782,10.0
2014-11-08T05:19:32.561782,1.35335283237
2014-11-08T05:19:33.561782,0.183156388887
2014-11-08T05:19:34.561782,0.0247875217667
2014-11-08T05:19:35.561782,0.00335462627903
2014-11-08T05:19:36.561782,0.000453999297625
2014-11-08T05:19:37.561782,6.14421235333e-05
2014-11-08T05:19:38.561782,8.31528719104e-06
2014-11-08T05:19:39.561782,1.12535174719e-06
2014-11-08T05:19:40.561782,1.52299797447e-07
2014-11-08T05:19:41.561782,2.06115362244e-08
2014-11-08T05:19:42.561782,2.78946809287e-09
2014-11-08T05:19:43.561782,3.77513454428e-10
2014-11-08T05:19:44.561782,5.10908902806e-11
2014-11-08T05:19:45.561782,6.91440010694e-12
2014-11-08T05:19:46.561782,9.35762296884e-13
2014-11-08T05:19:47.561782,2.000000000000000     ❶
2014-11-08T05:19:48.561782,2.000000000000000
2014-11-08T05:19:49.561782,2.000000000000000
2014-11-08T05:19:50.561782,2.000000000000000
2014-11-08T05:19:51.561782,2.000000000000000
2014-11-08T05:19:52.561782,2.000000000000000
2014-11-08T05:19:53.561782,2.000000000000000
```

```
2014-11-08T05:19:54.561782,2.000000000000000
2014-11-08T05:19:55.561782,2.000000000000000
2014-11-08T05:19:56.561782,1.92874984796e-21
2014-11-08T05:19:57.561782,2.61027906967e-22
2014-11-08T05:19:58.561782,3.5326285722e-23
2014-11-08T05:19:59.561782,4.78089288389e-24
2014-11-08T05:20:00.561782,6.47023492565e-25
2014-11-08T05:20:01.561782,8.7565107627e-26
2014-11-08T05:20:02.561782,1.18506486423e-26
2014-11-08T05:20:03.561782,1.60381089055e-27
2014-11-08T05:20:04.561782,2.1705220113e-28
2014-11-08T05:20:05.561782,2.93748211171e-29
2014-11-08T05:20:06.561782,3.97544973591e-30
2014-11-08T05:20:07.561782,5.38018616002e-31
2014-11-08T05:20:08.561782,7.28129017832e-32
2014-11-08T05:20:09.561782,9.85415468611e-33
2014-11-08T05:20:10.561782,1.3336148155e-33
2014-11-08T05:20:11.561782,1.80485138785e-34
2014-11-08T05:20:12.561782,NaN  ❷
2014-11-08T05:20:13.561782,NaN
2014-11-08T05:20:14.561782,NaN
2014-11-08T05:20:15.561782,NaN
2014-11-08T05:20:16.561782,8.19401262399e-39
2014-11-08T05:20:17.561782,1.10893901931e-39
2014-11-08T05:20:18.561782,1.50078576271e-40
2014-11-08T05:20:19.561782,2.03109266273e-41
2014-11-08T05:20:20.561782,2.74878500791e-42
```

❶ Uh oh, it looks like the reading was overwhelmed by another source moving through the room.

❷ At this point, it seems the detector was off, and no readings were made.

> NaN entries, as in this example, indicate that no number is stored in memory at the place where the data should be. NaN stands for "Not a Number."

Some experimentalists might see the NaN entries and immediately assume this data must be thrown away entirely. However, since experiments are often expensive and time-consuming to conduct, losing an entire run of data due to minor blips like this is often unacceptable. Concerns about data quality and inconsistencies are very common in science. Sometimes, dates are listed in a dozen different formats. Names are inconsistent across files. And sometimes data is erroneous. In this case, the section with too-high (2.0) counts due to the external radioactive source dwarfing the actual signal must be dealt with. How this section of the data is handled is a choice for the experimenter. Whatever the choice, however, tools exist to assist in the implementation.

The data from this run is ugly, but can it be saved with intelligent cleaning and modern tools? The following section will discuss one way to deal with missing data using Pandas.

Missing Data

Sometimes, data is missing. In some situations, a missing data point may be appropriate or expected, and can be handled gracefully. Often, however, it may need to be replaced with a default value, its effect on the statistical analysis of the results may need to be taken into consideration, or those data points may just need to be dropped.

Pandas is especially helpful in the event of missing data. In particular, Pandas has special methods for identifying, dropping, and replacing missing data.

With only a few lines in IPython, the NaN rows from the previous data can be dropped from the dataset entirely:

```
In [1]: import pandas as pd

In [2]: decay_df = pd.read_csv("many_decays.csv")

In [3]: decay_df.count()   ❶
Out[3]:
Time      50
Decays    46
dtype: int64

In [4]: decay_df.dropna()   ❷
Out[4]:
                         Time           Decays
0    2014-11-08T05:19:31.561782    1.000000e+01
1    2014-11-08T05:19:32.561782    1.353353e+00
2    2014-11-08T05:19:33.561782    1.831564e-01
3    2014-11-08T05:19:34.561782    2.478752e-02
4    2014-11-08T05:19:35.561782    3.354626e-03
5    2014-11-08T05:19:36.561782    4.539993e-04
6    2014-11-08T05:19:37.561782    6.144212e-05
7    2014-11-08T05:19:38.561782    8.315287e-06
8    2014-11-08T05:19:39.561782    1.125352e-06
9    2014-11-08T05:19:40.561782    1.522998e-07
10   2014-11-08T05:19:41.561782    2.061154e-08
11   2014-11-08T05:19:42.561782    2.789468e-09
12   2014-11-08T05:19:43.561782    3.775135e-10
13   2014-11-08T05:19:44.561782    5.109089e-11
14   2014-11-08T05:19:45.561782    6.914400e-12
15   2014-11-08T05:19:46.561782    9.357623e-13
```

```
16  2014-11-08T05:19:47.561782  2.000000e+00
17  2014-11-08T05:19:48.561782  2.000000e+00
18  2014-11-08T05:19:49.561782  2.000000e+00
19  2014-11-08T05:19:50.561782  2.000000e+00
20  2014-11-08T05:19:51.561782  2.000000e+00
21  2014-11-08T05:19:52.561782  2.000000e+00
22  2014-11-08T05:19:53.561782  2.000000e+00
23  2014-11-08T05:19:54.561782  2.000000e+00
24  2014-11-08T05:19:55.561782  2.000000e+00
25  2014-11-08T05:19:56.561782  1.928750e-21
26  2014-11-08T05:19:57.561782  2.610279e-22
27  2014-11-08T05:19:58.561782  3.532629e-23
28  2014-11-08T05:19:59.561782  4.780893e-24
29  2014-11-08T05:20:00.561782  6.470235e-25
30  2014-11-08T05:20:01.561782  8.756511e-26
31  2014-11-08T05:20:02.561782  1.185065e-26
32  2014-11-08T05:20:03.561782  1.603811e-27
33  2014-11-08T05:20:04.561782  2.170522e-28
34  2014-11-08T05:20:05.561782  2.937482e-29
35  2014-11-08T05:20:06.561782  3.975450e-30
36  2014-11-08T05:20:07.561782  5.380186e-31
37  2014-11-08T05:20:08.561782  7.281290e-32
38  2014-11-08T05:20:09.561782  9.854155e-33
39  2014-11-08T05:20:10.561782  1.333615e-33
40  2014-11-08T05:20:11.561782  1.804851e-34  ❸
45  2014-11-08T05:20:16.561782  8.194013e-39
46  2014-11-08T05:20:17.561782  1.108939e-39
47  2014-11-08T05:20:18.561782  1.500786e-40
48  2014-11-08T05:20:19.561782  2.031093e-41
49  2014-11-08T05:20:20.561782  2.748785e-42
```

❶ The data frame method count() successfully ignores the NaN rows automatically.

❷ The dropna() method returns the data excluding all rows containing a NaN value.

❸ Here, the time skips ahead 5 seconds, past the now-missing NaN rows.

Now the data is much cleaner, as the offending missing data has been dropped entirely. This automation of dropping NaN data is quite useful when you're preparing data for the next step: analysis.

Analysis

A fleet of tools is available for loading, processing, storing, and analyzing data computationally. In a Python data analysis environment, the numpy (*http://numpy.org*), scipy (*http://scipy.org*), and pandas (*http://pandas.pydata.org*) packages are the big hammers for numerical analysis. However, many packages within the SciPy and Sci-Kits ecosystems complement those hard-working tools. Some Python-based analysis toolkits to use, organized by discipline, can be found on the SciPy (*http://scipy.org*)

and SciKits (*https://scikits.appspot.com*) websites. There are too many to list here. However, some highlights include:

- Astronomy and astrophysics
    - Astropy (*http://astropy.org*): Core astronomy and astrophysics tools; includes FITS, ASCII, VOTable, and XML file format interfaces
    - PyRAF (*http://bit.ly/py-RAF*): Python-based interface to IRAF
    - SpacePy (*http://spacepy.lanl.gov/*): Data, analysis, and plotting tools for space sciences
    - SunPy (*http://sunpy.org/*): Solar data analysis environment
- Geophysics
    - OSGeo (*http://bit.ly/osgeo-lib*): GIS data import/export and analysis
    - Basemap (*http://matplotlib.org/basemap/*): 2D mapping
- Engineering
    - PyNE (*http://pyne.io*): Toolkit for nuclear engineering
    - `scikit-aero` (*https://scikits.appspot.com/scikit-aero*): Aeronautical engineering calculations in Python
- Mathematics
    - SymPy (*http://sympy.org*): Symbolic mathematics in Python
- Neuroscience
    - NIPY (*http://nipy.org*): Neuroimaging in Python
- Quantum physics and chemistry
    - QuTiP (*http://qutip.org/*): Quantum Toolbox in Python, simulating dynamics of open quantum systems
    - PyQuante (*http://pyquante.sourceforge.net/*): Python for Quantum Chemistry

The analysis step is very application specific and requires domain knowledge on the part of the physicist. A large part of analysis in the physical sciences, when models are derived, confirmed, or disproved based on experimental data, can be termed *inference* or *abstraction*. Abstraction can be an art as well as a science. It can be driven by, generally speaking, either side of the equation: the model or the data.

Model-Driven Analysis

In the case of the decay data, the model-driven analysis is very simple. To determine the decay constant of the isotope in question, we can fit an exponential to the data.

The well-known and accepted model for the decay of a radioactive isotope is $N = N_0 e^{-\lambda t}$.

Of course, that is a simple example. Most analysis in the physical sciences requires many steps of filtering and merging of data as well as integrations, interpolations, scaling, and so on.

> ## A Note on Floating-Point Arithmetic
>
> An excellent resource as you embark on the task of implementing your own numerical analysis algorithms is David Goldberg's paper, What Every Computer Scientist Should Know About Floating-Point Arithmetic. It sounds dry, but it truly is essential reading for any researcher who deals with algorithms that manipulate floating-point numbers.
>
> The paper contains a series of best practices for reducing numerical error obtained during even simple operations such as summation, multiplication, and division.
>
> As an example, the accuracy of simply summing a list of floating-point numbers varies strongly according to the order in which those numbers are summed. Errors, in general, are reduced when smaller numbers are added before larger ones, but more complex algorithms such as the Kahan summation algorithm improve greatly upon simple ordering.

While many of those techniques have now been encapsulated in numerical libraries, some algorithms for data analysis in physics have yet to be written. Furthermore, having a working knowledge of the implementation of the algorithms in the libraries being used for your analysis will help you to determine the applicability of an algorithm to your problem, or to differentiate two options when multiple algorithmic methods are available for a single problem.

Numerical Recipes: The Art of Scientific Computing, by William Press et al., is an excellent reference when implementing a common algorithm is necessary, and was once required reading for computational analysis. It is particularly relevant for model-driven analysis in the physical sciences, which often requires various methods for numerical integrations, solutions of differential equations, and evaluation of large systems of equations. This tome illuminates useful algorithms covering such elements of numerical analysis as:

- Interpolation and extrapolation
- Integration and derivation
- Evaluation of functions
- Inverse functions

- Fermi-Dirac integrals
- Random numbers
- Sorting
- Root finding

Beyond these algorithms, more elaborate methods exist. Many modern algorithms emphasize analysis that does not seek to measure a model based on data. Instead, it often seeks to *generate* models based on data. This is often termed *data-driven* analysis.

Data-Driven Analysis

In data-driven analysis, fancier methods are common. These include clustering algorithms, machine learning processes, and exotic statistical methods. Such algorithms are increasingly available in standard open source libraries and are increasingly common in physical sciences applications. They typically are used to infer models from the data (rather than confirm or deny models using the data). The Python ecosystem possesses many tools enabling such algorithms, including:

- Machine learning and clustering
    - scikit-learn (*https://scikits.appspot.com/scikit-learn*)
    - PyBrain (*http://www.pybrain.org/*)
    - Monte (*http://montepython.sourceforge.net/*)
    - PyPR (*http://pypr.sourceforge.net/*)
    - scipy-cluster (*http://bit.ly/scipy-cluster*)
- Statistics
    - Statsmodels (*http://statsmodels.sourceforge.net/*)
    - PyBayes (*https://github.com/strohel/PyBayes*)

Which algorithms and implementations of those algorithms to use and how to apply them to your data will be choices that are driven by your science. Whatever tools or techniques you use, however, data analysis results in conclusions that can, usually, be visualized.

Visualization

How you visualize your data is the first thing anyone will notice about your paper, and the last thing they'll forget. For this reason, visualization should be taken very seriously and should be regarded as a first-class element of any data analysis workflow.

A lot has been learned about how to present data and information most effectively. Much of this knowledge has emerged from business and marketing contexts.

In science—unlike in business, perhaps—visualization must not attempt to convince or persuade. Bias, obfuscation, and distortion are the mortal enemies of scientific visualization. Visualization in science must demonstrate, clarify, and explain.

Indeed, visualization best practices share many qualities with Python best practices. Python contains an easter egg: a poem on Python best practices, "The Zen of Python," by Tim Peters, is printed in response to the command `import this`. Though it was intended to illuminate guidelines for good Python programming, its first few lines also capture key rules that can be equally well applied to the display of information:

Code	Output
`import this`	The Zen of Python, by Tim Peters
	Beautiful is better than ugly.
	Explicit is better than implicit.
	Simple is better than complex.
	Complex is better than complicated.
	Flat is better than nested.
	Sparse is better than dense.
	Readability counts.
	Special cases aren't special enough to break the rules.
	Although practicality beats purity.

Combining these notions with insights gleaned from Edward Tufte's book *The Visual Display of Quantitative Information* and from Matthew Terry, physicist-turned-software-engineer, I hereby recommend the following "Zen of Scientific Visualization":

- Beautiful is better than ugly.
- Simple is better than complex.
- Complex is better than complicated.
- Clear is better than dense.
- Communicating is better than convincing.
- Text must not distract.
- People are not mantis shrimp.

The final recommendation may seem a bit odd. However, consider the biology of the mantis shrimp (*http://theoatmeal.com/comics/mantis_shrimp*). With vastly more color-perceiving cones in its eyes than a human, the mantis shrimp is able to distinguish colors with vastly greater resolution. People are not mantis shrimp. They are often confused and distracted by too many colors on a single plot. Additionally, many people in the physical sciences are colorblind,[1] so plots that rely too heavily on color to communicate may be somewhat discriminatory. Avoid complicated colormaps, and be sensitive to color blindness by avoiding heavy reliance on color.

Visualization Tools

Many libraries for plotting data exist. This section will introduce a few of the key libraries available for plotting publication-quality scientific data. The tools covered include:

Gnuplot
    Best for simple plots, as the syntax is peculiar

matplotlib
    A powerful plotting library in Python, robust and widely used

Bokeh
    Produces interactive plots appropriate for the Web, also interfaces with matplotlib

Inkscape
    Good for hand editing of scalable vector graphics

This section will introduce these tools by demonstrating their use to plot the decay data from our previous examples in a simple line plot. This introduction should be enough to help you get started with the right tool very quickly when you need to plot you work.

These tools are all available within the scientific Python ecosystem, with one exception: gnuplot. Gnuplot is not a Python tool, but it stands strong as a plotting option nonetheless.

Gnuplot

Gnuplot is a key tool in the physics arsenal. Though it is not a Python tool, it is sufficiently embedded in the physics community that we would be remiss were we to fail to address it here. While it has never had the most beautiful graphics, plentiful error messages, or pleasing syntax, physicists have loved it since its birth. Gnuplot is a workhorse in physics, for better or worse.

---

[1] While efforts are being made to improve the situation, it is an unfortunate fact that there is a long-standing gender imbalance in the physical sciences: even today, there are far more males than females in these fields. Because color vision deficiencies are more common in males than females, this imbalance means that color blindness is, in turn, more common in the physical sciences than in the general population.

The gnuplot interpreter is launched with the command gnuplot. In that interpreter, it is possible to enter commands to construct a plot. However, the most reproducible way to use gnuplot is by creating a script, which is then provided as an argument to the gnuplot command.

> **Exercise: Learn More About Gnuplot**
>
> Since gnuplot is a command-line tool, it is possible to learn more about it using the man command.
>
> 1. Open a terminal.
> 2. Find out how to use the gnuplot command via its man page.
>
> (Hint: For more on man pages, see Chapter 1.)

The following gnuplot script can be used to plot the decay data, along with a title and axis labels:

```
set title 'Decays' ❶
set ylabel 'Decays '
set xlabel 'Time (s)'
set grid  ❷
set term svg  ❸
set output 'plot_gnuplot.svg'  ❹
plot 'decays.csv' every ::1 using 1:2 with lines  ❺❻❼
```

❶ The set keyword defines variables like the title and axis labels.

❷ set can also add predefined customizations—in this case, grid lines.

❸ This sets the output terminal (file) to the SVG file format.

❹ This names the output file.

❺ The plot command accepts data from the input file.

❻ Of the rows in the input file, print all except the first.

❼ Of the columns in the input file, plot 1 and 2 against one another.

This script can be run with the gnuplot command on the command line. Try placing this code in a file called *decay_plot.gnuplot* and running the command gnuplot decay_plot.gnuplot. This script produces the visualization in Figure 7-2.

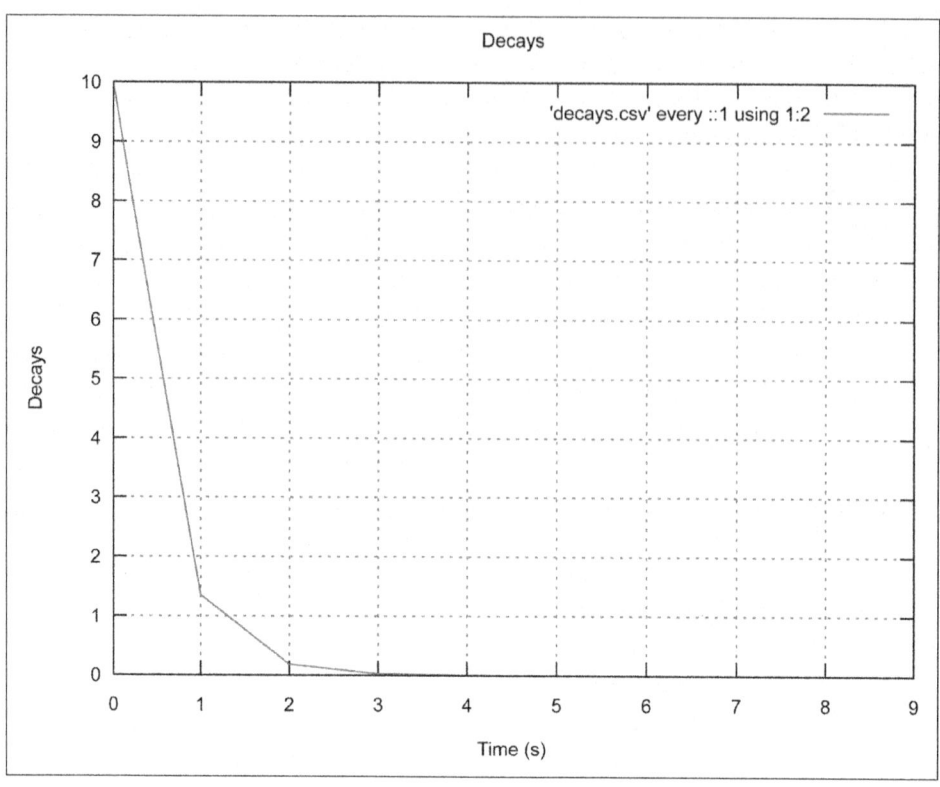

Figure 7-2. Gnuplot plot

By default, gnuplot uses red as the first color to plot in a line plot. Thus, the resulting plot line is red, though we did not dictate a line color. While this is handy, the second deployed default line color in gnuplot is green, which is unfortunate for color-blind people.

> **Exercise: Change the Color**
>
> 1. Open a terminal window.
> 2. Create the *decay_plot.gnuplot* file and Figure 7-2.
> 3. Modify the script to produce a blue line rather than red.
>
> (Hint: Use the man page and documentation online to determine how to change the color of the plot line.)

Though gnuplot is very simple and may be the first plotting library taught to a physicist on the command line, more effective physics can be accomplished if the plotting library is able to make direct use of the data preparation steps described in the previous sections. Additionally, graphics features implemented in more modern Python packages are somewhat superior aesthetically to the graphical capabilities in gnuplot. One such alternative is matplotlib.

matplotlib

The workhorse for scientific plotting in Python is matplotlib (*http://matplotlib.org*). We can reproduce our gnuplot plot with matplotlib by running the following Python script to create the new file:

```
import numpy as np  ❶

# as in the previous example, load decays.csv into a NumPy array
decaydata = np.loadtxt('decays.csv', delimiter=",", skiprows=1)

# provide handles for the x and y columns
time = decaydata[:,0]
decays = decaydata[:,1]

# import the matplotlib plotting functionality
import pylab as plt

plt.plot(time, decays)  ❷

plt.xlabel('Time (s)')
plt.ylabel('Decays')
plt.title('Decays')
plt.grid(True)  ❸
plt.savefig("decays_matplotlib.png")  ❹
```

❶ First we import numpy, so that we can load the data.

❷ This generates a plot of decays vs. time.

❸ This adds gridlines.

❹ This saves the figure as a PNG file (matplotlib guesses based on the extension).

This Python script can be run with the python command on the command line. To create the script on your own, place the code into a file called *decay_plot.py*. Running the command python decay_plot.py produces the plot in Figure 7-3.

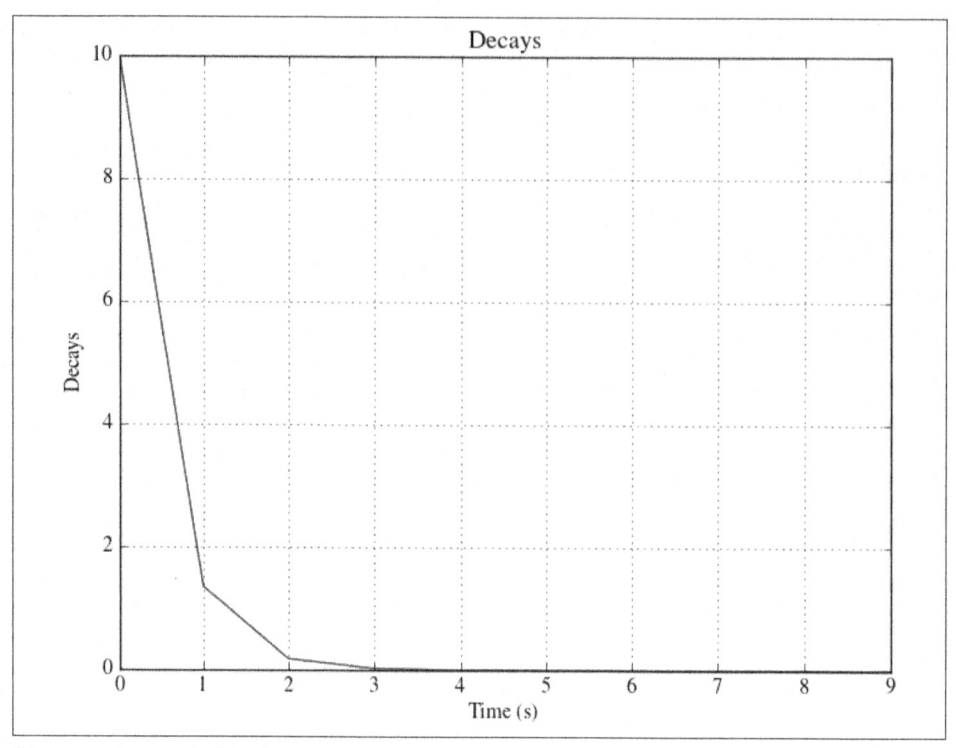

Figure 7-3. matplotlib plot

This plot is not very exciting, of course. When data is more complex—perhaps 2D or 3D, with multiple dependent parameters—matplotlib begins to show its true (very powerful) colors. Equipped with a plethora of plot types and aesthetic options, the power under the hood in matplotlib can be almost overwhelming. In such a situation, the *gallery* in matplotlib comes to the rescue.

The gallery

The best way to start with matplotlib is with the gallery (*http://matplotlib.org/ gallery.html*). Nearly every feature of the matplotlib plotting library is demonstrated by an example plot in the gallery, along with the source code that generated it. It contains a wealth of example scripts, lovingly created by the developers and users of the library.

In a physics context, the gallery is a powerful resource due to the speed with which it enables a researcher to identify desired features of a plot. With the source code for many features available, you can mix and match them to produce a compelling figure with your own scientific data, or, really, any customization at all.

Indeed, matplotlib plots can be used for nearly any purpose. One of the coolest examples in the gallery is certainly the polar plot used in the matplotlib logo (Figure 7-4).

Figure 7-4. matplotlib logo

In 2010, one of the authors of this book had the opportunity to help organize a talk by the creator of this extraordinary library, John D. Hunter.

When someone like this comes to give a talk about their plotting tool, the pressure is on: one must make an excellent flyer to advertise the event. The first step in making a flyer for the talk was to customize the script for the cool polar plot from the gallery. With matplotlib annotations, text boxes were added at specific *x, y* coordinates. To announce the event at the University of Wisconsin, both the Python script shown here and the resulting PDF plot Figure 7-5 were emailed to the students and staff:

```
#!/usr/bin/env python ❶

# Import various necessary Python and matplotlib packages
import numpy as np
import matplotlib.cm as cm ❷
from matplotlib.pyplot import figure, show, rc ❸
from matplotlib.patches import Ellipse ❹

# Create a square figure on which to place the plot
fig = figure(figsize=(8,8))

# Create square axes to hold the circular polar plot
ax = fig.add_axes([0.1, 0.1, 0.8, 0.8], polar=True)

# Generate 20 colored, angular wedges for the polar plot
N = 20
theta = np.arange(0.0, 2*np.pi, 2*np.pi/N)
radii = 10*np.random.rand(N)
width = np.pi/4*np.random.rand(N)
bars = ax.bar(theta, radii, width=width, bottom=0.0)
for r,bar in zip(radii, bars):
    bar.set_facecolor(cm.jet(r/10.))
    bar.set_alpha(0.5)

# Using dictionaries, create a color scheme for the text boxes
bbox_args = dict(boxstyle="round, pad=0.9", fc="green", alpha=0.5)
bbox_white = dict(boxstyle="round, pad=0.9", fc="1", alpha=0.9)
patch_white = dict(boxstyle="round, pad=1", fc="1", ec="1")

# Create various boxes with text annotations in them at specific
```

```python
    # x and y coordinates
    ax.annotate(" ",
        xy=(.5,.93),  ❺
        xycoords='figure fraction',  ❻
        ha="center", va="center",  ❼
        bbox=patch_white)  ❽

    ax.annotate('Matplotlib and the Python Ecosystem for Scientific Computing',
        xy=(.5,.95),
        xycoords='figure fraction',
        xytext=(0, 0), textcoords='offset points',
        size=15,
        ha="center", va="center",
        bbox=bbox_args)

    ax.annotate('Author and Lead Developer \n of Matplotlib ',
        xy=(.5,.82),
        xycoords='figure fraction',
        xytext=(0, 0), textcoords='offset points',
        ha="center", va="center",
        bbox=bbox_args)

    ax.annotate('John D. Hunter',
        xy=(.5,.89),
        xycoords='figure fraction',
        xytext=(0, 0), textcoords='offset points',
        size=15,
        ha="center", va="center",
        bbox=bbox_white)

    ax.annotate('Friday November 5th \n 2:00 pm \n1106ME ',
        xy=(.5,.25),
        xycoords='figure fraction',
        xytext=(0, 0), textcoords='offset points',
        size=15,
        ha="center", va="center",
        bbox=bbox_args)

    ax.annotate('Sponsored by: \n The Hacker Within, \n'
        'The University Lectures Committee, \n The Department of '
        'Medical Physics\n and \n The American Nuclear Society',
        xy=(.78,.1),
        xycoords='figure fraction',
        xytext=(0, 0), textcoords='offset points',
        size=9,
        ha="center", va="center",
        bbox=bbox_args)

    fig.savefig("plot.pdf")
```

❶ This common feature of executable Python scripts alerts the computer which Python to use.

❷ This imports the `colormaps` library from matplotlib.

❸ This imports other libraries (`color`, `figure`, `show`, `rc`) from matplotlib.

❹ This imports ellipse shapes from matplotlib (to be used as text boxes).

❺ This creates an annotation box at certain *x* and *y* coordinates.

❻ Those coordinates should be read as fractions of the figure height and width.

❼ The horizontal and vertical text should be aligned to the center of the box.

❽ The box being placed here should be white.

By executing the script (with `python scriptname.py`), everyone who received the email could produce the flyer shown in Figure 7-5 using matplotlib, the topic of the seminar.

It was a very proud moment for this author when John said he liked the flyer in Figure 7-5. After all, matplotlib was a key ingredient at that time not only in many dissertations, but also in the success of scientific Python. When John passed away in 2012, Fernando Perez described his contribution to the scientific computing community this way:

> In 2002, John was a postdoc at the University of Chicago hospital working on the analysis of epilepsy seizure data in children. Frustrated with the state of the existing proprietary solutions for this class of problems, he started using Python for his work, back when the scientific Python ecosystem was much, much smaller than it is today and this could have been seen as a crazy risk. Furthermore, he found that there were many half-baked solutions for data visualization in Python at the time, but none that truly met his needs. Undeterred, he went on to create matplotlib and thus overcome one of the key obstacles for Python to become the best solution for open source scientific and technical computing.

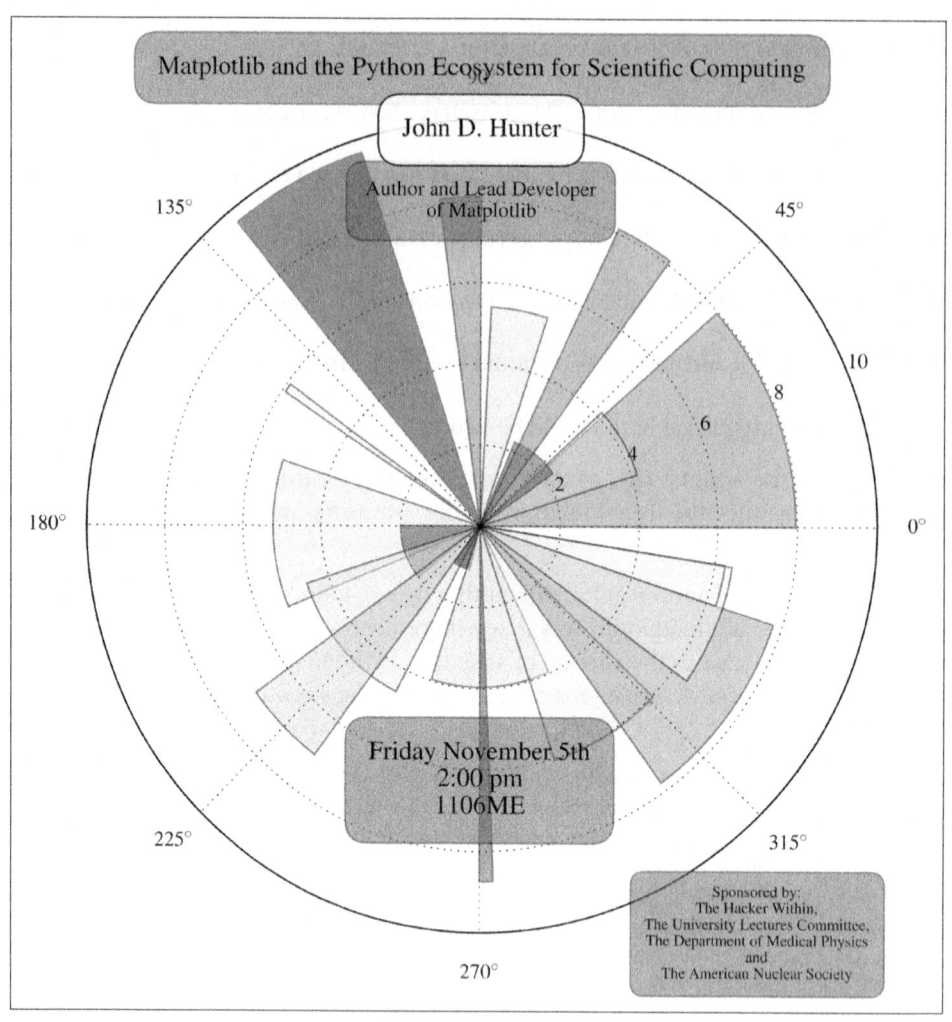

Figure 7-5. Announcement of 2010 John D. Hunter talk

Despite the loss of its creator, matplotlib continues to be improved by a vibrant team of extraordinary people, including Michael Droetboom, who now leads matplotlib development.

Additionally, matplotlib has provided a framework that other tools are capable of linking with. One such tool that is compatible with matplotlib is Bokeh.

Bokeh

Bokeh is a very simple, matplotlib-friendly API that can generate interactive plots appropriate for the Web. It abstracts somewhat from the matplotlib syntax, making the user experience cleaner. The following is a script to plot the decay data as an HTML file using Bokeh:

decay_bokeh.py

```
import numpy as np
# import the Bokeh plotting tools
from bokeh import plotting as bp

# as in the matplotlib example, load decays.csv into a NumPy array
decaydata = np.loadtxt('decays.csv',delimiter=",",skiprows=1)

# provide handles for the x and y columns
time = decaydata[:,0]
decays = decaydata[:,1]

# define some output file metadata
bp.output_file("decays.html", title="Experiment 1 Radioactivity")

# create a figure with fun Internet-friendly features (optional)
bp.figure(tools="pan,wheel_zoom,box_zoom,reset,previewsave")

# on that figure, create a line plot
bp.line(time, decays, x_axis_label="Time (s)", y_axis_label="Decays (#)",
    color='#1F78B4', legend='Decays per second')

# additional customization to the figure can be specified separately
bp.curplot().title = "Decays"
bp.grid().grid_line_alpha=0.3

# open a browser
bp.show()
```

While Bokeh can produce plots in many formats, it was intended to produce interactive plots for viewing in a browser. Thus, when this script is run with `python decay_bokeh.py`, a browser opens automatically to display the interactive, pannable, zoomable plot in Figure 7-6.

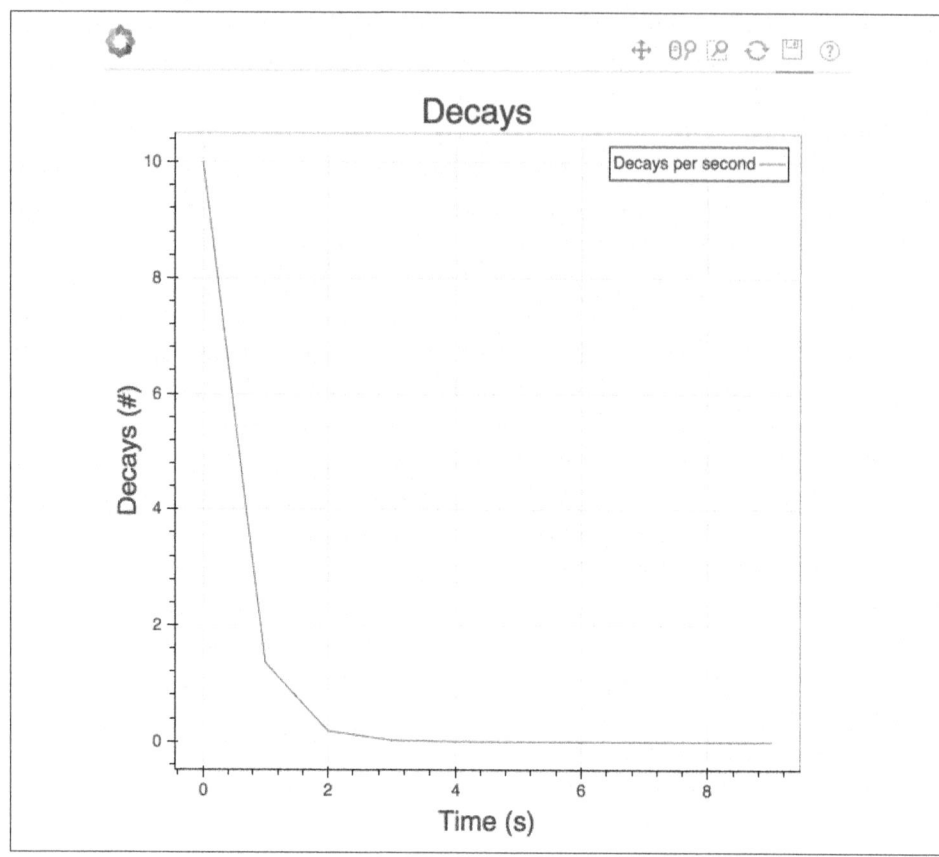

Figure 7-6. Decay plot with Bokeh

Bokeh is an excellent, easy-to-use tool for plots intended for publication on the Web. However, if the plot is complex or is intended for print media, matplotlib or gnuplot will serve that need better.

A key benefit of Bokeh, matplotlib, and gnuplot is their scriptable reproducibility. These tools are therefore the right choice for creating plots that do not yet exist. However, sometimes it is necessary to edit, crop, annotate, or otherwise manipulate an image file that already exists. The scripting tools in this section are all fully capable of handling these tasks. For cases when the original plot or diagram was not made with one of these tools, Inkscape is a good option for the physicist who needs to quickly tweak an image.

Inkscape

Inkscape is an open source project for producing and editing scalable vector graphics. Vector graphics are incredibly powerful. Rather than being pixelated, in a fixed-resolution manner, scalable vector graphics are just that—scalable. Inkscape can be difficult to automate, however, because it is a tool that relies on a graphical user interface for manipulating plots by hand. Of course, this should be done only as a last resort because doing anything by hand does not scale, even if your resulting graphics do.

Analysis and Visualization Wrap-up

The vastness of the fleet of data analysis and visualization tools available today is staggering, and new tools appear at a blinding rate. By walking through a common workflow (Figure 7-1) that plotted the course from raw data to visualization, this chapter briefly introduced some of the tools available for:

- Loading data
- Cleaning data
- Wrangling data
- Analyzing data
- Plotting data

However, we have only covered the bare minimum of available analysis and visualization tools. The intention here was simply to provide you with a notion of the landscape of tasks at hand and tools available to perform them. The following chapters will go into much more depth concerning NumPy (Chapter 9), storing data (Chapter 10), and data structures (Chapter 11).

This chapter covered a few general guidelines that can be followed concerning data loading and preparation:

- Plain-text numerical data of a reasonable size should be loaded with NumPy.
- HDF5 data should be loaded with PyTables or Pandas.
- Pandas is useful for most everything else, especially munging.
- Data can be converted between formats using Blaze.

Visualization is key to presenting the results of your analyses, and before making a decision about what visualization tools to use we recommend that you closely observe the publication-quality plots in your particular subfield of the physical sciences, to get an idea of what tools are typically used and what features are included. More complex data visualizations were out of scope for this chapter. However, many tools for complex visualizations exist. In particular, for physicists with volumetric or higher-dimensional datasets, we recommend sophisticated Python tools for volumetric data such as yt and `mayavi`.

While the ideal tool varies depending on characteristics of the data, workflow, and goals, keep in mind that some things are universal:

- Beautiful is better than ugly.
- Simple is better than complex.
- Complex is better than complicated.
- Clear is better than dense.
- Communicating is better than convincing.
- Text must not distract.
- People are not mantis shrimp.

CHAPTER 8
Regular Expressions

The limits of my language mean the limits of my world.

—Ludwig Wittgenstein

The world's first computers were women. They were the employees of research centers and national labs, where they inspected data, executed algorithms, and reorganized data. Their job title was "computer" because they computed. In the early days, computing meant evaluating raw data by hand for a variety of applications and experiments, including, famously, the Manhattan Project.

You, too, may have raw data. However, today's data should not be processed by hand. Today's data is usually too big, the risk of carpal tunnel is too high, and computers are too powerful to justify that. Processing raw textual physics data may require:

- Searching for and correcting irregularities
- Finding and replacing text across hundreds of files
- Evaluating mathematical expressions
- Manipulating number formatting
- Rearranging column-formatted data

This chapter will discuss regular expressions, a common syntax for matching patterns of characters in text files, data files, filenames, and other sequences of characters. This syntax is ubiquitous in the programming world because it can turn an enormous, tedious file cleanup task into a tiny one-line command. Additionally, it can help with day-to-day command-line navigation, file parsing, and text editing.

In the shell, regular expressions can be used to clean up and analyze raw data in conjunction with the search and print programs that will be discussed in this chapter. These, like *grep*, *sed*, and *awk*, were designed for finding, displaying, editing, and doing calculations with information in files. The regular expression module in Python (re) can be used in the same way, and this chapter will help to demonstrate how it boosts the already awesome power of Python.

In particular, this chapter will discuss basic regular expression syntax in the context of how it can help the physicist:

- Navigate the command line more efficiently.
- Quickly find files on the command line based on their content (*grep*).
- Find and replace a complex expression in many files at once (*sed*).
- Quickly do math on plain-text columns of data from the command line (*awk*).

We'll explore how you can accomplish some of these things in Python, as well.

> ### A Short History of Regular Expressions
>
> Because regular expressions are one of the oldest and most powerful tools in computing, let us take a brief foray into their history.
>
> With inspiration from two neurologists, Stephen Kleene, a 1940s mathematician, developed a formal notation for a logical classification of typeset characters. He called them regular sets. Combining elements of these *regular sets*, he created regular expressions. Ken Thompson, a programmer in the late 1960s, developed a text editor called *ed* with the ability to search a file for patterns defined with this notation. To search for a pattern within a document in *ed*, the user deployed the command to search *g*lobally for a *r*egular *e*xpression and *p*rint the modifications. It looked like `:g/re/p`. That command would become what is now known as *grep*.

Now that we have covered the basics, we are going to dive in. The following sections are a condensed, example-driven introduction that will help you get comfortable with the syntax of regular expressions.

Messy Magnetism

Regular expressions are a tool for text matching. Command-line scripting tools like *grep*, *sed*, *awk*, Python, and Perl make use of regular expressions. Using them (combined with a scripting language), a physicist can automate alteration of large files and sets of files. Additionally, the syntax of regular expressions is often deployable from within text editors like *vim*, and it can accordingly speed up code development enormously. As a concrete example, this chapter will follow the plight of a new graduate student in a lab studying the Earth's magnetism.

Imagine you have just joined a research group that analyzes data gathered by hundreds of satellites all across Earth's orbit. These satellites are being used to take simultaneous magnetic field measurements around the Earth. However, they are telecommunications satellites owned by private companies, so the researchers have access to the data only through agreements with the companies that own them. Accordingly, the data is sent to the researchers in many different formats. After a few months of data collection, many gigabytes of readings are now stored haphazardly on one of the computers in the lab. Before being analyzed, it will have to be organized and the various formats will need to be standardized.

As the newest member of the research group, you have therefore been tasked to:

- Find the files from each company and collect them into a single directory.
- Standardize the formats of dates, times, and labels across all the files.
- If the flux energy is in units of gauss (G), convert it to teslas (T).

All of this could take a very long time by hand. In fact, with over 400 files and thousands of necessary changes per file, it could take a very long time indeed. This chapter will show how, with regular expressions, you can spend the morning setting up a few scripts and leave the lab while the sun still shines in the sky.

To get started, we need to start navigating the filesystem to identify the files we are looking for. In the next section, we will use *metacharacters* that we are already familiar with as well as more complex regular expressions to find the data files.

Metacharacters on the Command Line

Day-to-day tasks, like finding files on the command line, can be sped up with regular expressions. By employing *metacharacters*, commands can be made to operate on many files at once.

Metacharacters are characters that have a special meaning aside from their literal meaning. *Literal characters*, on the other hand, are those that are being taken at face value. Regular expressions are strings made of metacharacters and literal characters.

We have already seen a number of metacharacters on the command line. In particular, in Chapter 1 we saw the use of the wildcard asterisk (*) and the backslash escape character (\). However, there are many more, and they follow a few rules:

- Alphanumeric characters match themselves.
- A dot (.) matches any character.
- Repeating patterns are matched with *, +, and ?.
- Character sets ([ ]) and the *or* operator (|) can match alternatives.

- The position markers ^ and $ match the beginning and end of a line, respectively.
- Parentheses can group things and extract information from matches.

Beginning with these basic rules, this section will show how to use the command-line tools ls and find to locate files whose filenames match a pattern.

Listing Files with Simple Patterns

We don't know a lot about the satellite data files we are looking for, but we do suspect that some of them end in the *.dat* extension. Ideally, we could at least start the search by browsing the directories and listing all files with names that end in ".dat".

In Chapter 1 we saw how * is used to match all of the filenames in a directory. That asterisk, or *wildcard* character, is actually capable of much more and can solve this problem in a flash. In the next section, we'll learn how.

The wildcard (*)

As discussed in Chapter 1, wildcard characters can be used to find files more effectively. Recall the asterisk. It allows the user to list all the text files in a directory by typing ls *.txt, because * means "zero or more of any character." This is because the asterisk is a metacharacter in the shell. Regular expressions are a language of metacharacters used for the purpose of pattern matching.

In a regular expression, the wildcard * matches the preceding item zero or more times. Some regular expressions and the text that they match within target strings are listed in Table 8-1. The first column shows the regular expression and the second shows the string to which it is applied. The bold elements of the string are those that were matched by the regular expression. For example, the regular expression zo* matches *zoooo* in "He zoooomed" and the first *z* in "motoguzzi."

Table 8-1. Matching with the wildcard

Expression	Matches
zo*	**zooo**med
zo*	moto**g**uzzi
zo*	proto**zo**an
p*i	mississ**ippi**

To find all of the .dat files in the current working directory, then, we can execute a simple ls command using the asterisk syntax. This command and its results are shown in Table 8-2.

Table 8-2. The wildcard on the command line

Code	Output
ls *.dat	atmos_sat_42.dat ... ... ❶ ... siriuscybernet_21.dat siriuscybernet_22.dat siriuscybernet_23.dat siriuscybernet_24.dat siriuscybernet_68.dat siriuscybernet_92.dat Telecomm99_2014.5.1.dat Telecomm99_2014.5.2.dat Telecomm99_2014.5.3.dat Telecomm99_2014.5.4.dat Telecomm99_2014.5.5.dat Telecomm99_2014.5.6.dat zorbital34l.2014.05.01.dat zorbital34l.2014.05.02.dat zorbital34l.2014.05.03.dat zorbital34l.2014.05.04.dat zorbital34l.2014.05.05.dat zorbital34l.2014.05.06.dat

❶ For brevity, not all of the files are listed; there are many.

This syntax means "List all files beginning with zero or more of any character, followed by the .dat string." While this will certainly uncover all of the .dat files in the current directory, it's not powerful enough for our task, for two reasons:

- The ls command only goes one level deep into the directory tree. Upon inspection, it becomes clear that the filesystem is riddled with directories and subdirectories of the data. For example, executing ls (with no argument) illuminates a few dozen directories named things like *MegaCorp* and *Oscorp*, which certainly contain more satellite data from the satellites owned by those megacorporations.

- Further inspection also indicates that a few files in the home directory weren't caught because they had a different file extension. Observed extensions so far include *.txt*, *.data*, and an all-caps *.DAT*.

Does this mean we have to enter every directory and execute multiple forms of the ls command? Traversing the filesystem one directory at a time, repeating a few ls *.dat commands in each directory, is not the way to spend an afternoon.

Globally Finding Filenames with Patterns (find)

A better way to spend the afternoon might be at the beach. To get out of the lab, we'll have to find a tool that can recursively search a lot of directories at once. That would improve this situation mightily. Thankfully, that tool exists, and it is called find.

The find command can be used in many ways. One option is to use it with regular expressions. In that mode, find is executed on the command line using the format:

    find [path] -regex "<expression>"

With this syntax, the top level of the search will be the indicated <path>. find will begin at that location and recursively parse filenames using regular expressions (-regex). The expression for which it will seek matches is provided between the double quotes.

To find all data files in the home directory *and all subdirectories of the home directory*, find can be used thus, in our case:

    find ~ -regex ".*\.dat"

This finds all files in ~ and all subdirectories of ~ whose names end in ".dat". But why are those dots and slashes needed? They weren't necessary when we used the ls command.

Irregular Expressions

Importantly, regular expressions are not as regular as one might imagine. That is, syntactic use of regular expressions will occasionally vary from tool to tool. *grep*, *awk*, *sed*, Python, and Perl have slightly varying implementations of certain characters. The find command can actually parse many flavors of regular expressions. You specify the flavor by using flags. Available flavors include:

- Basic regular expressions (BRE)
- Extended regular expressions (ERE)
- Perl-compatible regular expressions (PCRE)

Of course, use of each flavor will require investigation into its documentation. In this chapter, we'll introduce the basic metacharacters and character class treatments recognizable in most tools.

Fundamentally, this is because the set of metacharacters available to the `ls` command is a different flavor. While the wildcard is available on the command line, it doesn't mean the same thing on the command line that it does in proper regular expression syntax. On the command line, `.*` means "one dot (.), then zero or more of any character." In a regex, it means "zero or more of any character (.)."

So, we need the extra characters because the dot character (.) is not a metacharacter on the command line. On the command line, it signifies the current working directory or, when it appears at the beginning of a filename, a hidden file, as discussed in Chapter 1. However, the dot character (.) *is* a metacharacter in proper regular expressions. For this reason, the backslash is used before the *real* dot in ".dat" to indicate it should be taken literally.

That all may be a bit confusing. Let's take a step back and look at those two metacharacters (dot and backslash) more closely, so that we can digest this syntax.

Representing any character (.)

The dot character in a regular expression is very simple: it represents exactly one of *any* character. Note that the dot means something different on the command line, as just described.

> This bears repeating. On the command line, the dot does not mean "any character." In proper regular expressions, however, it does. On the command line, the dot already has a special meaning; to represent "any character," the ? can be used instead.

Since the dot means *any* character, it can be used to help match many files whose names share most, but not all, characters. Table 8-3 demonstrates the dot character in action.

Table 8-3. Matching with the dot character

Expression	Matches
r.d.	rads
r.d.	rodeo
r.d.	rider
r.d.	red!
r.d.	r2d2

In our example with the satellite *.dat* files, many of the Sirius Cybernetics satellite files had very similar names:

```
siriuscybernet_21.dat
siriuscybernet_22.dat
siriuscybernet_23.dat
siriuscybernet_24.dat
siriuscybernet_68.dat
siriuscybernet_92.dat
```

Indeed, many of those filenames include numbers in the 20s. Those files with numbers that start with 2 could be matched exactly with the regular expression:

```
siriuscybernet_2.\.dat
```

The first part of the expression, `siriuscybernet_2`, matches that string exactly. This pattern therefore matches the first four filenames in the preceding list, but not the last two (*siriuscybernet_68.dat* and *siriuscybernet_92.dat*).

The next character in the expression is the "one of any character" represented by the dot (.). In the case of the first file, the dot matches the *1*. In the case of the second file, it matches the *2*, and so on.

After that, however, there again is the backslash (\) character. Why does it keep popping up? We'll learn about this in the next section.

Escaping metacharacters (\)

Sometimes, as in the previous example, the pattern that you would like to match contains an actual dot (.) or an actual asterisk (*). To indicate the literal . character rather than the . metacharacter, it must be *escaped*. To escape a metacharacter is to escape its meta-meaning. We *escape* metacharacters by using the backslash (\). The expression . means "exactly one of any character." However, the expression \. means "exactly one dot." The difference is illustrated in Table 8-4.

Table 8-4. Escaping metacharacters

Expression	Matches
deat.*	deathstar
deat.*	death*

A common need for this escape character arises in the shell, where many commands require arguments separated by spaces. That is, in the shell, the space is a metacharacter for separating arguments, as in:

```
~ $ mv data.txt old.txt
```

Since some of those arguments are filenames, the question arises: how does the computer know the difference between a filename containing a space and a pair of arguments? The answer is that it doesn't know the difference:

```
~ $ mv data from yesterday.txt old.txt
usage: mv [-f | -i | -n] [-v] source target
       mv [-f | -i | -n] [-v] source ... directory
```

Just as with the *, to make the space a literal space, you must use the backslash \.

```
~ $ mv data\ from\ yesterday.txt old.txt
```

So, metacharacters really have two meanings: their special meaning and their literal meaning. The backslash can help you switch between the two.

> ## Escaping Literals
>
> Indeed, in addition to transforming metacharacters into literals, the ubiquitous backslash transforms many literals into metacharacters. Some examples include the end-of-line characters \n (linefeed) and \r (carriage return), as well as the tab metacharacter \t, NULL metacharacter \0, and whitespace metacharacter \s. Myriad other metacharacters exist, including the parentheses used in "Finding and Replacing Patterns in Files (sed)" on page 190.

Exercise: Escaping the Escape Character

1. Open a terminal.
2. Try to create a file that has a backslash in the filename with a command like `touch file\name`.
3. Use `ls` to examine the file you've just created. Did it work? Where is the slash?
4. Use what you've just learned to escape the escape character. Can you successfully make a file called *file\name*?

So, with this information, it is possible to execute the `find` command in such a way as to find all files in all subdirectories of the home directory with names that end in *.dat*. We can just use the pattern `.*\.dat`.

While that is a huge improvement over spending all afternoon traversing the filesystem, what about the files whose names end in *.data*, *.txt*, or *.DAT*? For that, we will have to proceed to the following section.

Finding either one pattern or another (|)

In order to match files with various extensions (*.txt*, *.dat*, *.data*, *.DAT*), we need to have an *or* Boolean available. With regular expressions, this is called *alternation* and is accomplished with the | syntax. That is, to search for any appearance of the word *proton* or the word *neutron*, you would separate the two with a vertical bar: `proton|neutron`. For more options, continue to expand the list: `proton|neutron|electron|boson`.

Exercise: Reverse-Engineer a Regex

The following string will find either *.dat* or *.DAT* extended files:

```
~ $ find . -regextype posix-extended -regex ".*\(\.dat\|\.DAT\)"
```

1. Can you tell why?
2. What are the slashes there for?
3. What about the extra specification of `-regextype posix-extended`?
4. Can you find out what that means from the `man` page for `find`?

Unfortunately, this "or" syntax is notoriously hard to read. There is a more readable way to specify something similar with character sets. The next section will discuss these.

Character sets ([...])

The syntax that allows matching of a set of characters is [] or [{}], depending on the tool being used. A character set matches any one of the enclosed characters. A few examples are given in Table 8-5.

Table 8-5. Character sets

Expression	Matches
`Col[ou]mbia`	**Columbia, Colombia**
`[Dd]ata`	**Data, data**
`[Dd][Aa][Tt][Aa]`	**DATA, data, Data**
`2014[-.]05[-.]10`	**2014.05.10, 2014-05-10**

This makes it easy to avoid worrying about capitalization, varying spellings across the pond, and many other matching issues. In our case, it means that *.DAT*, *.dat*, and *.data* files can all be found with one expression:

```
~ $ find . -regex ".*\.[Dd][Aa][Tt][Aa]*"
```

> ### Key Features of Character Sets
>
> - A character set will match any one character from the set of characters between the brackets. `[Dd]ad` will match *Dad* or *dad*.
> - Character sets can be specified in shorthand over a range using a hyphen. For letters, `[lmnop]` is identical to `[l-p]`. Each set will match any one of the characters "l" through "p" (like the "p" in "zap" or the "l" in "laser"). This works for both numbers and letters. Thus, `[0-9]` matches any digit.
> - A character set can indicate characters that we want to avoid. A caret in the square brackets (`[^{...}]`) denotes a negated character set: it will match anything *not* in the square brackets. For instance, `[\^aeiou]` matches only consonants.
> - Character sets can be combined. `[a-zA-Z]` is valid notation matching all letters, irrespective of case.
> - These sets also can be referred to by nicknames, such as `[:alpha:]` (letters), `[:alnum:]` (letters and numbers), `[:lower:]` (lowercase letters), `[:digit:]` (numbers), etc.

We may make it to the beach this afternoon after all! Now that we have been able to find all of the necessary files, we can go organize them. To write a shell script utilizing the output of the `find` command, go back to Chapter 1 and bash scripting to determine how to move all of the files we have found. It should take about six lines of code.

Next, the task moves on into the content of the files themselves. The `find` command on the command line only addresses the filenames. For more in-depth pattern searching, we will need the tried and true family of tools *grep*, *sed*, and *awk*.

grep, sed, and awk

We said before that *grep*, *sed*, and *awk* are a family of tools that use regular expressions and are available on the command line. They each have different capabilities:

- The `grep` command has the basic syntax `grep <pattern> <inputfile>`. *grep* grabs matched patterns and prints them.

- The sed command has the basic syntax `sed "s/<pattern>/<substitution>/" <inputfile>`. *Sed* combines *grep* with a substitution command.
- The awk command has the basic syntax `awk pattern [action]`. *awk* handles columns.

This chapter will touch on how each tool can help to accomplish the goals in the satellite data problem. *grep* will help us to investigate the contents of the files, *sed* will help us to make substitutions where formatting varies across the files, and *awk* will allow us to do simple calculations on the columns of data.

Now that we have rearranged all of the found files into a directory, our next task should be to learn some more about them. For this, we will use *grep*.

Finding Patterns in Files (grep)

grep is as essential a tool for programming as Google is for navigating the Internet. It is useful for everything from inspecting files to debugging. Indeed, *grep* works a lot like Google. It searches, globally, for regular expressions inside files, based on their content.

Day-to-day scientific computing, accordingly, relies on *grep* to make sense of the files encountered. For example, when I receive an error message generated from source code, I immediately search the code files for the phrase given by the exception. From this I can quickly find the filename and the line number at which the error was thrown.

In the satellite data example, we want to know a few things before we start fixing the files:

- How many and which of the files use Gs?
- Which files use the dash and which use the dot in date formats?

grep can help answer these questions. To answer the first question, we simply want to tell *grep* "search for all instances of Gs among the files in this directory." The syntax is simply `grep Gs *`. It's the simplest possible command, and yet it is so powerful! An example of the use of this tool is seen in Table 8-6.

Table 8-6. The wildcard with grep

Code	Output
grep Gs *	Oscorp.DAT_1:2000-1-1,481.983486734Gs
	Oscorp.DAT_1:2000-1-2,254.229864682Gs
	Oscorp.DAT_1:2000-1-3,57.4087373703Gs
	Oscorp.DAT_1:2000-1-4,425.027959432Gs
	Oscorp.DAT_1:2000-1-5,175.497450766Gs
	Oscorp.DAT_1:2000-1-6,304.130011333Gs
	Oscorp.DAT_1:2000-1-7,365.090569435Gs
	Oscorp.DAT_1:2000-1-8,357.834192688Gs
	Oscorp.DAT_1:2000-1-9,378.059846154Gs
	Oscorp.DAT_1:2000-1-10,179.401350076Gs
	...

If *grep* piques your curiosity, you may be interested in some of its extra features. Some common options for use with *grep* include:

-r
: The recursive flag. Do you recall from Chapter 1 what that means?

-c
: Counts matched lines.

-i
: Ignores capitalization. Can you think of an equivalent expression using brackets?

Additionally, further experience with *grep* can be had very safely within the system files on your computer. On a Unix or Linux platform, all of the words that the default spell-checker uses are stored in a plain-text file. The *grep* exercises in the following sidebar can be performed safely on that file.

Extended Exercises with grep

Enter the directory */usr/share/dict*, and investigate the document called *words*. Use wc -l. Use -c. On the command line, try the following exercises to familiarize yourself with grep:

- Look for the word *hacker* in the *words* document, using the command grep "hacker" words.
- Compare grep -c "within" words and grep -c "\within" words.
- Compare grep -c ".ithin" words to grep -c "\within" words.
- Try grep -c ".*ithin" words, etc.
- Try grep -c "pre.+" words and grep -c ".+pre.+" words.
- Compare grep "cat" words and grep "^cat" words.

- Compare grep "cat" words and grep "cat" words.
- Find blank lines with grep "^$" words.
- Compare grep ^[rstu]+$ words and grep ^[r-u]+$ words.
- Compare "grep \^[r-u]{3}" words to grep "[r-u]\{3,\}" words.

Now try the following challenges with the *words* file:

- Find all three-letter words.
- Find all three-letter words involving the letter *r*.
- Find the words with two consecutive *a*s.
- Find the words ending in *'s*.

As you can see, *grep* is indispensable to the efficient use of a computer. That you will need such a tool should not be surprising, as the importance of being able to find stuff efficiently scales directly with how much stuff is available. The speed with which these two things grow in the modern world is, arguably, the reason a worldwide megacompany was capable of bootstrapping itself out of no more than a search algorithm (Google).

All that said, it is often the case that even finding stuff is not powerful enough. Usually, the scientist needs to find patterns primarily *to replace them with something else*. The tool for this, on the command line, is *sed*. The next section will discuss *sed* in more detail.

Finding and Replacing Patterns in Files (sed)

sed is a tool for substitution. It is essentially the same as *grep*, but has one major extension: once a character string is matched, it can be substituted with something else. Thus, *sed* differs from *grep*, but can duplicate grep as well.

> Additionally, *sed* is enabled natively in the *vim* text editor. There, it can be accessed with <ESC>:s. For more help on *sed* within *vim*, type <ESC>:help sed and press Enter.

The syntax for *sed* substitution through a whole file is:

```
sed "s/<expression>/<substitution>/g" <inputfile>
```

This allows very quick substitution of a simple string in a file. In this example, the s (before the first slash) indicates "substitution" and the g (following the final slash) indicates substituting "globally" on every line of the file. Without the g, *sed* will only replace the first instance of the matched pattern in each line.

> Take note! *sed* can be run with the syntax `sed "s/<expression>/<substitution>/" <inputfile>`. However, without the g after the final slash, *sed* only changes the first match in each line. This is rarely the desired default behavior. Specify global substitution to capture all matches in a *line*!

The result of this syntax is simple: *sed* outputs the altered text of the file to the command line. With the default syntax, *sed* does not edit the file in-place. Rather, the changed text is sent to the standard output of the terminal. To capture this output, a temporary file is usually made.

Exercise: Redirect sed Output to a File

1. Execute a `sed` command on a file in your filesystem (try something simple like `"s/the/THE/g"`).
2. Note that the altered file text has appeared on the command line.
3. Using your knowledge of redirection (from Chapter 1), re-execute the command, this time sending the output to a temporary file.

If there were only one day's worth of data in the satellite data, it might make sense to directly substitute the badly formatted date type with the better-formatted date type, like so:

    sed "s/2014\.05\.01/2014-05-01/g" atmos_sat_42.dat

However, since we have many files, this will have to be scripted. Thankfully, it is possible for *sed* to save output as a new file, using this syntax:

    sed "s/<expression>/<substitution>/" <oldfile> > <newfile>

Also, for the brave, *sed* has a flag (`-i`) that causes it to edit the file in-place (no temporary file necessary):

    sed -i "s/<expression>/<substitution>/" <oldfile>

sed substituion can be made even more efficient if multiple replacement tasks are necessary per file. In that case, you can give multiple commands by piping *sed* output back into *sed* (recall the pipe from Chapter 1):

    sed "s/a/A/" oldfile.txt | sed "s/b/B/"

This is equivalent to including the -e flag before each substitution:

```
sed -e "s/a/A/" -e "s/b/B/" oldfile.txt
```

That task was easy enough, but the more challenging task for the satellite data will be to replace ill-formed date stamps with better-formed date stamps. In particular, a complex expression will need to be formed in order to match all of the known ill-formed date formats.

Finding and Replacing a Complex Pattern

To find and replace *all* dates, we must:

- Match the pattern for any ill-formatted date.
- Retrieve the date stamp and save the values.
- Substitute the format, using the saved date values.

As saving the values is necessary here, let us take a brief sidestep into capturing patterns during *sed* searches.

Since, sometimes, you'll need to reuse part of the pattern you matched, *sed* has syntax to hold the match in memory. It uses parentheses. Specifically, the following syntax matches x and remembers the match:

```
\(x\)
```

These are *capturing parentheses*. With these parentheses, (20[01][0-9]) matches and remembers the "2014" in "2014-05-01". That string can then be retrieved and reused during the substitution.

This remembering happens in a list that *sed* stores during the substitution. *sed* can store many of these snippets per substitution task, and they are retrieved in the order that they were created with a simple \N, where N is the index of the stored snippet.

We know that, in our example with the date formats, the pattern we would like to match is:

```
"20[01][0-9].*[0-9][0-9].*[0-9][0-9]"
```

The types of strings that will be matched by this include:

```
2014-05-01
2014-09-10
2015-10-30
2014.06.24
2014/09/23
2010/12/29
```

.
.
.

From this matched pattern, we would like to retrieve the year, month, and date separately so that the dates can be reformatted as "YYYY-MM-DD". With *sed* and its memory, the year is saved first and retrieved as \1. The month is saved second and retrieved as \2, and so on.

The `sed` command that fixes all of the malformed dates is then:

```
sed "s/\(20[01][0-9]\).*\([0-9][0-9]\).*\([0-9][0-9]\)/\1-\2-\3/g" <filename.dat>
```

Take some time to digest that. It's ugly, but should be comprehensible with some dissection. Note that regular expressions, like Perl, are a written language. These are not truly meant to be read.

Once that has been digested, we can allow the syntax to cement with some extra exposure in the next section.

sed Extras

Many useful things are possible with *sed*. Let's take the *phone.txt* file (in the code repository (*https://github.com/physics-codes/examples*) for this book) as an example.

sed is best for editing files, and it will do so globally with only a few keystrokes. For safety, we should try not to change important system files (woe betide he who damages the hex color code for peach puff). Let's relocate to the *regex* directory in the code files associated with this book.

With *sed*, we can use the d character to delete all blank lines in the file of phone numbers:

```
sed '/^$/d' phone.txt
```

It can also help us automatically double-space the file. We can do this in a brute-force way by deleting blank lines and appending carriage returns (\n) to each line:

```
sed -e '/^$/d' -e 's/^\(.\+\)$/\1\n/' phone.txt
```

But there's an easier way. Try G:

```
sed -e '/^$/d' -e G phone.txt
```

Or even just:

```
sed '/^$/d;G' phone.txt
```

Similar to our date exercise, it is possible to reformat the phone numbers in *phone.txt*. Note that this task uses the tool for remembering strings, \(\), discussed earlier in the chapter:

```
sed 's/.*\([0-9]\{3\}\).*\([0-9]\{3\}\).*\([0-9]\{3\}\).*/(\1)\2-\3/' phone.txt
```

You can number the lines of a file for a more readable display. The syntax for this one is somewhat complex. See if you can figure it out:

```
sed '/./=' wordlist.txt | sed '/./N; s/\n/ /'
```

Also, if you only want to modify a small section of the file, you can specify a line number to modify by indicating the line number at the beginning of the command:

```
sed '4 s/r/R/' wordlist.txt
```

You can even specify a range of line numbers to modify by specifying two numbers, separated by a comma:

```
sed 'sed '4,6 s/r/T/' wordlist.txt
```

You can also select lines to modify by pattern matching:

```
sed '/^z/ s/$/zzzzzzzzzz$/' wordlist.txt
sed '/wade/,/salt/ s/m/PPPPPPPPPP/' wordlist.txt
```

Before we move forward (to fix the units in our satellite data files), let's take a step back and reflect in Table 8-7 on what we have learned is possible with regular expressions (and a couple of extra nuggets to whet your appetite for more).

Table 8-7. Regular expression examples

Expression	Matches
uvwxyz	uvwxyz
[u-z]	One of either u, v, q, x, y, or z
[^]	One of any character except a space
p*i	Zero or more p characters followed by one i, such as pi or ppppi or i
.*	Zero or more of any character, such as super califragilisticexpialidocious or 42
\^spelunking.*(.*)	A line starting with spelunking, followed by an opening and closing parenthesis with any string in them
\\$	A line ending with just one backslash
\$	A (literal) dollar sign
.\{4\}z	Any four characters followed by a z

Now that we have successfully shown that regular expressions can be used to:

- Find files based on their names (*find*)
- Find files based on their content (*grep*)
- Replace content based on found patterns (*sed*)

The only remaining task is to show how to manipulate columns of data within files using patterns. For this, we will introduce *awk*.

Manipulating Columns of Data (awk)

A lot of data in physics begins in a simple format: columns of numbers in plain-text documents. Fortunately for us, a command-line tool called *awk* was invented long ago to quickly and efficiently sort, modify, and evaluate such files. This tool, a sibling to *sed* and *grep*, uses regular expressions to get the job done.

It's not elegant or modern—indeed, it may be the oldest tool in this particular book—but *awk* is always there. One day, you too will find yourself working on a high-performance computing cluster that holds all of your data, where you don't have permission to install a new version of Python. *awk* will be there to help you manipulate your columns of data.

Before we fix the units in our satellite data example, it is worth taking some time to understand *awk*. As an introductory example, we can investigate the files in the filesystem. On a Linux platform, a list of colors available to the system is found in the */usr/share/X11* directory. On a Unix (Mac OS X) platform, it is made available in */usr/X11/share/X11*.

The *rgb.txt* file in that directory looks like this:

```
255 250 250     snow
248 248 255     ghost white
248 248 255     GhostWhite
245 245 245     white smoke
245 245 245     WhiteSmoke
220 220 220     gainsboro
255 250 240     floral white
255 250 240     FloralWhite
253 245 230     old lace
253 245 230     OldLace
...
```

To get started investigating and manipulating this columnar data, note that *awk* can somewhat replicate what *sed* and *grep* do. Given a regular expression to match, *awk* will return the matching lines of a file. To get a feel for this, observe the results of the *awk* commands in Table 8-8.

Table 8-8. Using awk to find matching rows

Code	Output	
awk '/puff/' rgb.txt	255 218 185	peach puff
awk '/144/' rgb.txt	112 128 144	slate gray
	112 128 144	SlateGray
	112 128 144	slate grey
	112 128 144	SlateGrey
	30 144 255	dodger blue
	30 144 255	DodgerBlue
	208 32 144	violet red
	208 32 144	VioletRed
	30 144 255	DodgerBlue1
	144 238 144	PaleGreen2
	205 96 144	HotPink3
	205 41 144	maroon3
	144 238 144	light green
	144 238 144	LightGreen

We can select the colors that have 144 in the beginning of their hex strings as in Table 8-9—the caret tells *awk* to search for a match at the start of each line in the input file.

Table 8-9. awk and the caret

Code	Output	
awk '/^144/' rgb.txt	144 238 144	PaleGreen2
	144 238 144	light green
	144 238 144	LightGreen

We can even pick out the only color with 144 in the middle, as in Table 8-10.

Table 8-10. awk extended pattern

Code	Output	
awk '/^.*\ 144\ .*/' rgb.txt	30 144 255	dodger blue

In addition to just replicating some of *grep*'s capabilities, *awk* can add an action. However, it can only do actions on a column-wise basis. Note that dollar signs ($) indicate columns:

```
awk '{print $1$2$3}' rgb.txt
awk '/^255/{print $1$2$3}' rgb.txt
awk '/^.+ +.+144/{print $1$2$3}' rgb.txt
```

The column order can also be switched:

```
awk '{print $1," ",$2," ",$2," "$1}' rgb.txt
```

Or we can modify just one line:

```
awk NR==11'{print $1 $2 $3,"\t",$4}' rgb.txt
```

And finally, we can do math with *awk*:

```
awk NR==11'{print $1,"+",$2,"+",$3,"=",$1+$2+$3}' rgb.txt
```

The final task for the satellite data can be accomplished in an exactly analogous fashion.

Exercise: Convert Gs to Tesla

1. Use *sed* to parse one of the data files for instances of "Gs".
2. When found, use *awk* to multiply one column with another.
3. Finally, use *sed* to change the "Gs" string to "T".

It's okay to do this in multiple commands. However, it is possible to do it in one command.

Now that this complex task is complete, we can take the afternoon off! Or, we can spend it learning a little more Python. If you're in the latter camp, read on!

Python Regular Expressions

Everything we've seen how to do so far in this chapter is also possible in Python. Alternatives to all of these tools exist in the Python regular expression module re, which comes as part of the Python standard library. The re module allows Python-flavored regular expression pattern matching. Combined with the power of other Python modules, the features of *grep*, *sed*, and *awk* can be replicated in Python in an arguably cleaner and more robust syntax, ready for linking to other workflow process subparts.

We've emphasized the command-line tools due to their day-to-day importance in the life of a programmer. Very often, regular expression searches are one-offs, such that they are most swiftly executed on the command line. However, the power of regular expressions combined with the power of Python results in something quite formidable.

grep's capabilities can be replaced with:

- re.match(<pattern>, <string>) to match a regular expression pattern to the beginning of a string
- re.search(<pattern>, <string>) to search a string for the presence of a pattern

- `re.findall(<pattern>, <string>)` to find all occurrences of a pattern in a string

Similarly, the capabilities of *sed* can be replaced with:

- `re.sub(<pattern>, <replacement>, <string>)` to substitute all occurrences of a pattern found in a string
- `re.subn(<pattern>, <replacement>, <string>)` to substitute all occurrences of a pattern found in a string and return the number of substitutions made

The `re` model provides a few more powerful utilities as well. Namely:

- `re.split(pattern, string)` splits a string by the occurrences of a pattern.
- `re.finditer(pattern, string)` returns an iterator yielding a match object for each match.
- `re.compile(pattern)` precompiles a regular expression so the subsequent matches occur much more quickly.

In all of these functions, if a match to a regular expression is not found, then `None` is returned. If a match is found, then a special `MatchObject` is returned. `MatchObjects` have methods and attributes that allow you to determine the position in the string of the match, the original regular expression pattern, and the values captured by any parentheses with the `MatchObject.groups()` method. For example, let's try to match a date regular expression to some actual dates:

```
In [1]: import re ❶

In [2]: re.match("20[01][0-9].*[0-9][0-9].*[0-9][0-9]", '2015-12-16') ❷
Out[2]: <_sre.SRE_Match object; span=(0, 10), match='2015-12-16'>

In [3]: m = re.match("20[01][0-9].*[0-9][0-9].*[0-9][0-9]", '2015-12-16') ❸

In [4]: m.pos ❹
Out[4]: 0

In [5]: m.groups() ❺
Out[5]: ()

In [6]: m = re.match("20[01][0-9].*[0-9][0-9].*[0-9][0-9]", 'not-a-date') ❻

In [7]: m is None ❼
Out[7]: True
```

❶ First, import the regular expression module.

❷ The string matches the pattern, so a match is returned.

❸ Assign the match to a variable name for later use.

❹ Find the index in the string of the start of the match.

❺ Report all captured groups. This regular expression pattern had no capturing parentheses, so no substrings are reported.

❻ Try to match the date pattern against something that is not a date.

❼ Note how None is returned when the match fails.

To speed up matching multiple strings against a common pattern, it is always a good idea to compile() the pattern. Compiling takes much longer than matching. However, once you have a compiled pattern, all of the same functions are available as methods of the pattern. Since the pattern is already known, you don't need to pass it in when you call match() or search() or the other methods. Let's compile a version of the date regular expression that has capturing parentheses around the actual date values:

```
In [8]: re_date = re.compile("(20[01][0-9]).*([0-9][0-9]).*([0-9][0-9])") ❶

In [9]: re_date.match('2014-28-01') ❷
Out[9]: <_sre.SRE_Match object; span=(0, 10), match='2014-28-01'>

In [10]: m = re_date.match('2014-28-01') ❸

In [11]: m.groups() ❹
Out[11]: ('2014', '28', '01')
```

❶ Compile the regular expression and store it as the re_date variable.

❷ Use this variable to match against a string.

❸ Assign the match to a variable m for later use.

❹ Since the regular expression uses capturing parentheses, you can obtain the values within them using the groups() method. A tuple that has the same length as the number of capturing parentheses is returned.

More information on the re module can be found in the Python documentation (*http://bit.ly/relibrary*).

Regular Expressions Wrap-up

At this point, your regular expressions skills should include:

- How to speed up command-line use with metacharacters

- How to find files based on patterns in their names (`find`)
- How to find lines in files based on patterns in their content (`grep`)
- How to replace text patterns in files (`sed`)
- How to manipulate columns of data based on patterns (`awk`)

This chapter, along with its descriptions and examples, has been adapted from Software Carpentry (*http://software-carpentry.org*) material, the official Perl documentation (*http://perldoc.perl.org/perlre.html*), Python's `re` module documentation (*http://bit.ly/relibrary*), *Mastering Regular Expressions* by Jeffrey Friedl (O'Reilly), and the Mozilla Developer Network (*http://mzl.la/1xQ5yBs*). Those are all excellent resources and should be utilized for further information.

CHAPTER 9
NumPy: Thinking in Arrays

At the core of most computational physics problems lives an *array*. From the physics side, arrays are a natural way to describe numerical and discretized problems. This is because geometry may be chopped up into tetrahedrons (pyramids) or hexahedrons (cubes), and arrays may be used to represent scalar or vector values that live at every point in three-dimensional space. Furthermore, operations on arrays can be used to represent or approximate calculus operations, such as integration or derivatives. From the software side, an array is a contiguous block of memory where every element has the same type and layout. From both a physical and a computational perspective, arrays are concise, beautiful, and useful.

Every programing language that is serious about scientific computing has a notion of an *array data language*, either built into the language's core utilities or available as a third-party package. Since an array is essentially just a sequence of bits, the array data language adds semantics for handling arrays that are native to the host language while taking advantage of the fact that you know you have an unbroken sequence of bits.

Some languages, such as MATLAB and IDL, are centered around the array data language concept. Other general-purpose languages, such as Fortran, are meant for everyday programming but support powerful array constructs natively. In other generic languages, the reference array implementations come as projects external to the languages themselves. For example, Perl has PDL and Python has NumPy.

NumPy (pronounced *numb-pie*) is ubiquitous in the world of scientific Python. A great many packages are written using it as a base. This is in part because NumPy came out of the merger of two earlier competing array data languages in Python, called Numeric and Numarray. NumPy also understands both C- and Fortran-style arrays. It therefore provides a common exchange format for data coming from outside of Python as well.

What really makes NumPy popular, though, is that it is easy to learn, intuitive to use, and orders of magnitude more performant than using pure Python for array-based operations. If you are coming to NumPy from another array data language, you'll see a lot of Pythonic implementations of familiar concepts. If you are new to the world of numerics, NumPy provides a beautiful learning curve for you to climb and master.

Arrays

The basic type that NumPy provides is the N-dimensional array class `ndarray`. Rather than being created directly, ndarrays are often instantiated via the `array()` function that NumPy also provides. To create an array, import numpy and call `array()` on a sequence:

```
>>> import numpy as np
>>> np.array([6, 28, 496, 8128])
array([   6,   28,  496, 8128])
```

A common abbreviation for numpy is np, and you will almost always see the following import statement:

```
import numpy as np
```

This is so prevalent that we will use the np abbreviation from here on out.

NumPy provides a number of ways to create arrays in addition to the normal `array()` function. The four most common convenience functions are `arange()`, `zeros()`, `ones()`, and `empty()`. The `arange()` function takes a start, stop, and step and works exactly like Python's `range()` function, except that it returns an `ndarray`. The `zeros()` and `ones()` functions take an integer or tuple of integers and return an ndarray whose shape matches that of the tuple and whose elements are all either zero or one. The `empty()` function, on the other hand, will simply allocate memory without assigning it any values. This means that the contents of an empty array will be whatever happened to be in memory at the time. Often this looks like random noise, though sometimes you might get a single consistent number (often zero). Empty arrays are therefore most useful if you have existing data you want to load into an array, and you do not want to pay the cost of setting all the values to zero if you are just going to overwrite them. Here are some examples of how to create new arrays using the `arange()`, `zeros()`, `ones()`, and `empty()` functions:

Code	Returns
np.arange(6)	array([0, 1, 2, 3, 4, 5])
np.zeros(4)	array([0., 0., 0., 0.])
np.ones((2, 3))	array([[1., 1., 1.], [1., 1., 1.]])
np.empty(4)	array([1.28506949e-316, 6.95226953e-310, 8.30132260e-317, 6.95226842e-310])

The linspace() and logspace() functions are also important to know. These create an even linearly or logarithmically spaced grid of points between a lower and upper bound that is inclusive on both ends. Note that logspace() may also take a base keyword argument, which defaults to 10. The lower and upper bounds are then interpreted as the base to these powers.

Code	Returns
np.linspace(1, 2, 5)	array([1. , 1.25, 1.5 , 1.75, 2.])
np.logspace(1, -1, 3)	array([10. , 1. , 0.1])

You can also create arrays from iterators, HDF5 files, CSV files, and a special NumPy binary file format called *.npy*. Please refer to the NumPy documentation (*http://www.numpy.org/*) for more information on how to perform some of these more advanced tasks.

For all of these creation mechanisms, the ndarray class effectively represents a fixed-sized block of memory and accompanying metadata that defines the features of the array. The attributes of ndarray that define the layout of memory are listed in Table 9-1. You can manipulate most of these attributes directly to change the way the array functions without needing to create a whole new block of memory.

Table 9-1. Important ndarray attributes

Attribute	Description
data	Buffer to the raw array data
dtype	Type information about what is in data
base	Pointer to another array where data is stored, or None if data is stored here
ndim	Number of dimensions (int)
shape	Tuple of integers that represents the rank along each dimension; has length of ndim
size	Total number of elements (int), equal to the product of all of the elements of shape

Attribute	Description
itemsize	Number of bytes per element (int)
nbytes	Total number of bytes (int), equal to size times itemsize
strides	Number of bytes between the *i*th element of each axis and the *i*+1th element along the same axis (tuple of ints, length ndim)
flags	Low-level memory layout information

Modifying the attributes in an allowable way will automatically update the values of the other attributes. Since the data buffer is fixed-length, all modifications must preserve the size of the array. This fixed size restriction also implies that you cannot append to an existing array without copying memory.

A common method of reshaping an existing array is to assign a new tuple of integers to the shape attribute. This will change the shape in-place. For example:

Code	Returns
a = np.arange(4)	array([0, 1, 2, 3])
a.shape = (2, 2)	array([[0, 1], [2, 3]])

NumPy also provides a np.reshape() function that you can call with an array argument. However, this returns a reshaped copy of the original array. This is one of the central patterns of NumPy: operations that involve attributes or methods of ndarray occur in-place, while functions that take an ndarray as an argument return a modified copy.

The array attributes matter because they describe and determine the allowed operations with the array. Chief among these is the dtype attribute, to which the whole next section is dedicated.

dtypes

The dtype or data type is the most important ndarray attribute. The data type determines the size and meaning of each element of the array. The default system of dtypes that NumPy provides is more precise and broader for basic types than the type system that the Python language implements. As you would expect, dtypes focus on numeric types. The dtypes have a hierarchy based on abstract types, such as integers and floating-point data. Each abstract type has a concrete default size in bits that is used when precision is otherwise unspecified. Unlike Python types, all dtypes must have a constant size in memory. Even strings must have a fixed size in an array! This is so that the array as a whole has predictable properties. Modifying other attributes, such as the shape and stride of an array, would not work if the length of a type were allowed to change from element to element. Some types may or may not be present, depending on the architecture of your system and how NumPy was built. The system architecture may also affect the size of the default types.

The `dtypes` all have string character codes, as a concise mechanism for specifying the type. These are useful for creating more complicated types, as will be seen later. Some dtypes are *flexible*. This means that while any given array must have a fixed size, the dtype length may be different for different arrays. This is used for strings, where one array may have strings of length 10 and another array may have strings of length 42. The lengths for flexible types may be given explicitly in the `dtype`. Otherwise, they are computed from the longest element of the array.

Table 9-2 describes the basic `dtypes` available, along with their character codes, number of bytes (f means flexible), and corresponding Python types.

Table 9-2. Basic NumPy dtypes

dtype	Code	Bytes	Python	Description
bool_	?	1	bool	Boolean data type. Note that this takes up a full byte (8 bits) and is somewhat inefficient at storing a large number of bools. For a memory-efficient Boolean array, please see Ilan Schnell's bitarray (*https://github.com/ilanschnell/bitarray*) package.
bool8	?	1	bool	Alias to bool_.
int_			int	Default integer type; alias to either int32 or int64.
int0			int	Same as int_.
int8	b	1	int	Single-byte (8-bit) integer ranging from -128 to 127. Interchangeable with the C/C++ char type.
byte	b	1	int	Alias of int8.
int16	h	2	int	16-bit integer ranging from -32768 to 32767. Convertible to the C/C++ short type.
int32	i	4	int	32-bit integer ranging from -2147483648 to 2147483647. Usually interchangeable with the C/C++ int type; always convertible to the C/C++ long type.

dtype	Code	Bytes	Python	Description
int64	l	8	int	64-bit integer ranging from -9223372036854775808 to 9223372036854775807. Usually interchangeable with the C/C++ `long` type; always convertible to the C/C++ `long long` type. This has the same byte width as the native Python `int` type.
uint_			int	Default unsigned integer type; alias to either `uint32` or `uint64`.
uint0			int	Same as `uint_`.
uint8	B	1	int	Single-byte (8-bit) unsigned integer ranging from 0 to 255. Interchangeable with the C/C++ `unsigned char` type.
ubyte	B	1	int	Alias of `uint8`.
uint16	H	2	int	16-bit unsigned integer ranging from 0 to 65535. Convertible to the C/C++ `unsigned short` type.
uint32	I	4	int	32-bit unsigned integer ranging from 0 to 4294967295. Usually interchangeable with the C/C++ `unsigned int` type always convertible to the C/C++ `unsigned long` type.
int64	L	8	int	64-bit unsigned integer ranging from 0 to 18446744073709551615. Usually interchangeable with the C/C++ `unsigned long` type; always convertible to the C/C++ `unsigned long long` type.
float_	d	8	float	Alias to `float64`.
float16	e	2	float	16-bit floating-point number.
float32	f	4	float	32-bit floating-point number. Usually compatible with the C/C++ `float` type.
float64	d	8	float	64-bit floating-point number. Usually compatible with the C/C++ `double` type.
float96		12		96-bit floating-point number. Sometimes compatible with the C/C++ `long double` type.
float128	g	16		128-bit floating-point number. Sometimes compatible with the C/C++ `long double` type.
complex_	D	16	complex	Alias to `complex128`.
complex64	F	8	complex	64-bit complex floating-point number.

dtype	Code	Bytes	Python	Description
complex128	D	16	complex	128-bit complex floating-point number. Equivalent to the Python complex type and composed of two floats.
complex256	G	32		256-bit complex floating-point number.
string_	S	f	bytes	Bytes (or str in Python 2) data type. This is a flexible dtype.
string0	S	f	str	Alias of string_.
str_	S	f	str	Alias of string_.
unicode_	U	f	str	String (or Unicode in Python 2) data type. This is a flexible dtype.
unicode0	U	f	str	Alias of unicode_.
void	V	f		A raw data type of presumably C/C++ pointers. Arrays with this type make no presumptions about their contents.
void0	V	f		Alias of void.
object_	O	1	object	Generic dtype for holding any Python object. Implemented as an array of PyObject*.
object0	O	1	object	Alias of object_.

When you are creating an array, the dtype that is automatically selected will always be that of the least precise element. Say you have a list that is entirely integers with the exception of a single float. An array created from this list will have the dtype np.float64, because floats are less precise than integers. The order of data types sorted from greatest to least precision is Boolean, unsigned integer, integer, float, complex, string, and object. An example of this downcasting follows, where 28 is an integer in the a array and a float in the b array:

Code	Returns
a = np.array([6, 28, 496, 8128]) a.dtype	array([6, 28, 496, 8128]) dtype('int64')
b = np.array([6, 28.0, 496, 8128]) b.dtype	array([6.00000000e+00, 2.80000000e+01, 4.96000000e+02, 8.12800000e+03]) dtype('float64')

You can always force an array to have a given data type by passing dtype=<type> as a keyword argument to the array creation function. This will convert all elements of the

dtypes | 207

array to the given dtype, rather than relying on precision to figure out the type. In some circumstances this can lead to a loss of information (float to integer conversion, for instance). However, it has the benefit of giving you exactly what you want. Providing an explicit dtype is a good idea in most cases because it makes the code more readable. The dtypes that are passed in as keyword arguments may be any NumPy dtype (see Table 9-2), any of the string character codes (f, i, etc.), or any Python type (float, int, object, etc.):

Code	Returns
a = np.array([6, 28.0, 496, 8128], dtype=np.int8)	array([6, 28, -16, -64], dtype=int8)
b = np.array([6, 28.0, 496, 8128], dtype='f')	array([6.00000000e+00, 2.80000000e+01, 4.96000000e+02, 8.12800000e+03], dtype=float32)

For flexible data types, when you're using the character code the length of the type is given after the character code, inside of the string—that is, the code for a string of length 6 is 'S6'. The following example in IPython shows the result when an array with this data type is initialized with strings longer than length 6:

```
In [1]: import numpy as np

In [2]: np.array(['I will have length six', 'and so will I!'], dtype='S6')
Out[2]:
array([b'I will', b'and so'], dtype='|S6')
```

Now that you know how to manipulate arrays through their dtypes and other attributes, you are well poised to learn about other array manipulations. In the following section we will tackle array slicing, which looks a lot like slicing other Python sequences.

Slicing and Views

NumPy arrays have the same slicing semantics as Python lists when it comes to accessing elements or subarrays. Python list slicing was discussed in "Lists" on page 66. As applied to NumPy arrays, we see:

Code	Returns
a = np.arange(8)	array([0, 1, 2, 3, 4, 5, 6, 7])
a[::-1]	array([7, 6, 5, 4, 3, 2, 1, 0])
a[2:6]	array([2, 3, 4, 5])
a[1::3]	array([1, 4, 7])

What is different about slicing here is that because NumPy arrays are N-dimensional, you may slice along any and all axes! In Python, if you wish to slice along multiple axes—say, in a list of lists—you must slice the inner list for every element in the slice of the outer list:

```
outer = [...]
selection = [inner[a:b:c] for inner in outer[x:y:z]]
```

The number of nested for loops that is needed to slice lists of lists is the number of dimensions minus one. In NumPy, rather than indexing by a slice you can index by a tuple of slices, which each act on their own dimensions:

```
outer = np.array([...])
selection = outer[x:y:z, a:b:c]
```

The for loops implied by multidimensional slicing are all implicitly handled by NumPy at the C layer. This makes executing complex slices much faster than writing the for loops explicitly in Python. This is only useful if the array has a dimensionality greater than 1, however. If an axis is left out of a multidimensional slice, all elements along that dimension are included. Also note that rows come before columns in NumPy. In the following multidimensional slicing examples, we first need to create a multidimensional array:

Code	Returns

a = np.arange(16)
a.shape = (4, 4) ❶
	array([[0, 1, 2, 3],
	       [4, 5, 6, 7],
	       [8, 9, 10, 11],
	       [12, 13, 14, 15]])

a[::2, 1::2] ❷	array([[1, 3], [9, 11]])
a[1:3, 1:3] ❸	array([[5, 6], [9, 10]])

a[2::-1, :3] ❹	array([[8, 9, 10],
	       [4, 5, 6],
	       [0, 1, 2]])

❶ Create a 1D array and reshape it to be 4x4.

❷ Slice the even rows and the odd columns.

❸ Slice the inner 2x2 array.

❹ Reverse the first 3 rows, taking the first 3 columns.

The most important feature of array slicing to understand is that slices are *views* into the original array. No data is copied when a slice is made, making NumPy especially fast for slicing operations. This is because slices are regular enough to always be interpreted as manipulations of the original array's metadata (`shape`, `strides`, etc.). Because slices do not contain any of their data, the `base` attribute of the slice array is a reference back to the array that holds the data. For instance, if you take a slice of an array that is itself a slice, the bases of both slice arrays will point back to the original array.

That slice arrays are views means that modifications to their elements are reflected back in the original arrays. This makes sense, as there is only one block of memory between them. As a demonstration, if you have two arrays a and b, where b is a slice of a, then you can tell that b is a view if its base is a. Furthermore, changes to the contents of either a or b will also affect the other array. You can see this in the following example:

Code	Returns
a = np.arange(6)	array([0, 1, 2, 3, 4, 5])
b = a[1::2]	array([1, 3, 5])
b[1] = 42 ❶	array([1, 42, 5])
a ❷	array([0, 1, 2, 42, 4, 5])
b.base is a ❸	True

❶ Changing an element of b…

❷ …changes the corresponding element in a

❸ b is a view of a

If you truly want a copy of a slice of an array, you can always create a new array from the slice:

```
a = np.arange(16)
b = np.array(a[1::11])
```

Slices are not the only way to create a view. The `ndarray` class has a `view()` method on it that will give you a view into the whole array. This method takes two keyword arguments. The `dtype` keyword argument allows you to reinterpret the memory to another type without copying the data. The `type` argument allows you to change the kind of array that is returned. For example, we can view an `int64` array as an `int32` array with twice as many elements:

Code	Returns
a = np.arange(6, dtype=np.int64)	array([0, 1, 2, 3, 4, 5])
a.view('i4')	array([0, 0, 1, 0, 2, 0, 3, 0, 4, 0, 5, 0], dtype=int32)

Slices and views are an essential part of using NumPy efficiently. Knowing how to use these everyday operations—which do not copy data—makes your code run faster. In the next section you will learn about other common operations that do create new arrays but are still indispensable.

Arithmetic and Broadcasting

A defining feature of all array data languages is the ability to perform arithmetic operations in an *element-wise* fashion. This allows for concise mathematical expressions to be evaluated over an arbitrarily large amount of data. This works equally well for scalars as it does for arrays with the same shape. In the following example, we see how simple arithmetic operations (addition, subtraction, multiplication, etc.) are evaluated with an array as a variable:

Code	Returns
a = np.arange(6)	array([0, 1, 2, 3, 4, 5])
a - 1	array([-1, 0, 1, 2, 3, 4])
a + a	array([0, 2, 4, 6, 8, 10])
2*a**2 + 3*a + 1	array([1, 6, 15, 28, 45, 66])

Though this is extremely expressive, it can also be subtly expensive. For each operation, a new array is created and all elements are looped over. For simple expressions (such as a - 1) this overhead is fine, because the result is typically assigned to a new variable name. However, for more complex operations (such as 2*a**2 + 3*a + 1) the allocation of new arrays is somewhat wasteful since they are discarded immediately after the next operation is completed. Why create a special array for a**2 if it is going to be deleted when you finish computing 2*a**2? These ephemeral arrays are called *temporaries*.

Furthermore, each operation iterates through all elements of the array on its own. Since loops are more expensive than other forms of flow control (even in C), NumPy is not the most efficient for dealing with complex expressions. This is because NumPy does not store the context within which an operation is executed. This is natural for how Python and most other languages work. What it ends up meaning is that you can make performance improvements just by doing some algebraic rearrangements to minimize the total number of operations. As a contrived example, 6*a would run about twice as fast and use half the memory as the expression 3*(2*a). For more rigorous and thorough resolution of the temporary issue, please investigate the excellent numexpr (*https://github.com/pydata/numexpr*) package.

NumPy remains incredibly expressive and powerful for higher-order concepts, even with temporaries being perpetually created and destroyed. Suppose you have two arrays of the same shape, x and y. The numerical derivative dy/dx is given by this simple expression:

```
(y[1:] - y[:-1]) / (x[1:] - x[:-1])
```

This method treats the points in x and y as bin boundaries and returns the derivative for the center points ((x[1:] + x[:-1])/2). This has the side effect that the length of the result is 1 shorter than the lengths of an original arrays. If instead you wish to treat the points of the array as the center points with proper upper and lower bound handling so that the result has the same length as the original arrays, you can use NumPy's a gradient() function. The numerical derivative is then just:

```
np.gradient(y) / np.gradient(x)
```

The process of performing element-wise operations on arrays is not limited to scalars and arrays of the same shape. NumPy is able to *broadcast* arrays of different shapes together as long as their shapes follow some simple compatibility rules. Two shapes are compatible if:

- For each axis, the dimensions are equal (a.shape[i] == b.shape[i]), or the dimension of one is 1 (a.shape[i] == 1 or b.shape[i] == 1).
- The rank (number of dimensions) of one is less than that of the other (a.ndim < i or b.ndim < i).

When the ranks of two axes of two arrays are equal, the operation between them is computed element-wise. This is what we have seen so far for cases like a + a. When the length of an axis is 1 on array a and the length of the same axis on array b is greater than 1, the value of a is virtually stretched along the entire length of b in this dimension. Every element of b sees the value of a for this operation. This is where the term *broadcasting* comes from: one element of a goes to all elements of b. Similarly, for axes of b that are greater than the rank of a, the entire array a is stretched along the remaining dimensions of b. We have also already seen this as scalars (which have rank 0) have been applied to 1D and 2D arrays. Consider a 2×2 matrix times a 2×1 vector that broadcasts the multiplication:

Code	Returns
a = np.arange(4) a.shape = (2, 2)	array([[0, 1], [2, 3]])
b = np.array([[42], [43]])	array([[42], [43]])
a * b	array([[0, 42], [86, 129]])

Here, every column of a is multiplied element-wise by the values in b. Notably, this does not perform the dot product, which instead requires the aptly named dot() function:

Code	Returns
np.dot(a, b)	array([[43], [213]])

Normal Python multiplication (*) on arrays is implemented with broadcasting rules. These rules stretch lower-dimensional data into a higher dimension for only long enough to perform the operations. This is one kind of multiplication that can represent an outer product, in some situations. Broadcasting, just like in mathematics, is distinct from the inner product operation, where you should instead use the np.dot() function. This distinction is necessary to understand. As a more sophisticated example with a different operator, broadcasting also applies to adding a 4×3 array and a length-3 array:

Code	Returns

a = np.arange(12)
a.shape = (4, 3)
	array([[0, 1, 2],
	       [3, 4, 5],
	       [6, 7, 8],
	       [9, 10, 11]])

b = np.array([16, 17, 18])	array([16, 17, 18])

a + b	array([[16, 18, 20],
	       [19, 21, 23],
	       [22, 24, 26],
	       [25, 27, 29]])

Here, b is stretched along all four elements of the first axis of a. If instead a were a 3×4 array, the shapes would not match and the operation would fail. We can see this if we transpose the shape of a, as shown here:

Arithmetic and Broadcasting | 213

Code	Returns

a.shape = (3, 4)	array([[0, 1, 2, 3],
	      [4, 5, 6, 7],
	      [8, 9, 10, 11]])

a + b	ValueError: operands could not be broadcast together with shapes (3,4) (3,)

If, however, b was a 3×1 vector, the shapes would be broadcastable and the operation would be successful. Still, the result would be different than we saw previously:

Code	Returns

b.shape = (3, 1)	array([[16],
	      [17],
	      [18]])

a + b	array([[16, 17, 18, 19],
	      [21, 22, 23, 24],
	      [26, 27, 28, 29]])

This demonstrates the important point that dimensions can always be added to an existing ndarry as long as the length of the added dimension is 1. This is because the total number of elements in the array does not change when length-1 dimensions are added. This is a particularly useful feature with broadcasting.

Briefly adding fake dimensions for computations is so useful that NumPy has a special newaxis variable that you can use in an index to add a length-1 dimension. This reduces the amount that you have to reshape explicitly. In the following example, the dimensions do not match until b has a newaxis added to it:

Code	Returns
a = np.arange(6) a.shape = (2, 3)	array([[0, 1, 2], [3, 4, 5]])
b = np.array([2, 3])	array([2, 3])
a - b	ValueError: operands could not be broadcast together with shapes (2,) (2,3)
b[:, np.newaxis] - a	array([[2, 1, 0], [0, -1, -2]])

The newaxis index may appear as many times as needed before or after real data axes. Note, though, that NumPy arrays have a maximum of 32 dimensions. Using newaxis, you can show this easily:

```
>>> b[(slice(None),) + 32 * (np.newaxis,)] - a
IndexError: number of dimensions must be within [0, 32],
        indexing result would have 33

>>> b[(slice(None),) + 31 * (np.newaxis,)] - a
array([[[[[[[[[[[[[[[[[[[[[[[[[[[[[[[ 2,  1,  0],
                                   [-1, -2, -3]]]]]]]]]]]]]]]]]]]]]]]]]]]]]]]],
       [[[[[[[[[[[[[[[[[[[[[[[[[[[[[[[ 3,  2,  1],
                                   [ 0, -1, -2]]]]]]]]]]]]]]]]]]]]]]]]]]]]]]]])
```

As is hopefully clear, dealing with 32 dimensions can be somewhat tedious. Since you have now seen the basics of array manipulations, it is time to move on to something fancier.

Fancy Indexing

Slicing is a great way to pull out data from an array when the indices follow a regularly gridded pattern. With some multidimensional upgrades, slicing in NumPy follows the same pattern as for the built-in Python types. However, what if you want to pull out many arbitrary indices? Or you wish to pull out indices that follow a pattern, but one that is not regular enough, like the Fibonacci sequence? NumPy arrays handle these cases via *fancy indexing*.

Fancy indexing is where you index by an integer array or a list of integers, instead of indexing by a slice or newaxis. The *fancy* part comes from the following qualities:

- You may provide arbitrary indices.
- You can have repeated indices.
- Indices may be out of order.
- The shape of the index does not need to match the shape of the array.
- The shape of the index may have more or fewer dimensions than the array.
- Indices may be used seamlessly with slices.

The drawback to fancy indexing is that it requires copying the data into a new block of memory. Fancy indexing cannot in general be a view into the original array, like with slicing. This is due to the fact that there is no way to reason about what indices will or won't be present, since they are assumed to be arbitrary. Suppose we have the array 2*a**2 + 1, where a is in the range 0 to 8. The following fancy indexes may be applied:

Code	Returns
`a = 2*np.arange(8)**2 + 1`	`array([ 1,  3,  9, 19, 33, 51, 73, 99])`

`# pull out the fourth, last, and`
`# second indices`
`a[[3, -1, 1]]`
	`array([19, 99,  3])`

`# pull out the Fibonacci sequence`
`fib = np.array([0, 1, 1, 2, 3, 5])`
`a[fib]` ❶
	`array([ 1,  3,  3,  9, 19, 51])`

`# pull out a 2x2 array` `a[[[2, 7], [4, 2]]]` ❷	`array([[ 9, 99],` `       [33,  9]])`

❶ Note that the 1 index is repeated.

❷ The shape of the fancy index determines the shape of the result.

When you are mixing slicing with fancy indexing, each dimension must either be a slice or a fancy index. There is no need to union a slice and a fancy index along a single dimension, because such an operation can be fully described by a single fancy index. Note that even when the slices are present, a single axis that uses a fancy index will trigger the whole result to be a copy. It is always better to use slices when you can, as mixing slices and fancy indexes requires a multidimensional array. The following examples creates a 4×4 array that is then indexed by both slices and fancy indexes:

Code	Returns

`a = np.arange(16) - 8`
`a.shape = (4, 4)`
	`array([[-8, -7, -6, -5],`
	`       [-4, -3, -2, -1],`
	`       [ 0,  1,  2,  3],`
	`       [ 4,  5,  6,  7]])`

`# pull out the third, last, and`
`# first columns`
`a[:, [2, -1, 0]]`
	`array([[-6, -5, -8],`
	`       [-2, -1, -4],`
	`       [ 2,  3,  0],`
	`       [ 6,  7,  4]])`

`# pull out a Fibonacci sequence of`
`# rows for every other column, starting`
`# from the back`
`fib = np.array([0, 1, 1, 2, 3])`
`a[fib, ::-2]`
	`array([[-5, -7],`
	`       [-1, -3],`
	`       [-1, -3],`
	`       [ 3,  1],`
	`       [ 7,  5]])`

Note that you may also use a one-dimensional fancy index on each of the multiple dimensions independently. Each index is then interpreted as the coordinate for that dimension. Using the 4×4 array a from the example and a new fancy index i, we can apply i or various slices of i to each axis of a:

Code	Returns

`# get the diagonal with a range`
`i = np.arange(4)`
`a[i, i]`
	`array([-8, -3,  2,  7])`

`# lower diagonal by subtracting one to`
`# part of the range`
`a[i[1:], i[1:] - 1]`
	`array([-4,  1,  6])`
	
	`array([-7, -2,  3])`

`# upper diagonal by adding one to part`
`# of the range`
`a[i[:3], i[:3] + 1]`
	`array([-5, -2,  1,  4])`

`# anti-diagonal by reversal` `a[i, i[::-1]]`	

Fancy indexing is the feature that allows you to dice up NumPy arrays as you see fit. The fact that it is so arbitrary is why it is so powerful. However, it is easy to overuse this power, because the performance cost of copying data can sometimes be quite high. In everyday NumPy usage, fancy indexes are used all of the time. Embrace them, but also know their effects.

Related to the notion of fancy indexing and copying arbitrary data out of an array, the idea of *masking* is discussed in the next section.

Masking

A mask is like a fancy index in many respects, except that it *must* be a Boolean array. Masks may be used to index other arrays that have the same shape or the same length along an axis. If the value of the mask at a given location is `True`, then the value from the array appears in the result. If the value is `False`, then the data does not appear. As with fancy indexing, the application of a mask to an array will produce a copy of the data, not a view. A mask cannot be a Python list of `bool`s; it must truly be a NumPy array of `bool`s. Here's an example using a 3×3 matrix and a one-dimensional mask:

Code	Returns

`# create an array`
`a = np.arange(9)`
`a.shape = (3,3)`
	`array([[0, 1, 2],`
	`       [3, 4, 5],`
	`       [6, 7, 8]])`

`# create an all True mask` `m = np.ones(3, dtype=bool)`	`array([ True,  True,  True], dtype=bool)`
`# take the diagonal` `a[m, m]`	`array([0, 4, 8])`

In computational physics, masks can be used to pick out a region of a problem to either focus on or ignore. They are very useful for isolating the domain of a problem that is truly of interest.

Masks may also be multidimensional themselves. In this case, the mask indexes the array element-wise. The result of masking is typically a flat array. This is because the true parts of the mask do not necessarily form a coherent shape. In the following, m is a 3×3 Boolean array with four true elements. When used on our a array, the four values at the true locations appear in the result:

Code	Returns

`# create a mask`
`m = np.array([[1, 0, 1],`
`              [False, True, False],`
`              [0, 0, 1]], dtype=bool)`
	`array([[ True, False,  True],`
	`       [False,  True, False],`
	`       [False, False,  True]],`
	`      dtype=bool)`

`a[m]`	`array([0, 2, 4, 8])`

Masks are useful for hiding data that you know to be bad, unacceptable, or outside of what you find interesting at the moment. NumPy makes it easy to generate masks. The return of any comparison operator is a Boolean array. Rather than just being True or False, comparisons act element-wise. Masks that are generated from comparisons can be saved and used on other arrays. In the following example, less-than and greater-than comparisons generate valid masks:

Code	Returns

`a < 5` ❶	`array([[ True,  True,  True],`
	`       [ True,  True, False],`
	`       [False, False, False]],`
	`      dtype=bool)`

`m = (a >= 7)` ❷	`array([[False, False, False],`
	`       [False, False, False],`
	`       [False,  True,  True]],`
	`      dtype=bool)`

`a[m]` ❸	`array([7, 8])`

❶ Create a mask array.

❷ Create a mask and store it as m.

❸ Apply m to the original array it was created from.

What is particularly beautiful about this is that the mask can be generated in the indexing operation itself. You can read the following code as "a[i] such that a[i] is less than 5 for all i":

Code	Returns
a[a < 5]	array([0, 1, 2, 3, 4])

It is also possible to combine or modify masks with certain Python literal operators or their NumPy function equivalents. Table 9-3 displays the bitwise operators that are helpful when manipulating masks.

Table 9-3. NumPy bitwise operators

Operator	Function	Description
~	bitwise_not(x)	True for elements where x is False and False for elements where x is True. This is an alias to the numpy invert() function.
|	bitwise_or(x, y)	True for elements where either x or y or both are True.
^	bitwise_xor(x, y)	True for elements where either x or y (but not both) is True.
&	bitwise_and(x, y)	True for elements where both x and y are True.

As an example, the following generates two masks and then uses the bitwise or operator to combine them. The combined mask is then used to index an array:

Code	Returns
a[(a < 5) | (a >= 7)]	array([0, 1, 2, 3, 4, 7, 8])

Masks can and should be used in conjunction with NumPy's where() function. If you are familiar with the WHERE clause in SQL, this is conceptually similar. This function takes a Boolean array and returns a tuple of fancy indices that are the coordinates for where the mask is True. This function always returns a tuple so that it can be used in an indexing operation itself:

Code	Returns
np.where(a < 5)	(array([0, 0, 0, 1, 1]), array([0, 1, 2, 0, 1]))
a[np.where(a >= 7)]	array([7, 8])

Masking | 219

Passing the fancy index results of `where()` right back into the indexing operation is not recommended because it will be slower and use more memory than just passing in the mask directly. Taking the results of `where()` and modifying them in some way is recommended. For example, the following takes every column from a where a has any value that is less than 2:

Code	Returns

`a[:, np.where(a < 2)[1]]`	`array([[0, 1],`
	`        [3, 4],`
	`        [6, 7]])`

So far in this chapter you have learned how to manipulate arrays of a single, basic data type. In the following section you will discover how to create richer `dtypes` of your own that may better represent your data.

Structured Arrays

In most real-world data analysis scenarios, it is useful to have a notion of a *table* that has named columns, where each column may have its own type. In NumPy, these are called *structured arrays* or sometimes *record arrays*. This is because NumPy views them as one-dimensional arrays of structs, like you would find in C or C++.

You can construct structured arrays by compounding `dtypes` together in the `dtype()` constructor. The constructor may take a list of 2- or 3-tuples that describe the columns in the table. These tuples have the following form:

```
# 2-tuple
("<col name>", <col dtype>)

# 3-tuple
("<col name>", <col dtype>, <num>)
```

The first element of these tuples is the column name as a string. The second element is the `dtype` for the column, which may itself be another compound `dtype`. Thus, you can have subtables as part of your table. The third element of the tuple is optional; if present, it is an integer representing the number of elements that the column should have. If the number is not provided, a default value of 1 is assumed. Compound `dtypes` are similar in nature to SQL schemas or a CSV file's header line. Here are some simple examples:

Code	Returns
| ```
# a simple flat dtype
fluid = np.dtype([
    ('x', int),
    ('y', np.int64),
    ('rho', 'f8'),
    ('vel', 'f8'),
    ])
``` | ```
dtype([('x', '<i8'),
       ('y', '<i8'),
       ('rho', '<f8'),
       ('vel', '<f8')])
``` |
| ```
# a dtype with a nested dtype
# and a subarray
particles = np.dtype([
    ('pos', [('x', int),
             ('y', int),
             ('z', int)]),
    ('mass', float),
    ('vel', 'f4', 3)
    ])
``` | ```
dtype([('pos', [('x', '<i8'),
                ('y', '<i8'),
                ('z', '<i8')]),
       ('mass', '<f8'),
       ('vel', '<f4', (3,))])
``` |

All compound `dtypes` are implemented as void `dtypes` under the hood. There are also two `dtype` attributes that are only useful in the context of compound `dtypes`. The first, `names`, is a tuple of strings that gives the column names and their order. You can rename columns by resetting this attribute. The second, `fields`, is a dict-like object that maps the column names to the `dtypes`. The values in `fields` are read-only. This ensures that the `dtypes` are immutable, which is important because changing their size would change how the corresponding memory block of the array was sized. The following examples demonstrate accessing these attributes:

Code	Returns
`particles.names`	`('pos', 'mass', 'vel')`
| `fluid.fields` | ```
<dictproxy
{'rho': (dtype('float64'), 16),
 'vel': (dtype('float64'), 24),
 'x': (dtype('int64'), 0),
 'y': (dtype('int64'), 8)}>
``` |

You can create structured arrays by passing these data types into the array creation functions as usual. Note that in some cases, such as for `arange()`, the `dtype` that you pass in may not make sense. In such cases, the operation will fail. Functions such as `zeros()`, `ones()`, and `empty()` can take all data types. For example:
```

Code	Returns

np.zeros(4, dtype=particles)	array([((0, 0, 0), 0.0, [0.0, 0.0, 0.0]),
	      ((0, 0, 0), 0.0, [0.0, 0.0, 0.0]),
	      ((0, 0, 0), 0.0, [0.0, 0.0, 0.0]),
	      ((0, 0, 0), 0.0, [0.0, 0.0, 0.0])],
	     dtype=[('pos', [('x', '<i8'),
	                           ('y', '<i8'),
	                           ('z', '<i8')]),
	           ('mass', '<f8'),
	           ('vel', '<f4', (3,))])

Here, the sub-dtype column pos is displayed as a tuple and the subarray column vel is displayed as a list. This is because pos is implemented as three different named components, while vel is a subarray. Also note that when you wish to use array() to create a structured array from existing Python data the rows must be given as tuples. This is to prevent NumPy from interpreting the data as having more than one dimension along the structured axis. You can provide data to the array() function as follows:

Code	Returns

# note that the rows are tuples 	
f = np.array([(42, 43, 6.0, 2.1),
            (65, 66, 128.0, 3.7),
            (127, 128, 3.0, 1.5)],
          dtype=fluid)
	array([(42, 43, 6.0, 2.1),
	      (65, 66, 128.0, 3.7),
	      (127, 128, 3.0, 1.5)],
	     dtype=[('x', '<i8'),
	            ('y', '<i8'),
	            ('rho', '<f8'),
	            ('vel', '<f8')])

Indexing a structured array by an integer will pull out a single row. Indexing by a slice will return that slice of rows. Unlike with normal arrays, though, indexing by a string will return the column whose name matches the string. This dictionary-like access is further extended such that indexing by a list of strings will pull out multiple columns in the order given. This *must* be a list—not a tuple—to distinguish it from regular multidimensional indexing. For example:

Code	Returns
f[1]	(65, 66, 128.0, 3.7)

f[::2]	array([(42, 43, 6.0, 2.1),
	       (127, 128, 3.0, 1.5)],
	      dtype=[('x', '<i8'),
	             ('y', '<i8'),
	             ('rho', '<f8'),
	             ('vel', '<f8')])

f['rho']	array([6., 128., 3.])

f[['vel', 'x', 'rho']]	array([(2.1, 42, 6.0),
	       (3.7, 65, 128.0),
	       (1.5, 127, 3.0)],
	      dtype=[('vel', '<f8'),
	             ('x', '<i8'),
	             ('rho', '<f8')])

Having studied structured arrays, you now have seen a wide variety of different techniques for creating and manipulating arrays. The next section offers a brief peek into how many of these operations are abstracted under NumPy's hood.

Universal Functions

Now that we have seen how to define and manipulate arrays, we can discuss how to transform them. NumPy has a notion of *universal functions*, or *ufuncs*, that provide an interface for transforming arrays. Roughly speaking, a ufunc is a special callable object that implements the reduce(), reduceat(), outer(), accumulate(), and at() methods, as well as a handful of attributes. Knowing the details of how ufuncs are implemented is not critical. Only the most advanced NumPy users and developers truly need to worry about how to create and modify ufuncs.

What is much more important is understanding the suite of ufuncs that come as standard with NumPy. A few of them have already been presented: the bitwise operators we saw in Table 9-3 are universal functions.

Not all ufuncs will operate on all arrays, and some ufuncs may change the shape or size of the result they return. However, the idea behind using ufuncs is to write data transformations as generically as possible. A ufunc should fail if and only if the operation that is being attempted is illogical or inconsistent. Table 9-4 displays some of the most important ufuncs. For more information, please see the ufuncs documentation (*http://bit.ly/ufunc*).

Table 9-4. Important NumPy universal functions

Function	Description
add(a, b)	Addition operator (+)
subtract(a, b)	Subtraction operator (-)
multiply(a, b)	Multiplication operator (*)
divide(a, b)	Division operator (/)
power(a, b)	Power operator (**)
mod(a, b)	Modulus (%)
abs(a)	Absolute value
sqrt(a)	Positive square root
conj(a)	Complex conjugate
exp(a)	Exponential (e**a)
exp2(a)	Exponential with base 2 (2**a)
log(a)	Natural log
log2(a)	Log base 2
log10(a)	Log base 10
sin(a)	Sine
cos(a)	Cosine
tan(a)	Tangent
bitwise_or(a, b)	Bitwise | operator
bitwise_xor(a, b)	Bitwise ^ operator
bitwise_and(a, b)	Bitwise & operator
invert(a)	Bitwise inversion (i.e., the ~ operator)
left_shift(a, b)	Left bit shift operator (<<)

Function	Description
right_shift(a, b)	Right bit shift operator (>>)
minimum(a, b)	Minimum (note that this is different from np.min())
maximum(a, b)	Maximum (note that this is different from np.max())
isreal(a)	Test for zero imaginary component
iscomplex(a)	Test for zero real component
isfinite(a)	Test for noninfinite value
isinf(a)	Test for infinite value
isnan(a)	Test for Not a Number
floor(a)	Next-lowest integer
ceil(a)	Next-highest integer
trunc(a)	Truncate, remove noninteger bits

For example, we can take the sine of the linear range from zero to pi as follows:

Code	Returns
x = np.linspace(0.0, np.pi, 5)	array([0. , 0.78539816, 1.57079633, 2.35619449, 3.14159265])
np.sin(x)	array([0.00000000e+00, 7.07106781e-01, 1.00000000e+00, 7.07106781e-01, 1.22464680e-16])

Universal functions are very significant in NumPy. One brilliant aspect of NumPy's design is that even though they are fundamental to many common operations, as a user, you will almost never even notice that you are calling a universal function. They just work.

It is common for new users of NumPy to use Python's standard `math` module instead of the corresponding universal functions. The `math` module should be avoided with NumPy because it is slower and less flexible. These deficiencies are primarily because universal functions are built around the idea of arrays while `math` is built around the Python `float` type.

However, not every operation can be expressed solely using universal functions. Up next is a section that teaches about the vital odds and ends that have yet to be detailed.

Other Valuable Functions

In addition to the suite of ufuncs, NumPy also provides some miscellaneous functions that are critical for day-to-day use. In most cases these are self-explanatory; for instance, the `sum()` function sums elements in an array. Many of these allow you to supply keyword arguments. A common keyword argument is `axis`, which is `None` by default, indicating that these functions will operate over the entire array. However, if `axis` is an integer or tuple of integers, the function will operate only over those dimensions. Using `sum()` as an example:

Code	Returns

`a = np.arange(9)`
`a.shape = (3, 3)`
	`array([[0, 1, 2],`
	`       [3, 4, 5],`
	`       [6, 7, 8]])`

`np.sum(a)`	`36`
`np.sum(a, axis=0)`	`array([ 9, 12, 15])`
`np.sum(a, axis=1)`	`array([ 3, 12, 21])`

Many of these functions appear as methods on the `ndarray` class as well. Table 9-5 shows some of the most important global functions that NumPy provides. Please refer to the NumPy documentation (*http://www.numpy.org/*) for more information.

Table 9-5. Important NumPy global functions

function	Description
`sum(a)`	Adds together all array elements.
`prod(a)`	Multiplies together all array elements.
`min(a)`	Returns the smallest element in the array.

function	Description
max(a)	Returns the largest element in the array.
argmin(a)	Returns the location (index) of the minimum element.
argmax(a)	Returns the location (index) of the maximum element.
dot(a, b)	Computes the dot product of two arrays.
cross(a, b)	Computes the cross product of two arrays.
einsum(subs, arrs)	Computes the Einstein summation over subscripts and a list of arrays.
mean(a)	Computes the mean value of the array elements.
median(a)	Computes the median value of the array elements.
average(a, weights=None)	Returns the weighted average of an array.
std(a)	Returns the standard deviation of an array.
var(a)	Computes the variance of an array.
unique(a)	Returns the sorted unique elements of an array.
asarray(a, dtype)	Ensures the array is of a given dtype. If the array is already in the specified dtype, no copy is made.
atleast_1d(a)	Ensures that the array is least one-dimensional.
atleast_2d(a)	Ensures that the array is least two-dimensional.
atleast_3d(a)	Ensures that the array is least three-dimensional.
append(a, b)	Glues the values of two arrays together in a new array.
save(file, a)	Saves an array to disk.
load(file)	Loads an array from disk.
memmap(file)	Loads an array from disk lazily.

These functions can and do help, and using NumPy to its fullest often requires knowing them. Some of them you will likely reach for very soon, like sum(). Others may only rear their heads once or twice in a project, like `save()`. However, in all cases you will be glad that they exist when you need them.

NumPy Wrap-up

Congratulations! You now have a breadth of understanding about NumPy. More importantly, you now have the basic skills required to approach *any* array data language. They all share common themes on how to think about and manipulate arrays of data. Though the particulars of the syntax may vary between languages, the underlying concepts are the same. For NumPy in particular, though, you should now be comfortable with the following ideas:

- Arrays have an associated data type or `dtype`.
- Arrays are fixed-length, though their contents are mutable.
- Manipulating array attributes changes how you view the array, not the data itself.
- Slices are views and do not copy data.
- Fancy indexes are more general than slices but do copy data.
- Comparison operators return masks.
- Broadcasting stretches an array along applicable dimensions.
- Structured arrays use compound `dtypes` to represent tables.
- Universal and other functions are helpful for day-to-day NumPy use.

Still, NumPy arrays typically live only in memory. This means while it is a good mental model for performing calculations and trying to solve problems, NumPy is typically not the right tool for storing data and sharing it with your friends and colleagues. For those tasks, we will need to explore the tools coming up in Chapter 10.

CHAPTER 10
Storing Data: Files and HDF5

HDF5 stands for *Hierarchical Data Format 5*, a free and open source binary file type specification. HDF5 is built and supported by the HDF Group, which is an organization that split off from the University of Illinois Champagne-Urbana. What makes HDF5 great is the numerous libraries written to interact with files of this type and its extremely rich feature set.

HDF5 has become the default binary database for scientific computing. Unlike other software developers, scientists tend not to be primarily concerned with variable-length strings, and our data is highly structured. What sets our data apart is the sheer quantity of it.

The *Big Data* regime often deals with tables that have millions to billions of rows. The cutting edge of computational science is trying to figure how out to deal with data on the order of 10^{16} to 10^{18}. HDF5 is at the forefront of tackling this quantity of data. At this volume, data earns the term *exascale* because the size is roughly 1 exabyte. An exabyte is almost unimaginably large. And at this scale, any improvements that can be made to the storage size per element are worth implementing.

The beauty of HDF5 is that it works equally well on gargantuan data as it does on tiny datasets. This allows users to play around with subsets of their data on their laptops and then seamlessly deploy to the largest computers ever built, and everything in between.

A contributing factor to the popularity of HDF5 is that it is accessible from almost anywhere. The HDF Group supports interfaces in C, Fortran, Java, and C++ (mostly deprecated; use the C interface instead). The C interface is the default and most fully featured API. Third-party packages that interface with HDF5 are available in MAT-LAB, Mathematica, Haskell, and others. Python has two packages for using HDF5:

h5py and PyTables. Here, we will use PyTables and occasionally reference aspects of the C interface.

> **PyTables Versus h5py**
>
> Note that we have chosen PyTables here because its adds further querying capabilities to the HDF5 interface. These are important to learn about, as advanced querying comes up frequently in database theory. On the other hand, h5py exposes HDF5's underlying parallelism features directly to the user. For general use where you may want to ask sophisticated questions of your data, go with PyTables. In cases where you have large amounts of data that you don't need to question too deeply, go with h5py. For an excellent book on h5py, please see Andrew Collette's *Python and HDF5* (O'Reilly).

Before we can shoot for storing astronomical datasets, we first have to learn how normal-sized files are handled. Python has a lot of great tools for handling files of various types, since they are how most data—not just physics data—is stored. Having an understanding of how Python handles most files is needed to fully intuit how large-scale databases like HDF5 work. Thus, this chapter starts out with an overview section on normal files before proceeding on to HDF5 and all of its fancy features.

Files in Python

So far in this book, the discussion has revolved around *in-memory* operations. However, a real computer typically has a hard drive for long-term persistent storage. The operating system abstracts this into a collection of files. There are many situations in which you may need to interact with a file on your hard drive:

- Your collaborator emails you raw data. You download the attachment and want to look at the results.
- You want to email your collaborators some of your data, quickly.
- You need to use external code that takes an input or data file. You may need to run the program thousands of times, so you automate the generation of input files from data that you have in-memory in Python.
- An external program that you use writes out one (or more) result files, and you want to read them and perform further analysis.
- You want to keep an intermediate calculation around for debugging or validation.

Reading and writing files is about interacting with the outside world. The senders and receivers in these interactions can be humans, other programs, or both. Files provide a common object that enables these interactions. That said, files are further specialized into *formats*, such as *.csv*, *.doc*, *.json*, *.mp3*, *.png*, and so on. These formats denote the internal structure of the file. The *.txt* extension is an exception in that it is not a format; it is traditionally used to flag that a file has no specific internal structure but contains plain, free-flowing text. We do not have the time or space in this book to fully describe even a fraction of the popular file formats. However, Python will be able to open all of them (if not necessarily make sense of their internal structure).

In Python, to save or load data you go through a special *file handle* object. The built-in open() function will return a file object for you. This takes as its argument the path to the file as a string. Suppose you have a file called *data.txt* in the current directory. You could get a handle, f, to this file in Python with the following:

```
f = open('data.txt')
```

The open() call implicitly performs the following actions:

1. Makes sure that *data.txt* exists.
2. Creates a new handle to this file.
3. Sets the cursor position (pos) to the start of the file, pos = 0.

The call to open() does not read into memory any part of the file, write anything out to the file, or close the file. All of these actions must be done separately and explicitly and are accomplished through the use of file handle methods.

> Methods are functions that are defined on a class and bound to an object, as seen in Chapter 6.

Table 10-1 lists the most important file methods.

Table 10-1. Important file handle methods

Method	Description
f.read(n=-1)	Reads in n bytes from the file. If n is not present or is -1, the entire rest of the file is read.
f.readline()	Reads in the next full line from the file and returns a string that includes the newline character at the end.
f.readlines()	Reads in the remainder of the file and returns a list of strings that end in newlines.
f.seek(pos)	Moves the file cursor to the specified position.
f.tell()	Returns the current position in the file.

Method	Description
f.write(s)	Inserts the string s at the current position in the file.
f.flush()	Performs all pending write operations, making sure that they are really on disk.
f.close()	Closes the file. No more reading or writing can occur.

Suppose that *matrix.txt* represents a 4×4 matrix of integers. Each line in the file represents a row in the matrix. Column values are separated by commas. Ideally, we would be able to read this into Python as a list of lists of integers, since that is the most Pythonic way to represent a matrix of integers. This is not the most efficient representation for a matrix—a NumPy array would be better—but it is fairly common to read in data in a native Python format before continuing to another data structure. The following snippet of code shows how to read in this matrix. To follow along, first make sure that you create a *matrix.txt* file on your computer. You can do this by copying the contents shown here into your favorite text editor and saving the file with the right name:

matrix.txt	Code	Python matrix

1,4,15,9
0,11,7,3
2,8,12,13
14,5,10,6
	`f = open('matrix.txt')`
	`matrix = []`
	`for line in f.readlines():`
	`    row = [int(x) for x in line.split(',')]`
	`    matrix.append(row)`
	`f.close()`
		[[1, 4, 15, 9],
		[0, 11, 7, 3],
		[2, 8, 12, 13],
		[14, 5, 10, 6]]

Notice that the lines from a file are always strings. This means that you have to convert the string versions of the values in `matrix` into integers yourself. Python doesn't know that you mean for the content of the file to be a matrix. In fact, the only assumption about data types that Python can make is that a file contains strings. You have to tell it how to interpret the contents of a file. Thus, any numbers in the file must be converted from string forms to integers or floats. For a reminder on how to convert between variable types, see "Variables" on page 42. Special file readers for particular formats, which we won't see here, may perform these conversions for you automatically. However, under the covers, everything is still a string originally. At the end of the preceding code snippet, also note that the file must be closed manually. Even when you have reached the end of a file, Python does not assume that you are done reading from it.

> **Files Should Always Be Closed!**
>
> A file that remains open unnecessarily can lead to accidental data loss as well as being a security hazard. It is always better to close a file prematurely and open a new file handle than it is to leave one open and lingering. File handles are cheap to create, so performance should not be a concern.

Files are opened in one of multiple *modes*. The mode a file was opened with determines the methods that can be used on the handle. Invalid methods are still present on the handle, but trying to use them will raise an exception.

So far, we've only opened files in the default *read-only* mode. To change this, mode flags may be passed into the `open()` call after the filename. The mode is specified as a string of one or more characters with the special meanings listed in Table 10-2. A common example is to open the file for writing and erase the existing contents. This uses the `'w'` flag:

```
f = open('data.txt', 'w')
```

Table 10-2. Useful file modes

Mode	Meaning
`'r'`	Read-only. No writing possible. Starting `pos = 0`.
`'w'`	Write. If the file does not exist, it is created; if the file does exist, the current contents are deleted (*be careful!*). Starting `pos = 0`.
`'a'`	Append. Opens the file for writing but does not delete the current contents; creates the file if it does not exist. Starting `pos` is at the end of the file.
`'+'`	Update. Opens the file for both reading and writing; may be combined with other flags; does not delete the current contents. Starting `pos = 0`.

As a more sophisticated example, the following adds a row of zeros to the top of our matrix and a row of ones to the end:

Old matrix.txt	Code	New matrix.txt
1,4,15,9 0,11,7,3 2,8,12,13 14,5,10,6	`f = open('matrix.txt', 'r+')` ❶ `orig = f.read()` ❷ `f.seek(0)` ❸ `f.write('0,0,0,0\n')` ❹ `f.write(orig)` ❺ `f.write('\n1,1,1,1')` ❻ `f.close()` ❼	0,0,0,0 1,4,15,9 0,11,7,3 2,8,12,13 14,5,10,6 1,1,1,1

❶ Open the file in read and write mode, without overwriting the contents.

❷ Read the entire file into a single string.

❸ Go back to the start of the file.

❹ Write a new line, clobbering what was there.

❺ Write the original contents back to the file after the line that was just added.

❻ Write another new line after the original contents.

❼ Close the file now that we are done with it.

There are many times when no matter what happens in a block of code—success or failure, completion or exception—special safety code must be run at the end of the block. This is to prevent data loss, corruption, or even ending up in the wrong place on the filesystem. In Python, administering these potentially hazardous situations is known as *context management*. There are many context managers that perform defensive startup actions when the code block is *entered* (right before the first statement), and other cleanup actions when the block is *exited*. Code blocks may be exited either right after the last statement or following an uncaught exception. File handles are the most common context managers in Python. As we have mentioned before, files should always be closed. However, files can act as their own context managers. When using a file this way, the programmer does not need to remember to manually close the file; the call to the file's `close()` method happens automatically.

The `with` statement is how a context is entered and exited. The syntax for this statement introduces the `with` Python keyword and reuses the `as` keyword. The `with` statement has the following format:

```
with <context-manager> as <var>:
    <with-block>
```

Here, the <context-manager> is the actual context object, <var> is a local variable name that the context manager is assigned to, and the <with-block> is the code that is executed while the manager is open. The as <var> portion of this syntax is optional.

The *matrix.txt* file example from before can be expressed using a with statement as follows:

matrix.txt	Code	matrix

0,0,0,0
1,4,15,9
0,11,7,3
2,8,12,13
14,5,10,6
1,1,1,1
	`matrix = []`
	`with open('matrix.txt') as f:` ❶
	`    for line in f.readlines():`
	`        row = [int(x) for x in line.split(',')]`
	`        matrix.append(row)`
	❷
		[[0, 0, 0, 0],
		 [1, 4, 15, 9],
		 [0, 11, 7, 3],
		 [2, 8, 12, 13],
		 [14, 5, 10, 6],
		 [1, 1, 1, 1]]

❶ The file f is open directly following the colon (:).

❷ f is closed by the context manager once the indentation level returns to that of the with keyword.

Using with statements is the recommended way to use files, because not having to explicitly call f.close() all of the time makes your code much safer and more robust. Other kinds of context managers exist. You can write your own. However, files are the context managers that you will most frequently encounter. This is because it is easy to forget to close files, and the consequences can be relatively severe if you do.

Now that we have seen the basics of how to read and write normal Python files, we can start to discover HDF5. HDF5 is one of the richest and most useful file formats for scientific computing. HDF5 puts numeric data first. This helps distinguish it from other file formats where strings rule, like we have seen in this section. Before getting into the nitty-gritty, it will be helpful to first gain some perspective.

An Aside About Computer Architecture

Computers are physical tools, just like any other experimental device. So far, we have been able to ignore how they work and their internal structure. When it comes to data storage, though, there are enough subsystems simultaneously dancing that we need to understand what a computer *is* in order to effectively program it. As you will see later in this chapter, this knowledge can make the difference between your physics software taking a week to run or five minutes.

Overlooking other peripheral devices (keyboard, mouse, monitor), a basic computer has consisted of three main components since the 1980s: a central processing unit

(*CPU*), random-access memory (*RAM*), and a storage drive. Historically, the storage drive has gone by the name hard disk drive (*HDD*), because the device was made up of concentric spinning magnetic disks. More recent storage devices are built like a flash memory stick; these drives are called solid state drives (*SSDs*). The CPU, RAM, and storage can be thought of as living in series with each other, as seen in Figure 10-1.

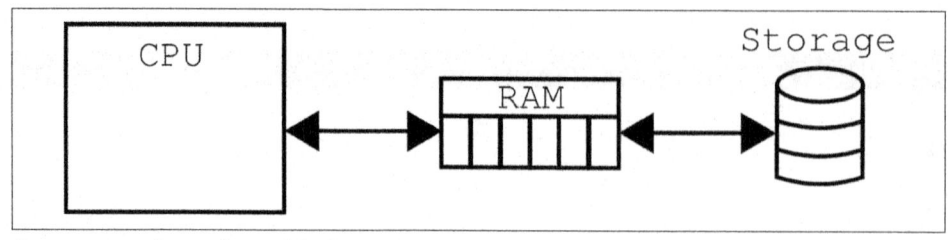

Figure 10-1. A simple model of a computer

In this simple model, the CPU can be thought of as a dumb calculator; RAM is what "remembers" what the CPU just did (it acts sort of like short-term memory), and the storage is what allows the computer to save data even when it is turned off (it's like long-term memory). In practice, when we talk about the computer "doing something," we really mean the CPU. When we talk about the filesystem, we really mean the storage. It is important to understand, therefore, that RAM is what shuffles bytes between these two components.

Of course, computer architectures have become much more complicated than this simple model. CPU caches are one major mainstream advancement. These caches are like small versions of RAM that live on the CPU. They contain copies of some of the data in RAM, but much closer to the processor. This prevents the computer from having to go out to main memory all of the time. For commonly accessed data, the caches can provide huge decreases in execution time. The caches are named after a hierarchy of level numbers, such *L1*, *L2*, and so on. In general, the higher the level number, the smaller the cache size, but the faster it is to access it. Currently, most processors come with L1 and L2 caches. Some processors are now also starting to come with L3 caches. Figure 10-2 represents a computer with CPU caches.

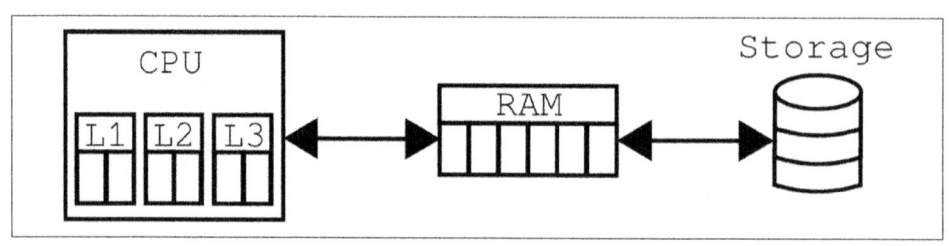

Figure 10-2. A computer with L1, L2, and L3 CPU caches

The other big innovation in computer architecture is graphics processing units, or *GPUs*. These are colloquially known as graphics cards. They are processors that live outside of the main CPU. A computer with a GPU is displayed in Figure 10-3.

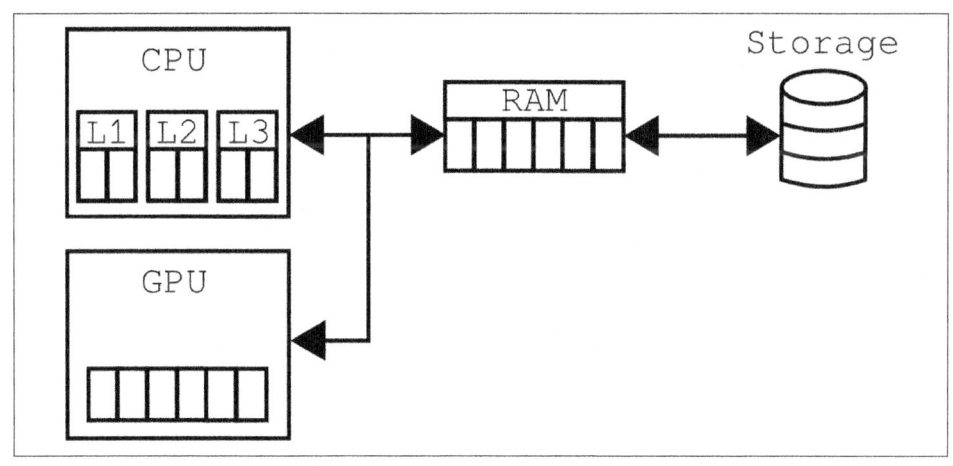

Figure 10-3. A computer with a GPU

Though there are many important differences between GPUs and CPUs, very roughly, you can think of GPUs as being really good at floating-point operations. CPUs, on the other hand, are much better at integer operations than GPUs (while still being pretty good with floating-point data). So, if you have an application that is primarily made up of floats, then GPUs may be a good mechanism to speed up your execution time.

Naturally, there is a lot more to computer engineering and architecture than what you have just seen. However, this gives you a good mental model of the internal structure of a computer. Keep this in mind as we proceed to talking about databases and HDF5. Many real-world programming trade-offs are made and balanced because of the physical performance of the underlying machine.

Big Ideas in HDF5

Persisting structured, numerical data to binary formats is superior to using plain-text ASCII files. This is because, by their nature, they are often smaller. Consider the following comparison between integers and floats in native and string representations:

```
# small ints          # medium ints
42     (4 bytes)      123456    (4 bytes)
'42'   (2 bytes)      '123456'  (6 bytes)

# near-int floats     # e-notation floats
12.34    (8 bytes)    42.424242E+42    (8 bytes)
'12.34'  (5 bytes)    '42.424242E+42'  (13 bytes)
```

In most cases, the native representation is smaller than the string version. Only by happenstance are small integers and near-integer floats smaller in their string forms. Such cases are relatively rare on average, so native formats almost always outperform the equivalent strings in terms of space.

Space is not the only concern for files. Speed also matters. Binary formats are always faster for I/O because in order to do real math with the numbers, if they are in a string form you have to convert them from strings to the native format. The Python conversion functions `int()` and `float()` are known to be relatively slow because the C conversion functions `atoi()` and `atof()` that they wrap around are expensive themselves.

Still, it is often desirable to have something more than a binary chunk of data in a file. HDF5 provides common database features such as the ability to store many datasets, user-defined metadata, optimized I/O, and the ability to query its contents. Unlike SQL, where every dataset lives in a single namespace, HDF5 allows datasets to live in a nested tree structure. In effect, HDF5 is a filesystem within a file—this is where the "hierarchical" in the name comes from.

PyTables provides the following basic dataset classes that serve as entry points for various HDF5 constructs:

`Array`
   The files of the filesystem

`CArray`
   Chunked arrays

`EArray`
   Extendable arrays

`VLArray`
   Variable-length arrays

`Table`
   Structured arrays

All of these must be composed of what are called *atomic types* in PyTables. The atomic types are roughly equivalent to the primitive NumPy types that were seen in "dtypes" on page 204. There are six kinds of atomic types supported by PyTables. Here are their names, descriptions, and supported sizes:

`bool`
   True or false type—8 bits

`int`
   Signed integer types—8, 16, 32 (default), and 64 bits

`uint`
: Unsigned integer types—8, 16, 32 (default), and 64 bits

`float`
: Floating-point types—16, 32, and 64 (default) bits

`complex`
: Complex floating-point types—64 and 128 (default) bits

`string`
: Fixed-length raw string type—8 bits times the length of the string

Other elements of the hierarchy may include:

Groups
: The directories of the filesystem; may contain other groups and datasets

Links
: Like soft links on the filesystem

Hidden nodes
: Like hidden files

These pieces together are the building blocks that you can use to richly describe and store your data. HDF5 has a lot of features and supports a wide variety of use cases. That said, simple operations are easy to implement. Let's start with basic file reading and writing.

File Manipulations

HDF5 files may be opened from Python via the PyTables interface. To get PyTables, first import `tables`. Like with `numpy` and `np`, it is common to abbreviate the `tables` import name to `tb`:

```
import tables as tb
f = tb.open_file('/path/to/file', 'a')
```

Files have modes that they may be opened in, similarly to how plain-text files are opened in Python. Table 10-3 displays the modes that are supported by PyTables.

Table 10-3. HDF5 file modes

Attribute	Description
r	Read-only—no data can be modified.
w	Write—a new file is created; if a file with that name exists, it is deleted.
a	Append—an existing file is opened for reading and writing, or if the file does not exist, it is created.
r+	Similar to a, but the file must already exist.

In HDF5, all nodes stem from a root node, "/" or f.root. In PyTables, you may access subnodes as attributes on nodes higher up in the hierarchy—e.g., f.root.a_group.some_data. This sort of access only works when all relevant nodes in the tree have names that are also valid Python variable names, however; this is known as *natural naming*.

Creating new nodes must be done on the file handle, not the nodes themselves. If we want to make a new group, we have to use the create_group() method on the file. This group may then be accessed via the location it was created in. For example, creating and accessing a group called a_group on the root node can be done as follows:

```
f.create_group('/', 'a_group', "My Group")
f.root.a_group
```

Possibly more important than groups, the meat of HDF5 comes from datasets. The two most common datasets are arrays and tables. These each have a corresponding create method that lives on the file handle, called create_array() and create_table(). Arrays are of fixed size, so you must create them with data. The type of the data in the HDF5 file will be interpreted via numpy. Tables, like NumPy structured arrays, have a set data type. Unlike arrays, tables are variable length, so we may append to them after they have been created. The following snippet shows how to create an array and a table and how to populate them using Python lists and NumPy arrays:

```
# integer array
f.create_array('/a_group', 'arthur_count', [1, 2, 5, 3])

# tables need descriptions
dt = np.dtype([('id', int), ('name', 'S10')])
knights = np.array([(42, 'Lancelot'), (12, 'Bedivere')], dtype=dt)
f.create_table('/', 'knights', dt)
f.root.knights.append(knights)
```

At this point, the hierarchy of groups and datasets in the file is represented by the following:

```
/
|-- a_group/
|    |-- arthur_count
|
|-- knights
```

Arrays and tables attempt to preserve the original *flavor*, or data structure, with which they were created. If a dataset was created with a Python list, then reading out the data will return a Python list. If a NumPy structured array was used to make the data, then a NumPy structured array will be returned. Note that you can read data from a dataset simply by slicing (described in Chapter 9). One great thing about PyTables and HDF5 is that only the sliced elements will be read in from disk. Parts of the dataset that are not included in the slice will not be touched. This speeds up reading by not making the computer do more work than it has to, and also allows you to read in portions of a dataset whose whole is much larger than the available memory. Using our sample `arthur_count` array, the following demonstrates flavor preservation. Also note that the type of the dataset comes from PyTables, and this is separate from the type of the data that is read in:

Code	Returns
f.root.a_group.arthur_count[:]	[1, 2, 5, 3]
type(f.root.a_group.arthur_count[:])	list
type(f.root.a_group.arthur_count)	tables.array.Array

Since the `arthur_count` array came from a Python list, only Python list slicing is available. However, if a dataset came from a NumPy array originally, then it can be accessed in a NumPy-like fashion. This includes slicing, fancy indexing, and masking. The following demonstrates this NumPy-like interface on our `knights` table:

Code	Returns
`f.root.knights[1]` ❶	`(12, 'Bedivere')`

`f.root.knights[:1]` ❷	`array([(42, 'Lancelot')],`
	`      dtype=[('id', '<i8'),`
	`             ('name', 'S10')])`

`mask = (f.root.knights.cols.id[:] < 28)`
`f.root.knights[mask]` ❸
	`array([(12, 'Bedivere')],`
	`      dtype=[('id', '<i8'),`
	`             ('name', 'S10')])`

`f.root.knights[([1, 0],)]` ❹	`array([(12, 'Bedivere'),`
	`       (42, 'Lancelot')],`
	`      dtype=[('id', '<i8'),`
	`             ('name', 'S10')])`

❶ Pull out just the second row.

❷ Slice the first row.

❸ Create a mask from the on-disk `id` column and apply this to the table.

❹ Fancy index the second and first rows, in that order.

Pulling in data from disk only as needed is known as *memory mapping*. HDF5 takes care of this for you automatically; this is one way that reading and writing to HDF5 files is optimized.

Now you know how to create and use the core node types: groups, arrays, and tables. However, you can use these elements of the hierarchy to even greater effect by combining them in meaningful ways. The following section discusses how to think about your information more broadly than within the confines of a single dataset.

Hierarchy Layout

Suppose you have a big table of similar objects. For example, consider a table of all of the particles that you have seen recently. This could be written as a list of tuples, as follows:

```
# particles: id, kind,     velocity
particles = [(42, 'electron', 72.0),
             (43, 'proton',   0.1),
             (44, 'electron', 76.8),
             (45, 'neutron',  0.39),
             (46, 'neutron',  0.72),
             (47, 'neutron',  0.55),
             (48, 'proton',   0.18),
             (49, 'neutron',  0.23),
             ...
            ]
```

Having a big table like this can be inefficient. If you know ahead of time that you normally want to look at all of the neutral and charged particles separately, then why search through all the particles all of the time? Instead, it would be better to split up the particles into multiple tables grouped by whether they are neutral or charged. The following shows such a split as applied to the original `particles` table:

```
neutral = [(45, 'neutron', 0.39),
           (46, 'neutron', 0.72),
           (47, 'neutron', 0.55),
           (49, 'neutron', 0.23),
           ...
          ]

charged = [(42, 'electron', 72.0),
           (43, 'proton',   0.1),
           (44, 'electron', 76.8),
           (48, 'proton',   0.18),
           ...
          ]
```

The `kind` column is now redundant in the `neutral` table because its value is the same for all rows. We can delete this whole column and rely on the structure of these two tables together to dictate that the `neutral` table always refers to neutrons. This space saving can be seen here:

```
neutral = [(45, 0.39),
           (46, 0.72),
           (47, 0.55),
           (49, 0.23),
           ...
          ]

charged = [(42, 'electron', 72.0),
           (43, 'proton',   0.1),
           (44, 'electron', 76.8),
           (48, 'proton',   0.18),
           ...
          ]
```

With these transformations we are embedding information directly into the semantics of the hierarchy. The deeper and broader your hierarchy is, the more information can be stored with it. For example, we could add another layer that distinguishes the particles based on detector:

```
/
|-- detector1/
|     |-- neutral
|     |-- charged
|
|-- detector2/
```

Hierarchy Layout | 243

```
|       |-- neutral
|       |-- charged
```

Data can and should be broken up like this to improve access time speeds. Such segregation based on available information is more efficient because there are:

- Fewer rows to search over
- Fewer rows to pull from disk
- Fewer columns in the description, which decreases the size of the rows

Dealing with less data is always faster than dealing with more. This is especially true because of how long it takes to read data from disk. Waiting around for access prior to computation is known as the starving CPU problem (Alted, 2010).

An analogy about access time with respect to different parts of a computer comes to us from Gustavo Duarte's article "What Your Computer Does While You Wait." This is helpful to keep in mind when making decisions about how to deal with large amounts of data. The metaphor goes as follows. If a processor's access of the L1 cache is analogous to you finding a word on a computer screen (3 seconds), then:

- Accessing the L2 cache is like getting a book from a bookshelf (15 s).
- Accessing main memory is like going to the break room, getting a snack, and chatting with your coworker (4 min).
- Accessing a (mechanical) HDD is like leaving your office, leaving your building, wandering the planet for a year and four months, then returning to your desk with the information finally made available.

From this analogy, you can see how important it is to minimize the number of times the computer has to go to the hard disk, minimize the amount of data that must be transferred from the hard disk, and keep the computer busy while it is waiting for information to be retrieved. Manipulating the layout of the hierarchy is a useful tool in this regard.

However, there are practical limits to hierarchy depth and breadth. For every new group or dataset, there is overhead. In HDF5 there is 64 KB of space reserved for metadata for every dataset. Thus, the practical limit is hit when the overhead size is comparable to the size of the underlying data being stored. Though your mileage and data may vary, a good rule of thumb is that three levels of hierarchy allowing any number of datasets at each level is enough structure for most problems.

So far in this chapter, you have learned how to read and write data in HDF5. You have even seen some manipulations to the file hierarchy that can be used to optimize your data. Up next, though, is a discussion of how HDF5 stores datasets. Understanding what actions HDF5 is actually performing opens up new avenues to data manipulation and optimization.

Chunking

Chunking is the ability to split up a dataset into smaller blocks of equal or lesser rank. Unlike the manipulations we saw in the previous section that occurred between multiple datasets, chunking happens within a single dataset and is implemented automatically. Chunks are how HDF5 stores data. This is a feature of HDF5 that has no direct analogy in NumPy.

Chunking requires that extra metadata be stored that points to the location of each chunk in the file. Thus, there is some overhead per chunk. The more chunks that exist, the more information must be stored about where those chunks live. However, chunking enables two key features that allow HDF5 to deal with huge amounts of data:

1. Sparse data may be stored efficiently.
2. Datasets may extend infinitely in all dimensions.

For example, suppose that you wanted to store a $10^9 \times 10^9$ item matrix. If this matrix was mostly made up of zeros, you could use HDF5 to store only the nonzero elements and obtain a gigantic savings in terms of execution time and database size. Currently, PyTables only allows one extendable dimension, typically rows. However, there is no such restriction in HDF5 itself, so if you use HDF5's C interface (or h5py), you may extend infinitely in every which way.

Consider that all datasets are composed of metadata (which describes the data type, shape, and more) and bytes that represent the data itself. In a contiguous dataset, all of the bytes are in a single array, right next to each other. This can be seen graphically in Figure 10-4. Alternatively, a chunked dataset splits up this array of bytes into many smaller arrays, each stored in a separate location. A representation of this may be seen in Figure 10-5.

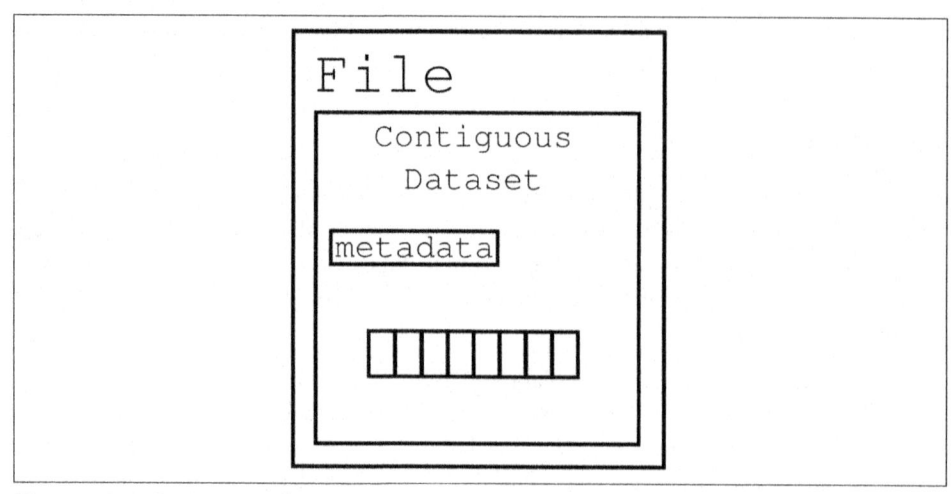

Figure 10-4. Contiguous dataset

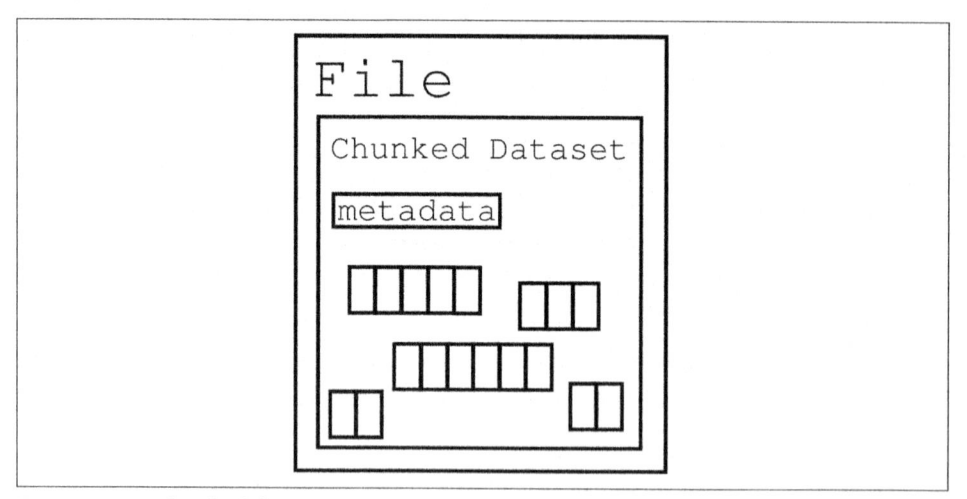

Figure 10-5. Chunked dataset

All reading and writing in HDF5 happens per-chunk. Contiguous datasets, like arrays, are the special case where there is only one chunk. Furthermore, the chunk shape and size is a property of the dataset itself and independent of the actual data. This has the following significant implications:

- Edge chunks may extend beyond the dataset.
- Default fill values are set in unallocated space, allowing for sparse datasets to only store the data they have.
- Reading and writing of chunks may happen in parallel.

- Small chunks are good for accessing only some of the data at a time.
- Large chunks are good for accessing lots of data at a time.

Chunks are not just a mechanism for laying out extensible datasets and efficiently implementing sparse data. Rather, chunks are *the* way that HDF5 reasons about data storage operations. HDF5 allows users to set the chunk size and shape on any chunked dataset. In PyTables, the chunkshape is a keyword argument to the creation methods. This is simply a tuple representing the chunk's shape and size along each dimension. In PyTables, we can make a new chunked array with the create_carray() method. The following is an example where we create a chunked array and explicitly set the chunk shape:

```
f.create_carray('/', 'omnomnom', data, chunkshape=(42,42))
```

If we had not explicitly set a chunk shape, PyTables would have picked one for us, using properties of the data and some heuristics to take a best guess as to the optimal chunking. In general, if you are unsure of what the chunk shape should be, just let PyTables pick one for you and do not set it manually.

> Tables, CArrays, and EArrays all accept the chunkshape keyword argument.

As a visual example of how chunking works, suppose you have a 3×3 matrix in memory. Storing this as a contiguous dataset would only require one flat array with nine elements. A diagram of this situation is displayed in Figure 10-6. However, we are free to choose any chunk shape that is less than or equal to 3×3. As an example, take 2×2 to be the chunk shape. The 3×3 matrix would then be represented by four 2×2 tiles overlaid onto the matrix. It is OK for the total shape of all of the tiles to extend beyond the bounds of the original data. Overextended chunk elements will simply go unused. Figure 10-7 shows this case, with shading used to denote the chunks.

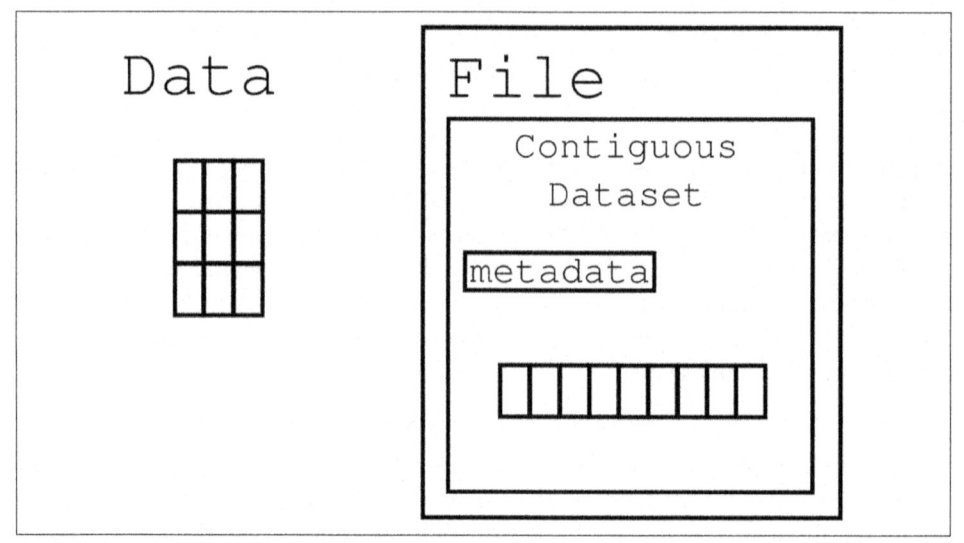

Figure 10-6. Contiguous 3x3 dataset, in memory and in an HDF5 file

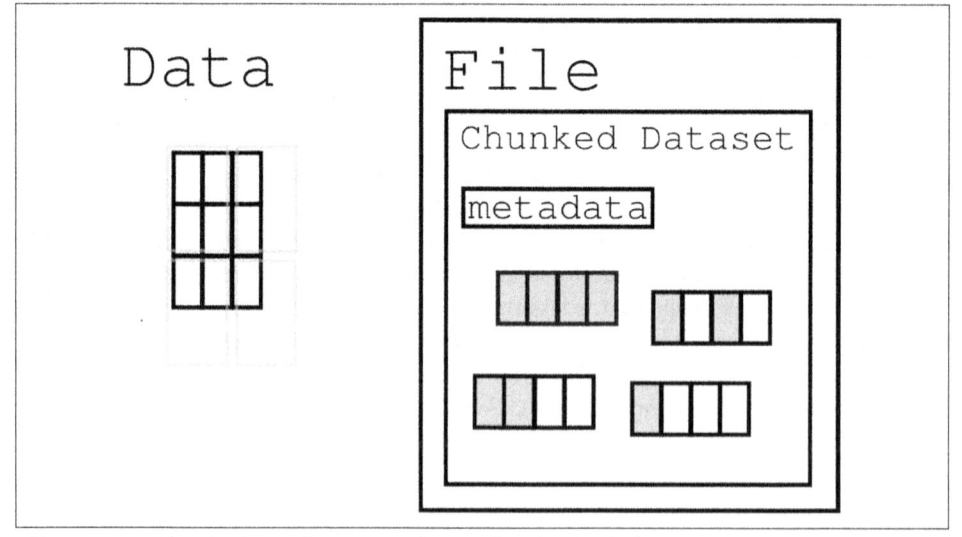

Figure 10-7. Chunked 3x3 dataset, with 2x2 chunks shaded

Manipulating the chunk shape is a great way to fine-tune the performance of an application. Typically, the best-guess chunk shape that PyTables provides is good enough for most cases. However, when you really need to squeeze every ounce of read and write speed out of your data, you can vary the chunk shape to try to find that perfect value.

While we are on the topic of performance, let's examine the different ways to perform calculations on our data. Sending and receiving data to and from the hard drive is not the only way to leverage PyTables, as we will see in the next section.

In-Core and Out-of-Core Operations

Calculations depend on the memory layout of the data. When implementing array operations, algorithms fall into two broad categories based on where you expect the data to be in memory. Operations that require all data to be in memory are called *in-core* and may be memory-bound. This means that the volume of data that may be processed is limited to the size of the computer's memory. For example, NumPy arrays are memory-bound because in order to exist, the whole array must be in memory. On the other hand, operations where the data can be external to memory are known as *out-of-core*. Such operations may be CPU-bound. Thus, "in-core" and "out-of-core" can be thought of in terms of which component (the RAM or the CPU) in Figure 10-1 is limiting the behavior of the machine. Unlike in NumPy, computations over HDF5 arrays may be performed in an out-of-core way. Let's explore in-core and out-of-core paradigms more deeply, before continuing. If you find any part of this discussion overwhelming or confusing, feel free to skip ahead to "Querying" on page 252.

In-Core

For a classic example of in-core operations, say that there are two NumPy arrays, a and b, sitting in memory. (The next section will discuss what to do if your data does not fit into memory.) The expression defined as c here is independent of the size and contents of a and b. As long as a and b are compatible, c may be computed:

```
a = np.array(...)
b = np.array(...)
c = 42 * a + 28 * b + 6
```

The expression for c creates three temporary arrays, which are discarded after they are used. Still, these temporaries take up space in memory. Thus, the free space needed in memory to compute c is at least six times the size of a or b. Here are the temporaries that would be computed when evaluating c:

```
temp1 = 42 * a
temp2 = 28 * b
temp3 = temp1 + temp2
c = temp3 + 6
```

In general, for N operations there are N-1 temporaries that have to be constructed. This wastes memory and is slower than if the temporaries were avoided altogether. Still, given how long reading from a file can take, pulling the data from disk is even slower than creating temporaries.

Alternatively, a less memory-intensive implementation would be to evaluate the whole expression element-wise. The following snippet shows how such a computation would be written:

```
c = np.empty(...)
for i in range(len(c)):
    c[i] = 42 * a[i] + 28 * b[i] + 6
```

This algorithm is the idiomatic way that you would write such an operation in C, C++, or Fortran. In those languages, this algorithm is very fast. However, in Python, this method is much slower than the simpler NumPy solution, even with the temporaries. This is in part because Python loops are slow and in part because of the very dynamic Python type system.

Slower still, suppose that a and b were HDF5 arrays on disk. The individual element access times would make this computation take thousands of times longer, or more. Repetitively reading small amounts of data from the hard drive is just that slow. Even with fast solid-state drives, this algorithm would still run like a turtle through a tar pit. In practice, idiomatic NumPy is good enough for almost all in-core operations. If you truly need something faster, you probably need to look to other programming languages.

Out-of-Core

Out-of-core operations combine the notions of element-wise evaluation and chunking. Even though this is most useful when access times are slow (i.e., when reading from disk), such operations can work on data that is fully inside of memory. Let's take again the example from the previous section, where c = 42 * a + 28 * b + 6. An out-of-core algorithm would apply a chunksize of 256 to the element-wise algorithm. The three possible solutions are seen here:

```
# setup
a = np.array(...)
b = np.array(...)

# how to compute c with numpy (in-core)
c = 42 * a + 28 * b + 6

# how to compute c element-wise (in-core)
c = np.empty(...)
for i in range(len(c)):
    c[i] = 42 * a[i] + 28 * b[i] + 6

# how to compute c with chunks (out-of-core)
c = np.empty(...)
for i in range(0, len(a), 256):
    r0, r1 = a[i:i+256], b[i:i+256]
    np.multiply(r0, 42, r2)
    np.multiply(r1, 28, r3)
```

```
np.add(r2, r3, r2)
np.add(r2,  6, r2)
c[i:i+256] = r2
```

Using the out-of-core strategy, no more than 256 elements need to be in memory at any time. While temporaries *are* created, their size is similarly limited to 256 elements. Feeding in only discrete chunks at a time allows out-of-core algorithms to function on infinite-length data while using only the finite memory resources of your computer. This problem lends itself nicely to parallelism, since every chunk can be assumed to be independent of every other chunk. We will learn more about parallelism in Chapter 12.

In Python, the numexpr library provides a way to perform chunked, element-wise computations on regular NumPy arrays. PyTables uses this library internally, but extends its application to HDF5 arrays on disk. The PyTables tb.Expr class implements the out-of-core interface that handles the expression evaluation. This has the distinct advantage that the full array is never in memory. The Expr class is critical because it is easy to have HDF5 datasets that are thousands, millions, or billions of times larger than the amount of memory on a laptop.

The expression for c can be calculated with the Expr class and with a and b as HDF5 arrays that live outside of memory. The following code represents this out-of-core strategy. Note that in this snippet even the c array lives on disk and is never fully brought into memory:

```
# open a new file
shape = (10, 10000)
f = tb.open_file("/tmp/expression.h5", "w")

# create the arrays
a = f.create_carray(f.root, 'a', tb.Float32Atom(dflt=1.), shape)
b = f.create_carray(f.root, 'b', tb.Float32Atom(dflt=2.), shape)
c = f.create_carray(f.root, 'c', tb.Float32Atom(dflt=3.), shape)

# evaluate the expression, using the c array as the output
expr = tb.Expr("42*a + 28*b + 6")
expr.set_output(c)
expr.eval()

# close the file
f.close()
```

Using tb.Expr or numexpr is much cleaner, simpler, and more robust than writing your own custom chunking algorithm. Thus, it is generally recommended to use the existing tools rather than trying to create your own.

In summary, in-core operations are fast and efficient when all of the data is already in memory. However, if the data needs to be pulled in from disk or if the data is too large to fit in memory, an out-of-core algorithm is best. For HDF5, PyTables provides the `Expr` class to perform out-of-core operations. PyTables uses this out-of-core capability to great effect by also giving users a mechanism to perform complex queries of their data. We'll look at this topic next.

Querying

One of the most common and most sophisticated operations is asking an existing dataset whether its elements satisfy some criteria. This is known as *querying*. PyTables defines special methods on tables to support fast and efficient querying. The following methods are defined for the `Table` class. Each takes a string condition, `cond`, that represents the expression to evaluate:

```
tb.Table.where(cond)                  # Returns an iterator of matches
tb.Table.get_where_list(cond)         # Returns a list of indices
tb.Table.read_where(cond)             # Returns a list of results
tb.Table.append_where(dest, cond)     # Appends matches to another table
```

The conditions used in these query calls are automatically evaluated as out-of-core operations for every row in the table. The `cond` expressions must return a Boolean. If the expression is `True`, then the row is included in the output. If the condition is `False`, then the row is skipped. These conditions are executed in the context of the column names and the other datasets in the HDF5 file. Variables in the current Python `locals()` and `globals()` functions are available in the expressions as well.

The `where()` method returns an iterator over all matched rows. It is very common to compare columns in a table to specific values or each other. The following example shows iterating over all rows where column 1 is less than 42 and column 2 is equal to column 3 for that row:

```
for row in tab.where('(col1 < 42) & (col2 == col3)'):
    # do something with row
```

Queries may be quite complex. This makes the benefits of out-of-core computation that much greater, since fewer temporaries are made and less data needs to be individually transferred from disk.

The `where()` method is a major reason why users choose PyTables over h5py or HDF5 on its own. The query interfaces are easy and intuitive to use, even though there are a lot of powerful tools beneath each method call. In the next section, we discuss another great feature that PyTables exposes in a pleasant interface. Unlike querying, this one is implemented by HDF5 itself.

Compression

Another approach to solving the starving CPU problem, where the processor is waiting around not doing anything while data comes in from disk, is through *compres-*

sion. Compression is when the dataset is piped through a zipping algorithm when it is written to disk and the inverse unzipping algorithm when data is read from disk. Compression happens on each chunk independently. Thus, compressed chunks end up on disk with a varying number bytes, even though they all share the same chunk shape. This strategy does have some storage overhead due to the zipping algorithm itself. Even with the compression overhead, however, zipping may drastically reduce file sizes *and* access times. For very regular data, it can reduce the data size nearly to zero.

That compressing and decompressing each chunk can reduce read and write times may, at first glance, seem counterintuitive. Compression and decompression clearly require more processing power than simply copying an array directly into memory from disk. However, because there is *less total information* transferred to and from disk, the time spent unzipping the array can be far less than what would be required to move the array around wholesale.

Compression is a feature of HDF5 itself. At a minimum, HDF5 is dependent on `zlib`, a generic zipping library. Therefore, some form of compression is always available. HDF5 compression capabilities are implemented with a plugin architecture. This allows for a variety of different algorithms to be used, including user-defined ones. PyTables supports the following compression routines: `zlib` (default), `lzo`, `bzip2`, and `blosc`.

Compression is enabled through a mechanism called *filters*. Filters get their name from the fact that they sit between the data in memory and the data on disk. Filters perform transformations on any data passing through them. PyTables has a `Filters` class that allows you to set both the compression algorithm name (as a string) and the compression level, an integer from 0 (no compression) to 9 (maximum compression). In PyTables, filters can be set on any group or dataset, or on the file itself. If unspecified, a dataset or group will inherit the filters from the node directly above it in the hierarchy. The following code demonstrates the various ways to apply filters:

```
# complevel goes from [0,9]
filters = tb.Filters(complevel=5, complib='blosc')  ❶

# filters may be set on the whole file
f = tb.open_file('/path/to/file', 'a', filters=filters)  ❷
f.filters = filters  ❸

# filters may also be set on most other nodes
f.create_table('/', 'table', desc, filters=filters)  ❹
f.root.group._v_filters = filters  ❺
```

❶ Create a `filters` object.

❷ Set filters when a file is created.

❸ Set filters on a file after it is opened.

❹ Set filters when a dataset is created.

❺ Set filters on an existing dataset.

Additionally, note that when you're choosing compression parameters, a mid-level (5) compression is usually sufficient. There is no need to go all the way up to the maximum compression (9), as it burns more processor time for only slight space improvements. It is also recommended that you use `zlib` if you must guarantee complete portability across all platforms. This is because it is the only compression algorithm guaranteed to be present by HDF5 itself. In general, if you are unsure of what to use for compression, just use `zlib`. It is the safest and most portable option. If you know you are going to stick to just using PyTables, however, it is ideal to use `blosc`. `blosc` comes from PyTables and is optimized for HDF5 and tabular data. The performance differences can be significant between `blosc` and `zlib`, which is a general-purpose library.

You have now learned all the essential aspects of creating HDF5 files and using them efficiently. However, you can interact with HDF5 in more ways than just through PyTables. The next section will go over some of the tools that make working with HDF5 even easier.

HDF5 Utilities

HDF5 is an excellent file format that supports many ways of accessing data from a wide variety of programming languages. However, it can be a fair amount of work to write a program to look into the contents of an HDF5 file every time you want to inspect one. For this reason, a number of utilities have been developed that can help you view or visualize the contents of these files. The three major command-line tools for looking into HDF5 files are as follows:

h5ls
: An `ls`-like tool for HDF5 files. This comes with HDF5. It displays metadata about the file, its hierarchy, and its datasets.

h5dump
: Prints the contents of an HDF5 table, in whole or in part, to the screen. This comes with HDF5.

ptdump
: A PyTables-based version of `h5dump`. This comes with PyTables.

These three programs all take an HDF5 file as an argument. For example, using h5ls on a file called *2srcs3rxts.h5* might display the following information about the datasets in this file:

```
$ h5ls 2srcs3rxts.h5
AgentDeaths           Dataset {7/Inf}    ❶
Agents                Dataset {7/Inf}
Compositions          Dataset {200/Inf}
InputFiles            Dataset {1/Inf}
ResCreators           Dataset {23/Inf}
Resources             Dataset {85/Inf}
SimulationTimeInfo    Dataset {1/Inf}
Transactions          Dataset {17/Inf}
```

❶ The first column contains the hierarchy node name. The second lists the node type—every node here is a dataset, since we do not have any groups. The third column contains the dataset size. The number before the slash is the current size and the number after the slash is the maximum size. Here, Inf indicates that all of the datasets are extensible and thus have infinite size.

For more information on how to use these utilities, type h5ls --help, h5dump --help, or man ptdump at the command line, respectively.

For the more visually inclined, there are also graphical tools that can be used to interactively explore HDF5 files. The two major viewers right now are *hdfview* and *ViTables*:

- *hdfview* is a Java-based HDF5 viewer and editor by the HDF group.
- *ViTables* is an independent PyTables- and Qt-based viewer for HDF5.

In general, *hdfview* is better at handling datasets that may extend infinitely in many dimensions. On the other hand, *ViTables* is better at displaying large amounts of data and also provides a graphical interface to the querying capabilities in PyTables. Both tools are valuable in their own right. However, if you are mainly using PyTables to interact with HDF5, *ViTables* should be all you need.

Storing Data Wrap-up

Having reached the end of this chapter, you should now be familiar with how to save and load data, both to regular files and HDF5 databases. Most physics problems can be cast into a form that is array-oriented. HDF5 is thus extremely helpful, since its fundamental abstractions are all based on arrays. Naturally, other file formats and databases exist and are useful for the cases they were designed around. Nothing quite matches the sublime beauty of HDF5 for computational physics, though.

At this point, you should be comfortable with the following tasks:

- Saving and loading plain-text files in Python
- Working with HDF5 tables, arrays, and groups
- Manipulating the hierarchy to enable more efficient data layouts

and familiar with these concepts:

- HDF5 tables are conceptually the same as NumPy structured arrays.
- All data reading and writing in HDF5 happens per-chunk.
- A contiguous dataset is a dataset with only one chunk.
- Computation can happen per-chunk, so only a small subset of the data ever has to be in memory at any given time.
- Going back and forth to and from disk can be very expensive.
- Querying allows you to efficiently read in only part of a table.
- Compressing datasets can speed up reading and writing, even though the processor does more work.
- HDF5 and other projects provide tools for inspecting HDF5 files from the command line or from a graphical interface.

What you have learned so far with respect to storing data is the beginning of a very deep topic. Sometimes as a computational scientist, you will need to dive in deeper and learn more. Most of the time, you probably won't. If you are already feeling overwhelmed, do not worry. Mastery of these topics will come through the natural course of experience and practice. If you feel like this chapter has only started to get you going, do not worry either. There is more to learn and more to enjoy. Understanding data storage issues goes hand in hand with understanding how computers are built and architected, and the trade-offs that are made in the process.

In the next chapter we shift gears, investigating data structures and their associated algorithms that seem to crop up over and over again in computational physics.

CHAPTER 11
Important Data Structures in Physics

In every domain there are algorithms and data structures that seem to pop up over and over again. This is because of how useful these data structures are, rather than the everything-is-a-nail-when-all-you-have-is-a-hammer phenomenon. This chapter will present some of the important and common data structures for physics.

This is by no means a complete listing. Nor is this chapter meant to take the place of a formal computer science data structures course—the canonical data structures in that domain tend not to be the same as those in physics, though there is some overlap. Rather, this is meant to whet your appetite. Many of the data structures here have important variants, which will be mentioned but not discussed in detail.

We have already seen the most important data structure: arrays. Since these received an entire chapter, we will forgo talking about them here. Instead, we will cover:

Hash tables
    Useful whenever you want to create associations between data

Data frames
    Similar to structured arrays or tables, but with added capabilities that make them invaluable for experimental data

B-trees
    Useful for managing array chunks and hierarchies

K-d trees
    Space-partitioning data structures that are useful when trying to reason about the closest neighboring points in space

Let's start at the top of this list and work our way down.

Hash Tables

Hash tables are *everywhere* in software development. The most common one that we have used so far is the Python dictionary. This data structure is so important that it forms the backbone of the Python language. Unlike in "Dictionaries" on page 73, when they were first introduced, we now get to answer the question, "How does a hash table work?"

A hash table is a special type of mapping from keys to values where the keys must be unique. In the simplest case there is a keys column and a values column in the table. The rows of this table are sorted according to the hash of the key modulus the length of the table. Missing or empty rows are allowed. Item insertion and value lookup all happen based on the hash value of the key. This means that their use presupposes a good mechanism for hashing the keys, which, as discussed in "Sets" on page 71, Python provides natively as the built-in `hash()` function.

For example, let's look at a sample hash table that maps names of subatomic particles to their mass in amu. We've set the length of the hash table to be eight even though there will only be four entries, so some rows are unused. This can be seen in Table 11-1.

Table 11-1. Sample hash table

i	Key	hash(key)	hash(key)%8	Value
0				
1	'neutrino'	-4886380829903577079	1	3.31656614e-9
2				
3				
4				
5	'proton'	-4127328116439630603	5	1.00727647
6	'electron'	4017007007162656526	6	0.000548579909
7	'neutron'	3690763999294691079	7	1.008664

Note that the row index i where data appears in the table is exactly the same as the result of the expression `hash(key)%table_len`. This lets us access elements based on the key very quickly by recomputing the index expression and then jumping to that point in the table. For example, the expression to find the value associated with the proton key, `tab['proton']`, would perform the following actions (in pseudocode) to actually get that value:

```
table['proton'] ->
    h = hash('proton')
    i = h % size(tab)
    return table[i].value
```

The main innovation of hash tables for associating data is that they prevent testing equality over all of the keys to find the corresponding value. In fact, even though the lookup mechanics are more complex, they perform on average much faster than if you were to search through an equivalent data structure, like a list of tuples. In the worst case, they still perform just as well as these alternatives.

For getting, setting, or deleting an item, hash tables are on average order-1. This means that whether you have an array with 10 elements or 50 billion elements, retrieving an item will take the same amount of time. In the worst case, getting, setting, and deleting items can take up to order-N, or O(N), where N is the size of the table. This incredibly unlikely situation is equivalent to looping over the whole table, which is what you would have to do in the worst case when using a list of tuples.

> **Big O Notation**
>
> Big O notation is a shorthand for describing the limiting behavior of an algorithm with respect to the size of the data. This looks like a function O() with one argument. It is meant to be read as "order of" the argument. For example, O(N) means "order-N" and O(n log(n)) is "order n log(n)." It is useful for quickly describing the relative speed of an algorithm without having to think about the specifics of the algorithm's implementation.

However, there are two big problems with hash tables as formulated previously. What happens when the table runs out of space and you want to add more items? And what happens when two different keys produce the same hash?

Resizing

When a hash table runs out of space to store new items, it must be resized. This typically involves allocating new space in memory, copying all of the data over, and reorganizing where each item lives in the array. This reorganization is needed because a hash table takes the modulus by the table size to determine the row index.

From Table 11-1, take the neutrino key, whose hash is -4886380829903577079. This hash value is the same no matter the table length. However, when you mod this value by 8, it produces an index of 1. If the table size were doubled, then the neutrino hash mod 16 would produce an index of 9. In general, each index will change as the result of a resize. Thus, a resize does more than just copy the data: it also rearranges it. For example, consider resizing Table 11-1 to length 12. This expanded table can be seen in Table 11-2.

Table 11-2. Longer sample hash table

i	Key	hash(key)	hash(key)%12	Value
0				
1				
2	'electron'	4017007007162656526	2	0.000548579909
3	'neutron'	3690763999294691079	3	1.008664
4				
5	'neutrino'	-4886380829903577079	5	3.31656614e-9
6				
7				
8				
9	'proton'	-4127328116439630603	9	1.00727647
10				
11				

The two hash tables seen in Table 11-1 and Table 11-2 contain the same information and are accessed in the same way. However, they have radically different layouts, solely based on their size.

The size of the table and the layout are handled automatically by the hash table itself. Users of a hash table should only have to worry about resizing insofar as to understand that multiple insertion operations that each emplace one item will almost always be more expensive than a single insertion that emplaces multiple items. This is because multiple insertions will force the hash table to go through all of the intermediate sizes. On insertion of multiple items, the hash table is allowed to jump directly to the final size.

For Python dictionaries, automatic resizing means that you should try to update() dicts when possible rather than assigning new entries (d[key] = value) over and over again. Given two dictionaries x and y, it is much better to use x.update(y) than to write:

```
for key, value in y.items():
    x[key] = value
```

Different hash table implementations choose different strategies for deciding when to resize (table is half full, three-quarters full, completely full, never) and by how much to resize (double the size, half the size, not at all). Resizing answers the question of what to do when a hash table runs out of space. However, we still need to address what to do when two keys accidentally produce the same index.

Collisions

A *hash collision* occurs when a new key hashes to the same value as an existing key in the table. Even though the space of hash values is huge, ranging from -9223372036854775808 to 9223372036854775807 on 64-bit systems, hash collisions are much more common than you might think. For simple data, it is easy to show that an empty string, the integer 0, the float 0.0, and False all hash to the same value (namely, zero):

```
hash('') == hash(0) == hash(0.0) == hash(False) == 0
```

However, even for random keys, hash collisions are an ever-present problem. Such collisions are an expression of the *birthday paradox*. Briefly stated, the likelihood that any two people in a room will have the same birthday is much higher than that of two people sharing a specific birthday (say, October 10th). In terms of hash tables, this can be restated as "the likelihood that any pair of keys will share a hash is much higher than the probability that a key will have a given hash (say, 42) and that another key will have that same hash (again, 42)."

Set the variable s as the size of the table and N as the number of distinct hash values possible for a given hash function. An approximate expression for the likelihood of a hash collision is given by $p_c(s)$:

$$p_c(s) = 1 - e^{\frac{-s(s-1)}{2N}}$$

This may be further approximated as:

$$p_c(s) = \frac{s^2}{2N}$$

For Python `dicts`, `N=2**64`, and Figure 11-1 shows this curve. After about a billion items, the probability of a hash collision in a Python `dict` starts going up dramatically. For greater than 10 billion items, a collision is effectively guaranteed. This is surprising, since the total range of the space is about 1.844e19 items. This means that a collision is likely to happen even though only one-billionth of the space is filled. It is important to note that the shape of this curve is the same for all values of `N`, though the location of the inflection point will change.

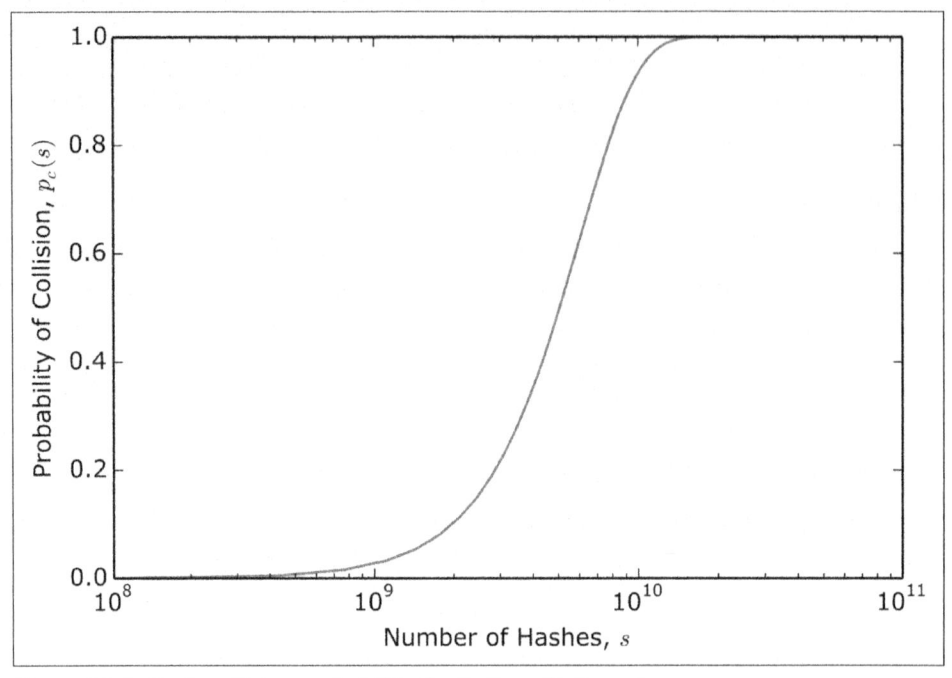

Figure 11-1. Hash collision probability for Python dictionaries

Given that hash collisions *will* happen in any real application, hash table implementations diverge in the way that they choose to handle such collisions. Some of the strategies follow the broad strokes presented here:

- Every index is a *bucket*, or list of key/value pairs. The hash will take you to the right bucket, and then a linear search through every item in the bucket will find the right value. This minimizes the number of resizes at the expense of a linear search and a more complex data structure. This is known as *separate chaining*.

- In the event of a collision, the hash is modified in a predictable and invertible way. Continued collisions will cause repeated modifications and hops. Searching for a key will first try the hash and the successive modifications. This is known as *open addressing* and is the strategy that Python dictionaries implement. This has the benefit that all items live flat in the table. There are no sublists to search

through. However, this comes at the cost that the index of a key is not computable from the hash of the key alone. Rather, the index depends on the full history of the hash table—when items were inserted and deleted.
- Always resize to the point of zero collisions. This works well where all of the keys must share the same type. This is sometimes called a *bidirectional* or *bijective hash map* because the keys are uniquely determined by their hashes, just like the hashes are uniquely determined by their keys.

As a user of hash tables, the details of how the collisions are handled are less important than the fact that collisions happen and they affect performance.

In summary, hash tables are beautiful and ubiquitous. On average they have incredible performance properties, being order-1 for all operations. This comes at the expense of significantly increased implementation complexity over more fundamental containers, such as arrays. Luckily, since almost every modern programming language supports hash tables natively, only in the rarest circumstances will you ever need to worry about writing one yourself.

Up next, we will talk about another data structure that you should never have to implement, but that is crucial to have in your analysis toolkit.

Data Frames

A relative newcomer, the *data frame* is a must-have data structure for data analysis. It is particularly useful to experimentalists. The data frame is an abstraction of tables and structured arrays. Given a collection of 1D arrays, the data frame allows you to form complex relationships between those arrays. Such relationships extend beyond being columns in a table, though that is possible too. One of the defining features of data frames that makes them invaluable for experimentalists is that they gracefully handle missing data. Anyone who has worked with real data before knows that this is a feature that should not be overlooked.

The data frame was popularized by the R language, which has a native data frame implementation and is largely (though not solely) used for statistics. More recently, the pandas package for Python implements a data frame and associated tools.

Data frames are effectively tables made up of named columns called *series*. Unlike in other table data structures we have seen, the series that make up a data frame may be dynamically added to or removed from the frame. A series can be thought of as a 1D NumPy array (it has a dtype) of values along with a corresponding *index* array. The index specifies how values of the series are located. If no index is provided, regular integers spanning from zero to one minus the length of the series are used. In this case, series are very similar to plain old NumPy arrays. The value of indexes, as we will see, is that they enable us to refer to data by more meaningful labels than zero through N-1. For example, if our values represented particle counts in a detector, the index could be made up of the strings 'proton', 'electron', and so on. The data frames themselves may also have one or more index arrays. If no index is provided, then zero to N-1 is assumed. In short, data frames are advanced in-memory tables that allow human-readable data access and manipulation through custom indexes.

The usage of data frames was first presented in Chapter 8. Here, we will cover their basic mechanics. For the following examples, you'll need to import the `pandas` package. Note that like `numpy` and `pytables`, the `pandas` package is almost always imported with an abbreviated alias, namely pd:

```
import pandas as pd
```

Let's start by diving into series.

Series

The `Series` class in `pandas` is effectively a one-dimensional NumPy array with an optional associated index. While this is not strictly accurate, much of the NumPy flavor has been transferred to `Series` objects. A series may be created using array-like mechanisms, and they share the same primitive `dtype` system that NumPy arrays use. The following example creates a series of 64-bit floats:

Code	Returns
`pd.Series([42, 43, 44], dtype='f8')`	0 42 ❶ 1 43 2 44 dtype: float64 ❷

❶ Index and values columns

❷ dtype of the values

Note that the column on the left is the index, while the column on the right displays the values. Alternatively, we could have passed in our own custom noninteger index. The following shows a series s with various particle names used as the index and the values representing the number of the associated particle that a detector has seen:

Code	Returns

`s = pd.Series([42, 43, 44],`
`       index=["electron",`
`              "proton",`
`              "neutron"])`
	electron 42
	proton 43
	neutron 44
	dtype: int64

The index itself is very important, because this dictates how the elements of the series are accessed. The index is an immutable ndarray that is composed of only hashable data types. As with the keys in a dictionary, the hashability ensures that the elements of an index may be used to safely retrieve values from a series. In the following code snippet, we see that we can index into a series in a dict-like fashion to pull out a single value. We can also slice by indices to create a subseries, because the index itself has an order. Finally, even though the series has an index, we can always go back to indexing with integers:

Code	Returns
s['electron']	42
# inclusive bounds s['electron':'proton']	electron 42 proton 43 dtype: int64
# integer indexing still OK s[1:]	proton 43 neutron 44 dtype: int64

Series may also be created from dictionaries. In this case, the keys become the index and the elements are sorted according to the keys. The following code demonstrates a series t being created from a dict with string particle keys and associated integer values:

Code	Returns

t = pd.Series({'electron': 6,
              'neutron': 28,
              'proton': 496,
              'neutrino': 8128})
	electron 6
	neutrino 8128
	neutron 28
	proton 496
	dtype: int64

Additionally, arithmetic and other operations may be performed on a series or a combination of series. When two series interact, if an element with a particular index exists in one series and that index does not appear in the other series, the result will contain all indices. However, the value will be NaN (Not a Number) for the missing index. This means that the datasets will only grow, and you will not lose an index. However, the presence of a NaN may not be desired. Reusing the s and t series from our previous examples, the following code adds these two series together. Since t has a neutrino element that s does not, the expression s + t will have a neutrino element, but its value will be NaN:

Code	Returns
s + t	electron 48 neutrino NaN neutron 72 proton 539 dtype: float64

The advantage of having NaN elements show up in the resulting series is that they make it very clear that the input series to the operation did not share a common basis. Sometimes this is OK, like when you do not care about neutrinos. At other times, like when you want to sum up the total number of counts, the NaN elements are problematic. This forces you to deal with them and adhere to best practices. There are two approaches for dealing with a NaN. The first is to go back to the original series and make sure that they all share a common index. The second is to filter or mask out the NaN elements after the other operations have completed. In general, it is probably best to go back to the original series and ensure a common basis for comparison.

Unless otherwise specified, for almost all operations pandas will return a copy of the data rather than a view, as was discussed in "Slicing and Views" on page 208. This is to prevent accidental data corruption. However, it comes at the cost of speed and memory efficiency.

Now that we know how to manipulate series on their own, we can combine many series into a single data frame.

The Data Frame Structure

The DataFrame object can be understood as a collection of series. These series need not share the same index, though in practice it is useful if they do because then all of the data will share a common basis. The data frame is a table-like structure akin to a NumPy structured array or a PyTables Table. The usefulness of data frames is the same as other table data structures. They make analyzing, visualizing, and storing complex heterogeneous data easier. Data frames, in particular, provide a lot of helpful semantics that other table data structures do not necessarily have. Data frames are distinct from other table-like structures, though, because their columns are series, not arrays. We can create a data frame from a dictionary of arrays, lists, or series. The keys of the dictionary become the column names. Reusing the definitions of s and t from before, we can create a data frame called df:

Code	Returns
| `df = pd.DataFrame({'S': s, 'T': t})` | ``` 
               S     T
electron      42     6
neutrino     NaN  8128
neutron       44    28
proton        43   496

[4 rows x 2 columns]
``` |

You can also create a data frame from a NumPy structured array or a list of tuples. Data frames may be saved and loaded from CSV files, HDF5 files (via PyTables), HTML tables, SQL, and a variety of other formats.

Data frames may be sliced, be appended to, or have rows removed from them, much like other table types. However, data frames also have the sophisticated indexing semantics that series do. The following code demonstrates some data frame manipulations:

Code	Returns
| `df[::2]` ❶ | ```
               S   T
electron      42   6
neutron       44  28

[2 rows x 2 columns]
``` |
| `dg = df.append(` ❷ <br> `        pd.DataFrame({'S': [-8128]},` <br> `            index=['antineutrino']))` | ```
                  S     T
electron         42     6
neutrino        NaN  8128
neutron          44    28
proton           43   496
antineutrino  -8128   NaN

[5 rows x 2 columns]
``` |
| `dh = dg.drop('neutron')` ❸ | ```
                  S     T
electron         42     6
neutrino        NaN  8128
proton           43   496
antineutrino  -8128   NaN

[4 rows x 2 columns]
``` |

❶ Slice every other element.

❷ Add a new index to the data frame and value to S.

❸ Delete the neutron index.

You may also easily transpose the rows and columns via the T attribute, as seen here:

Data Frames | 267

Code	Returns
df.T	<pre> electron neutrino neutron proton
S   42   NaN   44   43
T   6   8128   28   496

[2 rows x 4 columns]</pre> |

Arithmetic and other operations may be applied to the whole data frame. The following example creates a Boolean mask data frame by comparing whether the data is less than 42. Note that the Boolean mask is itself another data frame:

Code	Returns
df < 42	<pre> S T
electron   True   True
neutrino False False
neutron   False   True
proton   True False

[4 rows x 2 columns]</pre> |

A major innovation of the data frame is the ability to add and remove columns easily. With NumPy structured arrays, adding a new column to an existing array involves creating a new compound dtype to represent the new table, interleaving the new column data with the existing table, and copying all of the data into a new structured array. With data frames, the notion of a column is flexible and interchangeable with the notion of an index. Data frames are thus much more limber than traditional tables for representing and manipulating data. Column access and manipulation occurs via dict-like indexing. Such manipulation can be seen with the existing df data frame:

Code	Returns
| `# accessing a single column`<br>`# will return a series`<br>`df['T']` | ```
electron      6
neutrino   8128
neutron      28
proton      496
Name: T, dtype: int64
``` |
| `# setting a name to a series`<br>`# or expression will add a`<br>`# column to the frame`<br>`df['small'] = df['T'] < 100` | ```
            S     T  small
electron   42     6   True
neutrino  NaN  8128  False
neutron    44    28   True
proton     43   496  False

[4 rows x 3 columns]
``` |
| `# deleting a column will`<br>`# remove it from the frame`<br>`del df['small']` | ```
            S     T
electron   42     6
neutrino  NaN  8128
neutron    44    28
proton     43   496

[4 rows x 2 columns]
``` |

These kinds of column operations, along with reindexing, `groupby()`, missing data handling, plotting, and a host of other features, make data frames an amazing data structure. We have only scratched the surface here. While they may not be able to handle the kinds of data volumes that parallel chunked arrays do, for everyday data analysis needs nothing beats the flexibility of data frames. As long as your data can nicely fit into memory, then data frames are a great choice.

The next section takes us to the other end of the data volume spectrum, with a detailed discussion of B-trees.

B-Trees

B-trees are one of the most common data structures for searching over big chunks of data. This makes them very useful for databases. It is not an understatement to say that HDF5 itself is largely based on the B-tree. Chunks in a dataset are stored and managed via B-trees. Furthermore, the hierarchy itself (the collection of groups and arrays and tables) is also managed through a B-tree.

A B-tree is a tree data structure where all of the nodes are sorted in a breadth-first manner. Each node may have many subnodes. This structure makes it easy to search for a specific element because, starting at the top root node, you can simply test whether the value you are looking for is greater or less than the values in the nodes that you are at currently. The following example represents a B-tree of the Fibonacci sequence. Here, the square brackets ([]) represent nodes in the B-tree. The numbers inside of the square brackets are the values that the B-tree is currently storing:
```

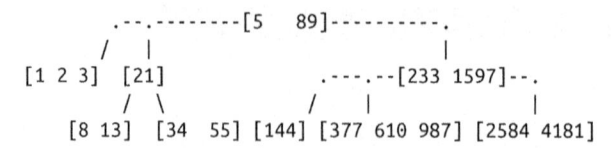

```
          .---.--------[5   89]----------.
         /    |                          |
   [1 2 3]  [21]            .----.--[233 1597]--.
     / \                   /    |                |
 [8 13] [34  55]    [144] [377 610 987]    [2584 4181]
```

It is important to note that each node in a B-tree may store many values, and the number of values in each node may vary. In the simplest case, where each node is constrained to have a single value, the B-tree becomes a *binary search tree* (not to be confused with a *binary tree*, which we won't discuss here). The following diagram shows a binary search tree specialization of a small B-tree:

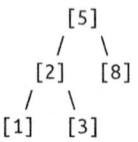

```
      [5]
      / \
    [2] [8]
    / \
  [1] [3]
```

B-trees (and binary search trees) may be *rotated*. This means that the nodes can be rearranged to have a different structure without breaking the search and ordering properties. For example, the tree above may be rotated to the following tree with equivalent average search properties:

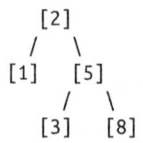

```
      [2]
      / \
    [1] [5]
        / \
      [3] [8]
```

B-trees are very effective for nonlinearly organizing array data. The index of the array determines on which node in the tree the array lives. The tree itself manages the locations of all of the nodes. The nodes manage the data chunks assigned to them. The ability for nodes to be inserted and removed at arbitrary indices allows for arrays to have missing chunks, be infinitely long, and be extendable.

In practice, B-trees tend to follow some additional simple rules as a way of reaping performance benefits and making the logic easier to understand:

- The height of the tree, h, is constant. All leaves (terminal nodes) exist at the same height.
- The root node has height 0.
- The maximum number of child nodes, m, is kept below a constant number across all nodes.
- Nodes should be split as evenly as possible over the tree in order to be *balanced*.

The size of a tree is measured by how many nodes (n) it has. Getting, setting, and deleting nodes in a B-tree are all order $\log(n)$ operations on average. The worst-case behavior for these operations is also order log(n), and it is possible to do better than this in the unlikely event that the node you are looking for is higher up the tree than being on a leaf. These properties make B-trees highly desirable from a reliability standpoint. Table 11-3 shows a comparison between B-tree and hash table performance.

Table 11-3. Performance comparison between B-trees and hash tables for common operations

Operation	Hash table average	Hash table worst	B-tree average	B-tree worst
Get: x[key]	O(1)	O(n)	O(log n)	O(log n)
Set: x[key] = value	O(1)	O(n)	O(log n)	O(log n)
Delete: del x[key]	O(1)	O(n)	O(log n)	O(log n)

In practice, you will want to use a B-tree whenever you need to quickly find an element in a large, regular, and potentially sparse dataset. If you happen to be writing a database, B-trees are probably what you want to use to index into the datasets on disk. B-trees were presented here because they are used frequently under the covers of the databases that we have already seen, such as HDF5. Furthermore, if you have a mapping that you want to ensure is always sorted by its keys, you will want to use a B-tree instead of a dictionary. Still, in most other circumstances, a dictionary, NumPy array, or data frame is typically a better choice than a B-tree.

B-trees as a data structure are too complex to give a full and working example here. While there is no accepted standard implementation in Python, there are many libraries that have support for B-trees. You can install and play around with them. A few that are worth looking into are:

btree
  A dict-like C extension module implementation

BTrees
  A generic B-tree implementation optimized for the Zope Object Database (ZODB)

blist
  A list-, tuple-, set-, and dict-like data structure implemented as a B+-tree (a variant of a strict B-tree)

Let's take a quick look at the blist package (*http://stutzbachenterprises.com/blist/*), since it has the best support for Python 2 and 3. This package has a sorteddict type that implements a traditional B-tree data structure. Creating a sorteddict is similar to creating a Python dictionary. The following code imports `sorteddict`, creates a new B-tree with some initial values, and adds a value after it has been created:

Code	Returns

`from blist import sorteddict`
 	
`b = sorteddict(first="Albert",`
`            last="Einstein",`
`            birthday=[1879, 3, 14])`
 	
`b['died'] = [1955, 4, 18]`
 	
`list(b.keys())`
	`sorteddict({'birthday': [1879, 3, 14],`
	`            'first': 'Albert',`
	`            'last': 'Einstein'})`
	
	`['birthday', 'died', 'first', 'last']`

The keys always appear sorted, because of how B-trees work. Even though the `sorted dict` implements a dictionary interface, its performance characteristics and underlying implementation are very different from that of a hash table.

> We have now seen three possible ways to store associations between keys and values: hash tables, series, and B-trees. Which one you should use depends on your needs and the properties of the data structure.

B-trees are great for data storage and for organizing array chunks. For this reason, they are used in databases quite a bit. The next section presents a different tree structure that excels at storing geometry.

K-D Trees

A *k-d tree*, or *k-dimensional tree*, is another tree data structure. This one excels at finding the nearest neighbor for points in a k-dimensional space. This is extraordinarily useful for many physics calculations. Often times, when solving geometric partial differential equations, the effects that matter the most in the volume at hand come from the directly surrounding cells. Splitting up the problem geometry into a k-d tree can make it much faster to find the nearest neighbor cells.

The big idea behind k-d trees is that any point (along with the problem bounds) defines a `k-1` dimensional hyperplane that partitions the remaining space into two sections. For example, in 1D a point p on a line l will split up the line into the space above p and the space below p. In two dimensions, a line will split up a box. In three dimensions, a plane will split a cube, and so on. The points in a k-d tree can then be placed into a structure similar to a binary search tree. The difference here is that sorting is based on the point itself along an axis `a` and that `a` is equal to the depth level of the point modulo the number of dimensions, k. Thus, a effectively defines an orientation for how points should partition the space.

K-d trees are not often modified once they are instantiated. They typically have get or query methods but do not have insert or delete methods. The structure of space is what it is, and if you want to restructure space you should create a whole new tree. For n points, k-d trees are order log(n) on average for all of their operations. In the worst case, they are order n.

Furthermore, k-d trees are most effective when k is small. This is ideal for physics calculations, where k is typically 3 for three spatial dimensions and rarely goes above 6. For simplicity, the examples here will use k=2. This makes the partitions simple line segments.

Since we do not need to worry about insertions or deletions, we can represent a sample k-d tree algorithm as follows:

```
class Node(object):  ❶

    def __init__(self, point, left=None, right=None):  ❷
        self.point = point
        self.left = left
        self.right = right

    def __repr__(self):  ❸
        isleaf = self.left is None and self.right is None
        s = repr(self.point)
        if not isleaf:
            s = "[ " + s + ":"
        if self.left is not None:
            s += "\n left = " + "\n ".join(repr(self.left).split('\n'))
        if self.right is not None:
            s += "\n right = " + "\n ".join(repr(self.right).split('\n'))
        if not isleaf:
            s += "\n ]"
        return s

def kdtree(points, depth=0):  ❹
    if len(points) == 0:
        return None
    k = len(points[0])
    a = depth % k
    points = sorted(points, key=lambda x: x[a])
    i = int(len(points) / 2)  # middle index, rounded down
    node_left = kdtree(points[:i], depth + 1)
    node_right = kdtree(points[i+1:], depth + 1)
    node = Node(points[i], node_left, node_right)
    return node
```

❶ A tree consists of nodes.

❷ A node is defined by its point. Since this is a binary search tree, it may have one node to the left and one node to the right.

❸ A string representation of the node given its relative location in the tree.

❹ A recursive function that returns the root node given a list of points. This will automatically be balanced.

As you can see, the Node class is very simple. It holds a point, a left child node, and a right child node. The heavy lifting is done by the kdtree() function, which takes a sequence of points and recursively sets up the nodes. Each node having only two children makes k-d trees much easier to manipulate than B-trees. As an example, consider the following random sequence of points internal to the range [0, 6] in 2-space. These can be placed into a k-d tree using code like the following:.

Code	Returns
| ```
points = [(1, 2), (3, 2),
          (5, 5), (2, 1),
          (4, 3), (1, 5)]
root = kdtree(points)
print(root)
``` | ```
[(3, 2):
 left = [(1, 2):
   left = (2, 1)
   right = (1, 5)
 ]
 right = [(5, 5):
   left = (4, 3)
 ]
]
``` |

The partitioning generated by this k-d tree is visualized in Figure 11-2.

For a rigorously tested, out-of-the-box k-d tree, you should use the KDTree class found in scipy.spatial. This is a NumPy-based formulation that has more impressive querying utilities than simply setting up the tree. The implementation of KDTree also fails over to a brute-force search when the search space goes above a user-definable parameter. Using the points list from our previous example, the following creates an instance of the KDTree class:

```
from scipy.spatial import KDTree
tree = KDTree(points)
```

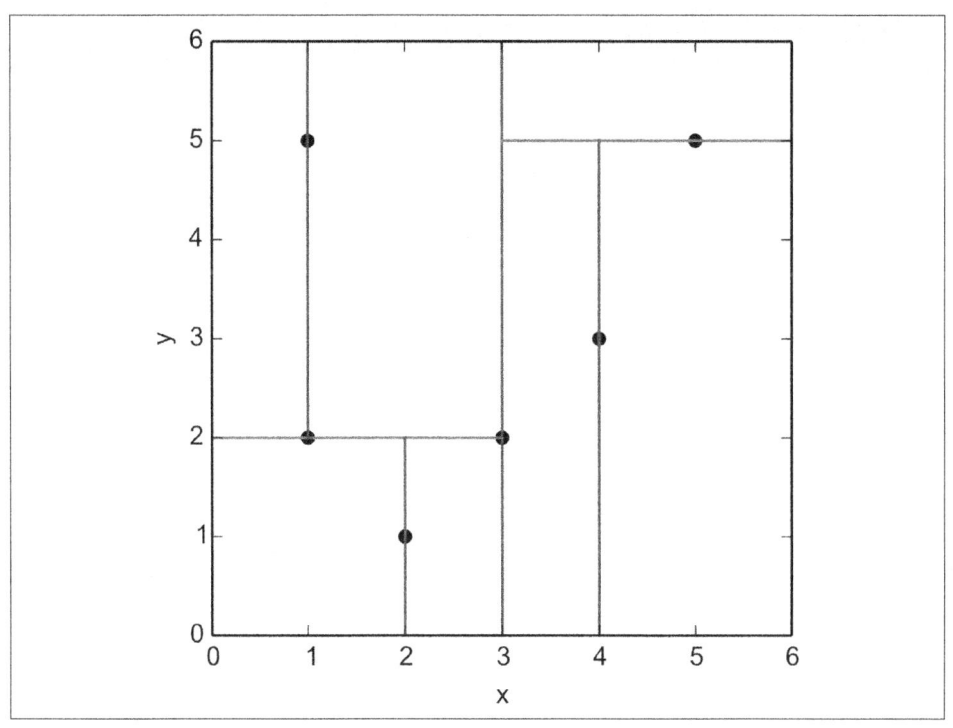

Figure 11-2. K-d tree partitioning example

This `tree` object has a `data` attribute that is a NumPy array representing the points. If the points were originally a NumPy array the `data` attribute would be a view, not a copy. The following shows that the tree's `data` retains the order of the original `points`:

Code	Returns
tree.data	array([[1, 2], [3, 2], [5, 5], [2, 1], [4, 3], [1, 5]])

The `query()` method on the `KDTree` class takes a sequence of points anywhere in space and returns information on the N nearest points. It returns an array of distances to these points as well as the indices into the data array of the points themselves. Note that `query()` does not return the cell in which a point lives. Using the `tree` we constructed previously again, let's find the nearest point in the tree to the location (4.5, 1.25):

Code	Returns
| ```
# query() defaults to only the closest point
dist, idx = tree.query([(4.5, 1.25)])

dist

idx

# fancy index by idx to get the point
tree.data[idx]
``` | ```
# results

array([ 1.67705098])

array([1])

array([[3, 2]])
``` |

The result of this query may be seen graphically in Figure 11-3.

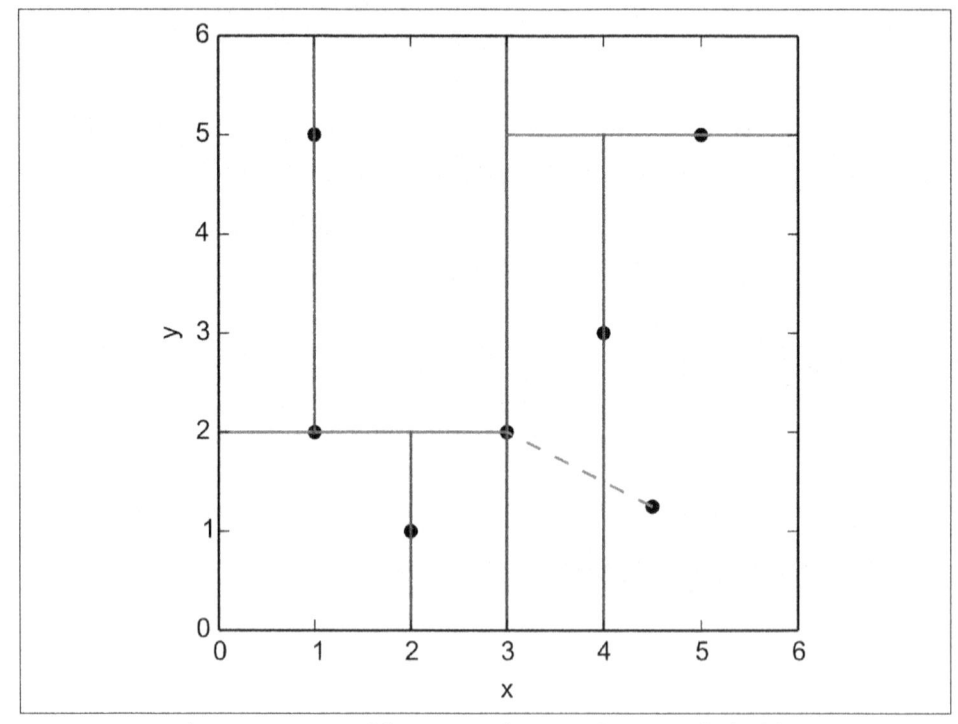

Figure 11-3. K-d tree nearest neighbor query, distance shown as dashed line

These querying capabilities are very useful whenever you have data for some points in space and want to compute the corresponding values for any other point based on the data that you have. For example, say each of your points has a corresponding measurement for magnitude of the electric field at that location. You could use a k-d tree to determine the nearest neighbor to any point in space. This then lets you approximate the electric field at your new point based on the distance from the closest measured value. In fact, this strategy may be used for any scalar or vector field.

The underlying physics may change, but finding the nearest neighbors via a k-d tree does not.

Due to their broad applicability, k-d trees come up over and over again in computational physics. For more information on the KDTree class, please refer to the SciPy documentation (*http://docs.scipy.org/doc/*).

Data Structures Wrap-up

As a scientist, organizing data is an integral part of daily life. However, there are many ways to represent the data you have. Familiarity with a variety of strategies gives you more flexibility to choose the option that best fits any given task. What you have learned here are some of the most significant to physics-based fields. Of these, hash tables are useful to just about everyone (and are not limited to physics). K-d trees, on the other hand, are most useful when you're thinking about problems spatially. Physicists do so a lot more than other folks. You should now be familiar with the following concepts:

- Hash tables require a good hash function, which is provided to you by Python.
- Resizing a hash table can be expensive.
- Hash collisions *will* happen.
- Series are like NumPy arrays but with more generic indexing capabilities.
- Data frames are like tables with series as columns.
- Data frames handle missing data through NaN values.
- B-trees can be used to organize chunks of an array.
- You may rotate B-trees to change their structure without altering their performance.
- A binary search tree is a B-tree with only one piece of data per node.
- K-d trees are a variant of the binary search tree that are organized by points in k-dimensional space.
- K-d trees are exceptionally useful for problems involving geometry.

Now that you know how to represent your data, the next chapter will teach you how to analyze and visualize it.

CHAPTER 12
Performing in Parallel

A natural way to approach parallel computing is to ask the question, "How do I do many things at once?" However, problems more often arise in the form of, "How do I solve my *one* problem faster?" From the standpoint of the computer or operating system, parallelism is about simultaneously performing tasks. From the user's perspective, parallelism is about determining dependencies between different pieces of code and data so that the code may be executed faster.

Programming in parallel can be fun. It represents a different way of thinking from the traditional "x then y then z" procedure that we have seen up until now. That said, parallelism typically makes problems faster for the computer to execute, but harder for you to program. Debugging, opening files, and even printing to the screen all become more difficult to reason about in the face of P processors. Parallel computing has its own set of rewards, challenges, and terminology. These are important to understand, because you must be more like a mechanic than a driver when programming in parallel.

Physicists often approach parallelism only when their problems finally demand it. They also tend to push back that need as far as possible. It is not yet easier to program in parallel than it is to program procedurally. Here are some typical reasons that parallel solutions are implemented:

- The problem creates or requires too much data for a normal machine.
- The sun would explode before the computation would finish.
- The algorithm is easy to parallelize.
- The physics itself cannot be simulated at all with smaller resources.

This chapter will focus on how to write parallel programs and make effective use of your computing resources. We will not be focusing on the underlying computer science that enables parallelism, although computer science topics in parallel computing are fascinating. You do not need to know how to build a bicycle to be able to ride one. By analogy, you can create computational physics programs without knowing how to build your own supercomputer. Here, we opt to cover what you need to know to be effective in your research. Let's start with some terminology.

Scale and Scalability

While parallelism is often necessary for large problems, it is also useful for smaller problems. This notion of size is referred to as *scale*. Computational scale is a somewhat ambiguous term that can be measured in a few different ways: the number of processes, the computation rate, and the data size.

A simple definition for scale is that it is proportional to the number of P processes that are used. This provides the maximum *degree of parallelism* that is possible from the point of view of the computer.

> It is better to talk about the number of *processes* used rather than the number of *processors*. This is because the number of processes is independent of the hardware.

Another popular measure of parallelism is the number of floating-point operations per second, or *FLOPS*. This is the rate at which arithmetic operations (addition, multiplication, modulus, etc.) happen on float data over the whole machine. Since a lot of scientific code mostly deals with floats, this is a reasonable measure of how fast the computer can do meaningful computations. FLOPS can sometimes be misleading, though. It makes no claims about integer operations, which are typically faster and important for number theory and cryptography. It also makes no claims about string operations, which are typically slower and important for genomics and biology. The use of graphics processing units (GPUs) is also a way to game the FLOPS system. GPUs are designed to pump through large quantities of floating-point operations, but they need special programming environments and are not great in high-data situations. FLOPS is a good measure of how fast a computer can work in an ideal situation, but bear in mind that obtaining that ideal can be tricky.

The final measure of scale that is commonly used is how much data is generated as a result of the computation. This is easy to measure; simply count how many bytes are written out. This tells you nothing of how long it took to generate those bytes, but the data size is still important for computer architects because it gives a scale of how much RAM and storage space will be required.

There are two important points to consider given the different scale metrics:

1. A metric tends to be proportional to the other metrics—a machine that is large on one scale is often large on the others.
2. Achieving a certain computation scale is preceded by having achieved lower scales first.

For example, point 1 states that when you have more processes available, you are capable of more FLOPS. Point 2 says that you cannot store a petabyte without storing a terabyte, you cannot store a terabyte without storing a gigabyte, and so on.

Together, these points imply that you should try to *scale up* your code slowly and methodically. Start with the trivial case of trying to run your code on one process. Then attempt 10 processes, and then 100, and up. It may seem obvious, but do not write code that is supposed to work on a million cores right out of the gate. You can't, and it won't. There will be bugs, and they will be hard to track down and resolve. Slowly scaling up allows you to address these issues one by one as they emerge, and on the scale at which they first appear.

Scalability is an indication of how easy or hard it is to go up in scale. There are many ways of measuring scalability, but the one that applies the most to computing is runtime performance. For a given machine, scalability is often measured in strong and weak forms.

Strong scaling is defined as how the runtime changes as a function of the number of processors for a fixed total problem size. Typically this is measured by the *speedup*, s. This is the ratio of time it takes to execute on one processor, t_1, to the time it takes to execute on P processors, t_P:

$$s(P) = \frac{t_1}{t_P}$$

In a perfectly efficient system, the strong scaling speedup is linear. Doubling the number of processors will cause the problem to run in half the time.

Weak scaling, on the other hand, is defined as how the runtime changes as a function of the number of processors for a fixed problem size per processor. This is typically measured in what is known as the *sizeup*, z. For a problem size N, the sizeup is defined as:

$$z(P) = \frac{t_1}{t_P} \times \frac{N_P}{N_1}$$

In a perfectly weak scaling system, the sizeup is linear. Doubling the number of processors will double the size of the problem that may be solved.

Both strong and weak scaling are constrained by *Amdahl's law*. This law follows from the observation that some fraction of an algorithm–call it α–cannot be parallelized. Thus, the maximum speedup or sizeup possible for P processors is given as:

$$\max(s(P)) = \frac{1}{\alpha - \frac{1-\alpha}{P}}$$

Taking this to the limit of an infinite number of processors, the maximum possible speedup is given as:

$$\max(s) = \lim_{P \to \infty} \frac{1}{\alpha - \frac{1-\alpha}{P}} = \frac{1}{\alpha}$$

Thus, Amdahl's law states that if, for example, 10% of your program is unparallelizable ($\alpha = 0.1$), then the best possible speedup you could ever achieve is a factor of 10. In practice, it is sometimes difficult to know what α is in more sophisticated algorithms. Additionally, α is usually much smaller than 0.1, so the speedups achievable are much greater. However, Amdahl's law is important because it points out that there *is* a limit.

What matters more than problem scale, though, is the algorithm. The next section discusses the types of parallel problems that exist.

Problem Classification

Certain algorithms lend themselves naturally to parallelism, while others do not. Consider summing a large array of numbers. Any part of this array may be summed up independently from any other part. The partial sums can then be summed together themselves and achieve the same result as if the array had been summed in series. Whether or not the partial sums are computed on the same processor or at the same time does not matter. Algorithms like this one with a high degree of independence are known as *embarrassingly parallel* problems. Other embarrassingly parallel examples include Monte Carlo simulations, rendering the Mandelbrot set, and certain optimization techniques such as stochastic and genetic algorithms. For embarrassingly parallel problems, the more processors that you throw at them, the faster they will run.

On the other hand, many problems do not fall into this category. This is often because there is an unavoidable bottleneck in the algorithm. A classic operation that is difficult to parallelize is inverting a matrix. In general, every element of the inverse matrix is dependent on every element in the original matrix. This makes it more difficult to write efficient parallel code, because every process must know about the entire original matrix. Furthermore, many operations are the same or similar between elements.

All hope is not lost for non-embarrassingly parallel problems. In most cases there are mathematical transformations you can make to temporarily decrease the degree of dependence in part of the problem. These transformations often come with the cost of a little excess computing elsewhere. On the whole though, the problem solution will have faster runtimes due to increased parallelism. In other cases, the more you know about your data the more parallelism you can eke out. Returning to the inverse matrix example, if you know the matrix is sparse or block diagonal there are more parallelizable algorithms that can be used instead of a fully generic solver. It is not worth your time or the computer's time to multiply a bunch of zeros together.

At large scales, these classifications have names based on the architecture of the machines that run them rather than the properties of the algorithm. The better known of these is *high-performance computing* (HPC). Such machines are built to run non-embarrassingly parallel problems. When people talk about supercomputers, they are referring to HPC systems. As a rule, all of the *nodes* on an HPC system are the same. Each node has the same number of CPUs and GPUs and the same amount of memory, and each processor runs the same operating system. Though the nodes are connected together in a predetermined *topology* (rings, tauruses, etc.), node homogenization gives the illusion that any subset of nodes will act the same as any other subset of nodes of the same size.

You should consider using HPC systems when your problems are large and not embarrassingly parallel, or they are large and you need the node homogeneity. That said, RAM per node has been on a downward trend in HPC, so if you have a high-memory application, you have extra work to do in making the algorithm data parallel in addition to compute parallel. Because they are a little trickier to program, but not difficult conceptually, we will largely skip discussing data-parallel algorithms here.

HPC's lesser-known sibling is called *high-throughput computing* (HTC). As its name implies, HTC is designed to pump through as many operations as possible with little to no communication between them. This makes HTC ideally suited to embarrassingly parallel problems. Nodes in an HTC system need not be the same. In some cases, they do not even have to share the same operating system. It is incorrect to think of an HTC system as a single machine. Instead, it is a coordinated network of machines. When these machines are spread over a continent or the entire world, HTC is sometimes known as *distributed computing*.

In both HPC and HTC systems, the most expensive component of a parallel algorithm is communication. The more frequently you have to communicate between processes, and the more data that must be sent, the longer the calculation takes. This almost invariably means that one node will be waiting while another node finishes. In HPC, you can minimize communication time by trying to place communicating nodes topologically close to each other. In HTC systems, there are only two phases with communication—task initialization and return. For short tasks, though, this communication time can dominate the execution time. So, in HTC it is always a good idea to perform as much work as possible on a node before returning.

Now that you have an understanding of the problem space, you can learn how to solve actual problems in parallel.

Example: N-Body Problem

Throughout the rest of this chapter we will be exploring parallelism through the lens of a real-world problem. Namely, we will be looking at the *N-body problem*. This is a great problem for computational physics because there are no analytical solutions except when N=2, or in certain cases when N=3. Numerical approximations are the best that we can get. Furthermore, as we will see in the following sections, a major portion of this problem lends itself nicely to parallelism while another portion does not.

The N-body problem is a generalization of the classic 2-body problem that governs the equations of motion for two masses. As put forward by Newton's law of gravity, we see:

$$\frac{dp}{dt} = G\frac{m_1 m_2}{r^2}$$

where p is momentum, t is time, G is the gravitational constant, m_1 and m_2 are the masses, and r is the distance between them. In most cases, constant mass is a reasonable assumption. Thus, the force on the *i*th mass from N-1 other bodies is as follows:

$$m_i \frac{d^2 \mathbf{x}_i}{dt^2} = \sum_{j=1, i \neq j}^{N} G\frac{m_i m_j (\mathbf{x}_j - \mathbf{x}_i)}{\|\mathbf{x}_j - \mathbf{x}_i\|^3}$$

with $\mathbf{x}_i$ being the position of the *i*th body. Rearranging the masses, we can compute the acceleration, $\mathbf{a}_i$:

$$\mathbf{a}_i = G \sum_{j=1, i \neq j}^{N} \frac{m_j (\mathbf{x}_j - \mathbf{x}_i)}{\|\mathbf{x}_j - \mathbf{x}_i\|^3}$$

This term for the acceleration can then be plugged into the standard time-discretized equations of motion:

$$\mathbf{v}_{i,s} = \mathbf{a}_{i,s-1}\Delta t + \mathbf{v}_{i,s-1}$$

$$\mathbf{x}_{i,s} = \mathbf{a}_{i,s-1}\Delta t^2 + \mathbf{v}_{i,s-1}\Delta t + \mathbf{x}_{i,s-1}$$

where $\mathbf{v}$ is the velocity, s indexes time, and $\Delta t = t_s - t_{s-1}$ is the length of the time step. Given the initial conditions $\mathbf{x}_{i,0}$ and $\mathbf{v}_{i,0}$ for all N bodies and a time step Δt, these equations will numerically solve for the future positions, velocities, and accelerations.

Now that we have a statement of the N-body problem, the following sections solve this problem in their own illuminating ways.

No Parallelism

When you're implementing a parallel algorithm, it is almost always easier to implement and analyze a sequential version first. This is because you can study, debug, and get a general sense of the behavior of the algorithm without adding an extra variable: the number of processors. This version is also almost always easier to program and requires fewer lines of code. A common and reliable strategy is to write once in series and then rewrite in parallel. Related to this is the somewhat subtle point that a sequential algorithm is not the same as a parallel algorithm with P=1. Though the two both run on one processor, they may have very different performance characteristics.

The following collection of functions implements a non-parallel solution to a single time step of the N-body problem, as well as generating some initial conditions:

```
import numpy as np

def remove_i(x, i):
    """Drops the ith element of an array."""
    shape = (x.shape[0]-1,) + x.shape[1:]
    y = np.empty(shape, dtype=float)
    y[:i] = x[:i]
    y[i:] = x[i+1:]
    return y

def a(i, x, G, m):
    """The acceleration of the ith mass."""
    x_i = x[i]
    x_j = remove_i(x, i)    ❶
    m_j = remove_i(m, i)
    diff = x_j - x_i
    mag3 = np.sum(diff**2, axis=1)**1.5
    result = G * np.sum(diff * (m_j / mag3)[:,np.newaxis], axis=0)    ❷
    return result

def timestep(x0, v0, G, m, dt):
    """Computes the next position and velocity for all masses given
```

```
        initial conditions and a time step size.
        """
        N = len(x0)
        x1 = np.empty(x0.shape, dtype=float)
        v1 = np.empty(v0.shape, dtype=float)
        for i in range(N):  ❸
            a_i0 = a(i, x0, G, m)
            v1[i] = a_i0 * dt + v0[i]
            x1[i] = a_i0 * dt**2 + v0[i] * dt + x0[i]
        return x1, v1

    def initial_cond(N, D):
        """Generates initial conditions for N unity masses at rest
        starting at random positions in D-dimensional space.
        """
        x0 = np.random.rand(N, D)  ❹
        v0 = np.zeros((N, D), dtype=float)
        m = np.ones(N, dtype=float)
        return x0, v0, m
```

❶ We should not compute the acceleration from this mass to itself.

❷ Compute the acceleration on the *i*th mass.

❸ Update the locations for all masses for each time step.

❹ Random initial conditions are fine for our sample simulator. In a real simulator, you would be able to specify initial particle positions and velocities by hand.

Here, the function a() solves for the acceleration of the *i*th mass, timestep() advances the positions and velocities of all of the masses, remove_i() is a simple helper function, and initial_cond() creates the initial conditions for randomly placed unit masses at rest. The masses are all placed within the unit cube. All of these functions are parameterized for N masses in D dimensions (usually 2 or 3). The masses here are treated as points and do not collide by hitting one another.

In most initial configurations, the masses will start accelerating toward one another and thereby gain velocity. After this, their momentums are typically not aligned correctly for the masses to orbit one another. They leave the unit cube, flying off in whatever direction they were last pointed in.

Since this is for demonstration purposes, we will set the gravitational constant G=1. A reasonable time step in this case is dt=1e-3. An example of the initial conditions and the first time step may be seen in Figures 12-1 and 12-2.

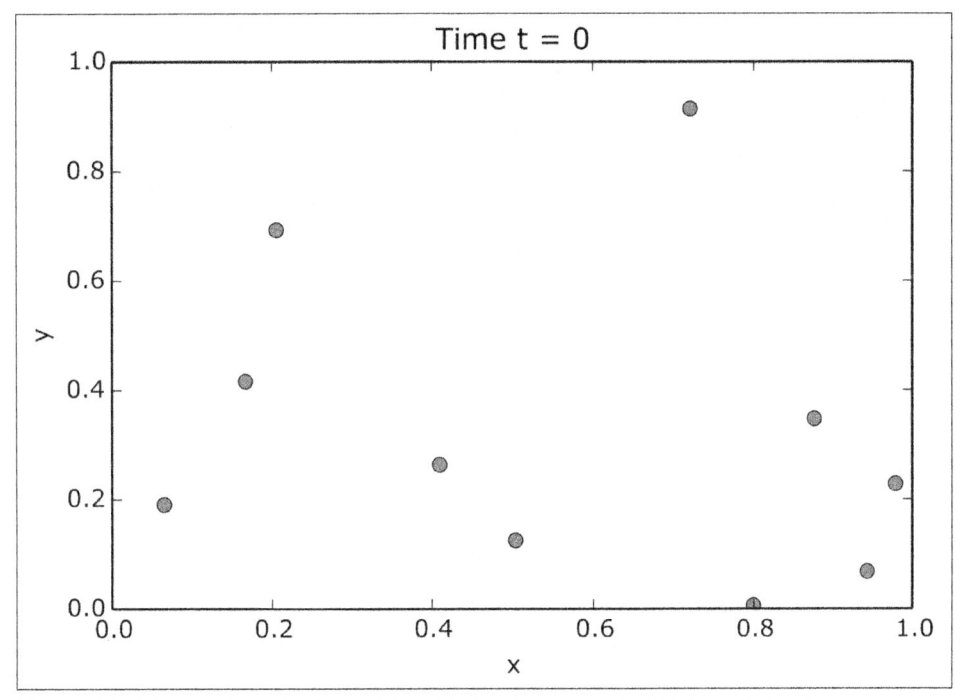

Figure 12-1. Example N-body positions for bodies at rest (initial conditions)

The data for these figures is generated by the following code:

```
x0, v0, m = initial_cond(10, 2)
x1, v1 = timestep(x0, v0, 1.0, m, 1.0e-3)
```

More generally, we can also write a simple driver function that simulates S time steps. The `simulate()` function takes the positions and velocities and updates them for each time step:

```
def simulate(N, D, S, G, dt):
    x0, v0, m = initial_cond(N, D)
    for s in range(S):
        x1, v1 = timestep(x0, v0, G, m, dt)
        x0, v0 = x1, v1
```

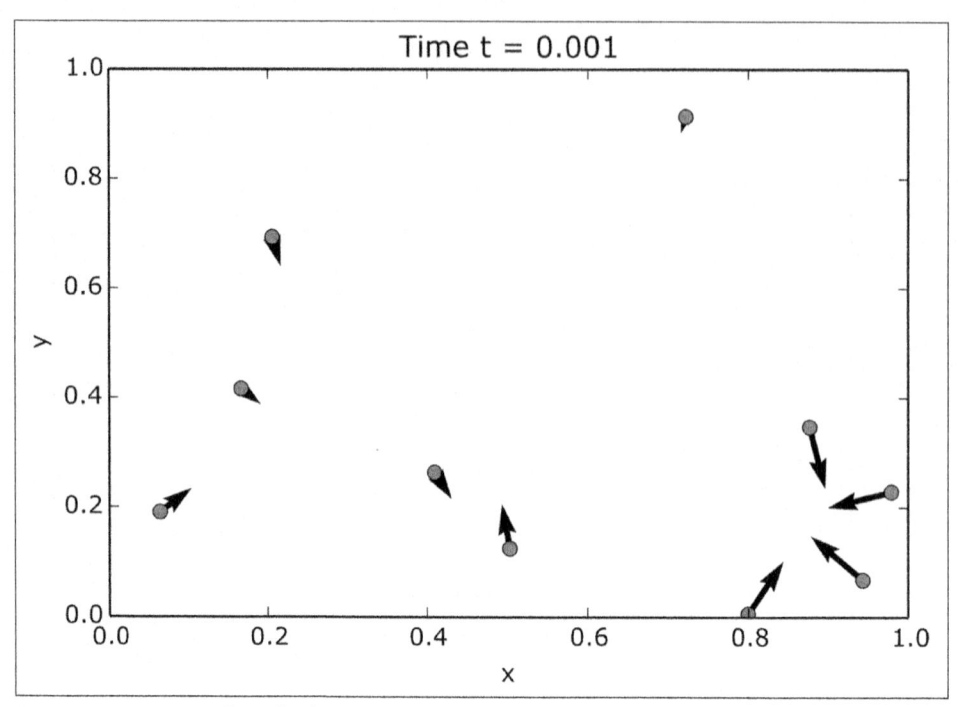

Figure 12-2. Example N-body positions and velocities after one time step

We can measure the performance of this non-parallel algorithm by calling `simulate()` for a variety of different N values and timing how long it takes. The following snippet shows how to automate these timings using Python's built-in `time` module. Here, we scale the particles by increasing powers of two:

```
import time
Ns = [2, 4, 8, 16, 32, 64, 128, 256, 512, 1024, 2048, 4096, 8192]
runtimes = []
for N in Ns:
    start = time.time()
    simulate(N, 3, 300, 1.0, 1e-3)
    stop = time.time()
    runtimes.append(stop - start)
```

Here, 300 time steps were chosen because after this point almost all of the bodies have left the original unit cube. The bodies are unlikely to engage in further interactions. Three hundred time steps is also large enough that the overhead that comes from just starting and running the simulations is minimized.

Intuition tells us that the N-body problem is order-N^2. This is because for every new body added, another iteration is added in the outer loop (the `timestep()` function) and another iteration is added in the inner loop (the `a()` function). Thus, we would expect the runtimes to be roughly parabolic as a function of N. Call the runtime of an

N-body simulation t_N. Figure 12-3 shows the ratio between the runtimes of t_N and t_2, the runtime of the 2-body problem.

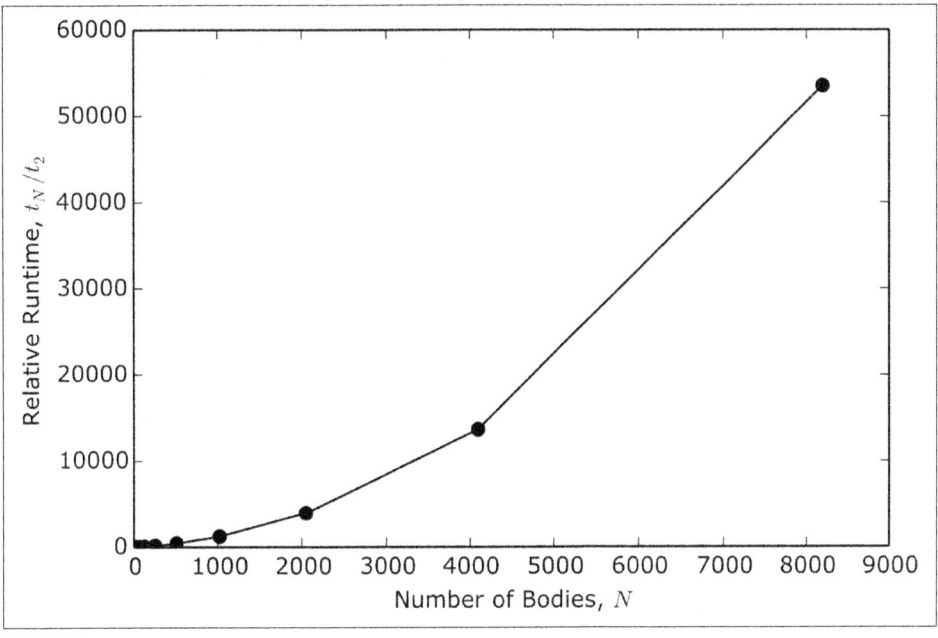

Figure 12-3. Relative runtimes of a non-parallel N-body simulation

Figure 12-3 is more or less parabolic, especially for large values of N. Since this simulation is done in serial, it is a little excessive to go much beyond 8,192 bodies. More telling, though, is the relative doubling time. This is defined as t_N divided by the runtime of half as many bodies, $t_{\frac{N}{2}}$. Since the problem is order-N^2, if you double the number of particles the runtime should go up by a factor of 4 no matter how many particles there are in the simulation. Figure 12-4 displays this quantity.

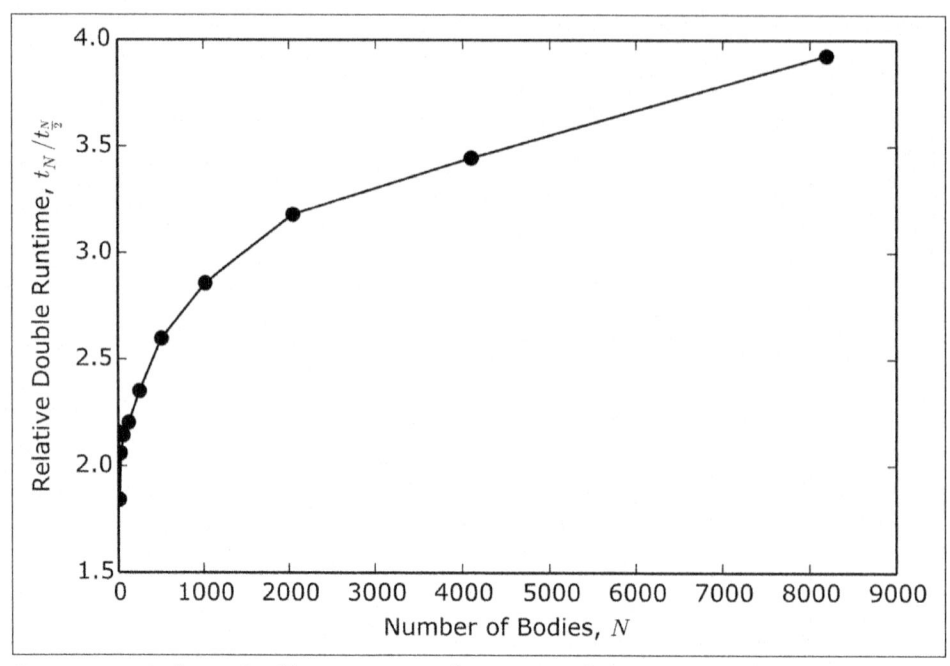

Figure 12-4. Relative doubling runtimes of a non-parallel N-body simulation

From this figure, there seem to be two distinct regimes in the relative doubling time. For large N, the relative doubling time does indeed seem to approach the magic 4 value. However for small N, this ratio is much closer to 2. In some cases it is even below 2, which implies that adding bodies decreases the amount of time spent on each body.

That Figure 12-4 is not a flat line close to 4 means that for low N, the overhead of the simulation dominates. Python, numpy, copying memory, and other facets of the implementation take up more of the runtime than the actual computation of the positions, velocities, and accelerations. It is not until about 2,000 bodies that the effects of N are really pertinent. Thus, for small N it probably is not worth parallelizing at all. The return on investment would be too low. For larger N, though, it will make a world of difference.

Let's now take a look at our first parallel N-body solver.

Threads

Threads are most people's first thought when they approach parallelism. However, threading in Python is almost certainly the wrong tool for the job. We cover threading here because it introduces some basic parallel concepts. You should also know why you should *not* use it.

> You should not use threads in Python for most scientific tasks! You should read this section anyway, though, because threading can be useful outside of Python.

Threads are objects that perform work and are not *blocked* by the work that other threads may be doing simultaneously. Executing code in a thread does not stop another thread from executing its own code. Threads may *communicate* with each other through their state. Threads may *spawn* child threads, and there is no limit to how many threads a program may run. There is always at least one thread in Python —the main thread.

Threads are one-time use. They have a `run()` method that, when it completes the thread, is no longer considered *alive*. This `run()` method is not directly called but is instead called implicitly when the `start()` method is called. This lets threads defer their work until a later point in time when the work has been properly set up. Threads cannot be killed externally; instead, the `run()` method must be allowed to complete. The exception to this rule is that if the thread is a *daemon*, the thread will die when the Python program itself exits.

Threads can be found in the `threading` module in the Python standard library. This exposes an interface that is similar to the threading interfaces found in other languages, such as Java or C.

Threading's big gotcha is that *all threads execute in the same process as Python itself*. This is because Python is an interpreted language, and the Python interpreter itself lives in only one process. This means that even though the threads may not block each other's execution, their combined speed is limited to that of a single processor on the machine. Number crunching for the most part is *CPU-bound*—that is, it is limited by the speed of the processor. If the computer is already running at top speed, adding more things for it to do does not make it go any faster. In fact, adding more tasks often hurts execution time. The processor becomes crowded.

The one-processor limit to threads is due to a detail of the way that the standard Python interpreter (called *CPython*, because it is written in C) is written. This detail is called the *global interpreter lock*, or *GIL*. The GIL determines which thread is currently running and when to switch from executing one thread to executing another. Several attempts to have Python without the GIL have been proposed over time. Most notable is the PyPy project (*http://pypy.org/*). None of these, however, have usurped the popularity of the classic CPython interpreter. CPython remains especially dominant in the scientific computing world.

So why do people use threads? For high-latency tasks, like reading a file from disk or downloading information over a network, most of the time the program is just waiting to get back the next chunk of data before it can proceed. Threads work well in these cases because the spare time can be put to good use executing other code.

However, in CPU-bound tasks, the processor is already as busy as it can be, and threads cause more problems than they solve. Except for reading in the occasional large dataset, scientific code is overwhelmingly CPU-bound.

Even though you probably should not use them, let's take a look at how the N-body problem is solved with threads. The first and easiest tactic for parallelization is to identify the parts of the code that do not depend on one another. With complete independence, the processes do not have to communicate with one another, so parallelism should come easy. In some algorithms this independence manifests as an inner loop and in some it is an outer loop. In some algorithms you can even switch which is the inner and which is the outer loop, to better fit your needs.

In the N-body problem, the loop in the `timestep()` function is the most easily parallelized. This allows us to only modify this one function. We can leave the implementations of the `remove_i()`, `a()`, and `initial_cond()` functions the same as they were in "No Parallelism" on page 285.

Before we can modify the `timestep()` function, we need to define threads that instead do the work of a single iteration. The following code implements a thread for use with the N-body problem:

```
from threading import Thread

class Worker(Thread):  ❶
    """Computes x, v, and a of the ith body."""
    def __init__(self, *args, **kwargs):
        super(Worker, self).__init__(*args, **kwargs)
        self.inputs = []   ❷
        self.results = []  ❸
        self.running = True  ❹
        self.daemon = True   ❺
        self.start()

    def run(self):
        while self.running:
            if len(self.inputs) == 0:  ❻
                continue
            i, x0, v0, G, m, dt = self.inputs.pop(0)
            a_i0 = a(i, x0, G, m)  ❼
            v_i1 = a_i0 * dt + v0[i]
            x_i1 = a_i0 * dt**2 + v0[i] * dt + x0[i]
            result = (i, x_i1, v_i1)
            self.results.append(result)
```

❶ To use threading, we need to subclass `Thread`.

❷ `inputs` is a buffer of work for this thread to churn through. We cannot call the `run()` method directly, so `inputs` provides a way to send data to the thread.

❸ A buffer of return values. Again, since we do not call the `run()` method ourselves, we have to have a place to store our `results`.

❹ Setting this to `False` will cause the `run()` method to end safely.

❺ Allow the thread to be terminated when Python exits.

❻ Check if there is work to do.

❼ The body of the original `timestep()` function.

First, we grab the `Thread` class from the `threading` module that comes with Python. To make this useful to the N-body problem, we subclass it under the name `Worker` and then override the constructor and the `run()` methods.

The constructor has two major purposes: setting up data structures used for communication and starting the thread. The `inputs` and `results` lists are used to load work onto the thread and get corresponding return values. The `running` flag is used to end the life of the thread cleanly—it will not stop midway while computing a result. The `daemon` flag is used to tell Python that this thread may be stopped when the program exits. Finally, the call to the `start()` method at the end of the constructor immediately makes the thread available for work when it is created.

While the `running` flag is true, the `run()` method will sit around waiting for work. Work comes in the form of a nonempty `inputs` list. The elements of this list are tuples that contain all of the arguments to `a()` as well as the time step value `dt`. When work arrives, the `run()` method proceeds to compute the next values of acceleration, velocity, and position for the *i*th body. The `run()` method then stores i, the next position, and the next velocity as a tuple in the `results` list, so that they can be read back by whoever put the arguments in the `inputs` list. It is important to keep i in the `results` list because it creates an unambiguous way of later determining which input goes with which result.

Managing individual one-time-use threads can be painful. Creating and starting threads can be expensive. It is best if thread objects can be reused for as long as possible. To handle this, a pattern known as *thread pools* is extremely common. The idea here is that the pool is created with some fixed size, and it is then responsible for creating, starting, farming out work to, stopping, and finally deleting the threads it manages. Python does not provide a thread pool to use off the shelf. However, they are fairly easy to implement. The following is a pool that manages instances of the `Worker` class:

```
class Pool(object):
    """A collection of P worker threads that distributes tasks
    evenly across them.
    """
    def __init__(self, size):
        self.size = size
        self.workers = [Worker() for p in range(size)]  ❶

    def do(self, tasks):
        for p in range(self.size):
            self.workers[p].inputs += tasks[p::self.size]  ❷
        while any([len(worker.inputs) != 0 for worker in self.workers]):
            pass  ❸
        results = []
        for worker in self.workers:  ❹
            results += worker.results
            worker.results.clear()
        return results  ❺

    def __del__(self):  ❻
        for worker in self.workers:
            worker.running = False
```

❶ Create new workers according to the size.

❷ Evenly distribute tasks across workers with slicing.

❸ Wait for all of the workers to finish all of their tasks.

❹ Get back the results from the workers and clean out the workers.

❺ Return the complete list of results for all inputs.

❻ Stop the workers when the pool itself is shut down.

Instantiating a Pool creates as many workers as were requested and keeps references to them for future use. The do( ) method takes a list of tasks—all of inputs that we want to run—and appends them to the workers' inputs lists. The tasks are appended using slicing to spread them out among all workers as evenly as possible. The pool then waits until the workers do not have anything left in their inputs lists. Once all of the work has been completed, the results are aggregated together, the workers' results lists are cleared for future work, and all of the results together are returned. Note that the results are not guaranteed to be in the order of the tasks that spawned them. In most cases, the order will be unpredictable. Lastly, the pool is responsible for stopping the workers from running once the pool itself is shut down in the delete method.

Using the `Worker` and `Pool` classes, our `timestep()` function can be rewritten to now accept a `pool` object to use for its computation. Note that since the results come back in an effectively random order, we must reconstruct them according to the i value. Doing so, as well as setting up the tasks, is fairly simple and is very cheap compared to the actual computation. The following reimplements the `timestep()` function:

```
def timestep(x0, v0, G, m, dt, pool):
    """Computes the next position and velocity for all masses given
    initial conditions and a time step size.
    """
    N = len(x0)
    tasks = [(i, x0, v0, G, m, dt) for i in range(N)]  ❶
    results = pool.do(tasks)  ❷
    x1 = np.empty(x0.shape, dtype=float)
    v1 = np.empty(v0.shape, dtype=float)
    for i, x_i1, v_i1 in results:  ❸
        x1[i] = x_i1
        v1[i] = v_i1
    return x1, v1
```

❶ Create a task for each body.

❷ Run the tasks.

❸ Rearrange the results since they are probably out of order.

Rewriting `timestep()` necessitates rewriting the `simulate()` function to create a Pool. To make scaling easy, `simulate()` should also be parameterized based on the number of workers, P. Parallel algorithms need to be told with what degree of parallelism they should be run. A pool-aware version of `simulate()` is shown here:

```
def simulate(P, N, D, S, G, dt):
    x0, v0, m = initial_cond(N, D)
    pool = Pool(P)
    for s in range(S):
        x1, v1 = timestep(x0, v0, G, m, dt, pool)
        x0, v0 = x1, v1
```

Unlike in "No Parallelism" on page 285, where we investigated how the N-body problem behaved as a function of the number of bodies, here we can determine how the simulation performs as a function of P. The following snippet has a fixed total problem size of 64 bodies over 300 time steps:

```
Ps = [1, 2, 4, 8]
runtimes = []
for P in Ps:
    start = time.time()
    simulate(P, 64, 3, 300, 1.0, 1e-3)
    stop = time.time()
    runtimes.append(stop - start)
```

In a perfect world, the strong scaling would have a speedup factor of 2x for every time P is doubled. However, the results here, shown in Figure 12-5, demonstrate that the speedup is actually a slowdown.

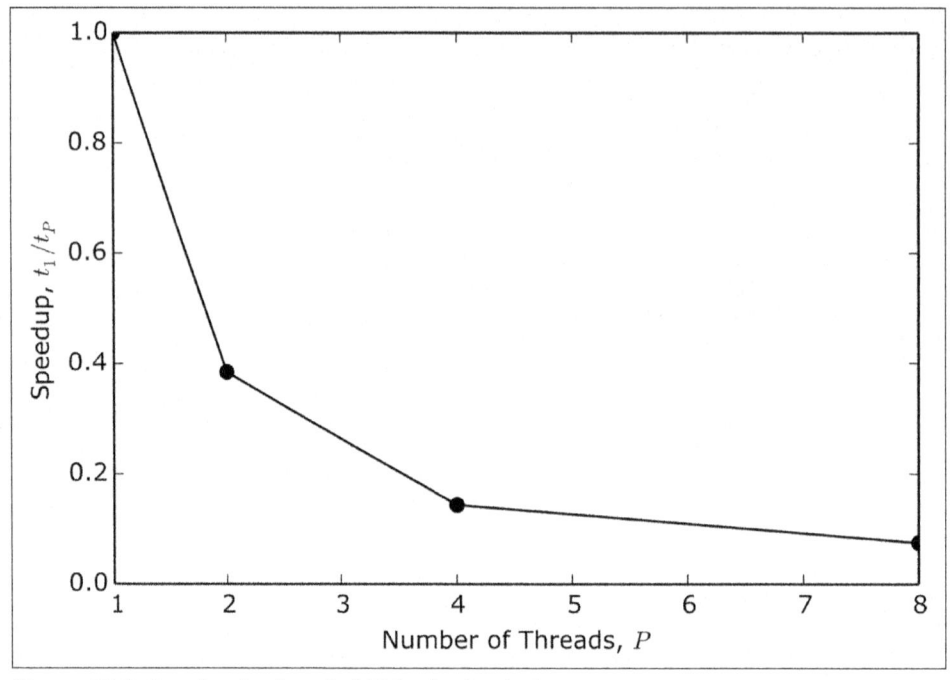

Figure 12-5. Speedup in threaded N-body simulation

Why does this happen? Every time we add threads, we add more load to the CPU. The single processor then has to spend more time figuring out which thread should be allowed to execute. As you can see, this gets very ugly very fast. Adding more threads to CPU-bound tasks does not enable more parallelism; it causes less. Even when there is only one worker thread, the main thread still exists. This causes the P=1 case that performs the best here to still be around 2.5 times *slower* than the equivalent non-parallel case seen in "No Parallelism" on page 285. These excessive slowdowns are why you should avoid threads.

In the next section, we'll see a strategy that actually does yield real speedups.

Multiprocessing

Multiprocessing is Python's way of handing off the responsibility of scheduling parallel tasks to the operating system. Modern operating systems are really good at *multitasking*. Multitasking allows for the sum of all processes to vastly exceed the resource limits of the computer. This is possible because multitasking does not necessarily allow for all processes to be simultaneously active. Any time you are browsing the Web, playing music, storing data on a flash drive, and running your fancy new USB toaster at the same time, you and your computer are multitasking.

The capability for multitasking is central to being able to use a computer for common tasks. On a single-CPU system, only one process can be physically executing at a time. A multitasking OS may pause this process and *switch contexts* to another process. Frequent, well-timed context switching gives the illusion of several processes running at once. When there are many processors available, multitasking can easily distribute a process across them, granting a degree of parallelism for free.

Forgetting filesystems, device drivers, and other human interactions, it is common for people to now view the main job of the operating system as scheduling tasks. At this point, most operating systems have extremely well-implemented schedulers. From Python's perspective, why not let someone else take care of this hard problem? It is, after all, a *solved* hard problem. For CPU-intensive tasks, multitasking is exactly what you should do.

> What is called "threading" in other languages is actually implemented more similarly to multiprocessing in Python.

Creating or *spawning* a new OS-level process is handled by Python. However, note that on POSIX (Unix-like—i.e., Linux and Mac OS X) machines, all processes are *forks* of the processes that spawned them. A fork inherits the environment of its parent. However, any modifications to the environment that the fork makes *will not* be reflected back in the parent process. Forks may spawn their own forks, which will get the modified environment. Killing a parent process will kill all of its forks. So, when Python spawns more processes, they all see the environment that the main Python process had. Multiprocessing interacts more directly with the OS, so other OS-level concepts that were seen in Chapter 1, such as piping, sometimes make an appearance here too.

Multiprocessing in Python is implemented via the standard library `multiprocessing` module. This provides a `threading`-like interface to handling processes. There are two major distinctions, though:

1. Multiprocessing cannot be used directly from an interactive interpreter. The main module (`__main__`) must be importable by the forks.
2. The `multiprocessing` module provides a `Pool` class for us. We do not need to write our own.

The `Pool` class has a number of different methods that implement parallelism in slightly different ways. However, the `map()` method works extraordinarily well for almost all problems.

`Pool.map()` has a similar interface to the built-in `map()` function. It takes two arguments—a function and an iterable of arguments to pass into that function—and returns a list of values in the same order that was given in the original iterable. `Pool.map()` blocks further execution of Python code until all of the results are ready. The major limitation of `Pool.map()` is that the function that it executes must have only one argument. You can overcome this easily by storing the arguments you need in a tuple or dictionary before calling it.

For the N-body problem, we no longer need the `Worker` and `Pool` classes that we used in "Threads" on page 290. Instead, we simply need a `timestep_i()` function that computes the time step evolution of the *i*th body. To make it easier to use with `multiprocessing`, `timestep_i()` should only have one argument. The following code defines this function:

```
from multiprocessing import Pool

def timestep_i(args):
    """Computes the next position and velocity for the ith mass."""
    i, x0, v0, G, m, dt = args   ❶
    a_i0 = a(i, x0, G, m)   ❷
    v_i1 = a_i0 * dt + v0[i]
    x_i1 = a_i0 * dt**2 + v0[i] * dt + x0[i]
    return i, x_i1, v_i1
```

❶ Unpack the arguments to the original `timestep()` function.

❷ The body of the original `timestep()` function.

Note that the first operation here is to unpack `args` into the variables we have come to know for this problem. Furthermore, the actual `timestep()` function from "Threads" on page 290 must be altered slightly to account for the new kind of process pool and the `timestep_i()` function. Mostly, we need to swap out the old `do()` call for the new `Pool.map()` call and pass it `timestep_i`, as well as the tasks:

```
def timestep(x0, v0, G, m, dt, pool):
    """Computes the next position and velocity for all masses given
    initial conditions and a time step size.
    """
    N = len(x0)
    tasks = [(i, x0, v0, G, m, dt) for i in range(N)]
    results = pool.map(timestep_i, tasks)   ❶
    x1 = np.empty(x0.shape, dtype=float)
    v1 = np.empty(v0.shape, dtype=float)
    for i, x_i1, v_i1 in results:
```

```
        x1[i] = x_i1
        v1[i] = v_i1
    return x1, v1
```

❶ Replace the old do() method with multiprocessing's Pool.map().

All of the other functions from "Threads" on page 290 remain the same, including simulate().

So how does multiprocessing scale? Let's take the fairly regular computing situation of programming on a dual-core laptop. This is not a whole lot of parallelism, but it is better than nothing. We would expect that a pool with two processes would run twice as fast as a pool with one process. At one process per physical processor, we would expect peak performance. For greater than two processes we would expect some performance degradation, as the operating system spends more time context switching. Example results can be seen in Figure 12-6. These were generated with the following code:

```
import time
Ps = [1, 2, 4, 8]
runtimes = []
for P in Ps:
    start = time.time()
    simulate(P, 256, 3, 300, 1.0, 1e-3)
    stop = time.time()
    runtimes.append(stop - start)
```

As you can see in Figure 12-6, it is not precisely 2x faster, but is in the neighborhood of 1.8x. The extra 0.2x that is missing goes to overhead in Python, processing forking, and the N-body algorithm. As expected, for pools of size 4 and 8 the speedup is worse than in the two processes case, though it is still a considerable improvement over the one processor case. Finally, it is worth noting that the one process case is 1.3x *slower* than the equivalent case with no parallelism. This is because of the overhead in setting up the parallel infrastructure. However, this initial 30% burden is quickly overcome in the two processor case. All of these results together indicate that multiprocessing is certainly worth it, even if you overestimate the number of processors that you have by a bit.

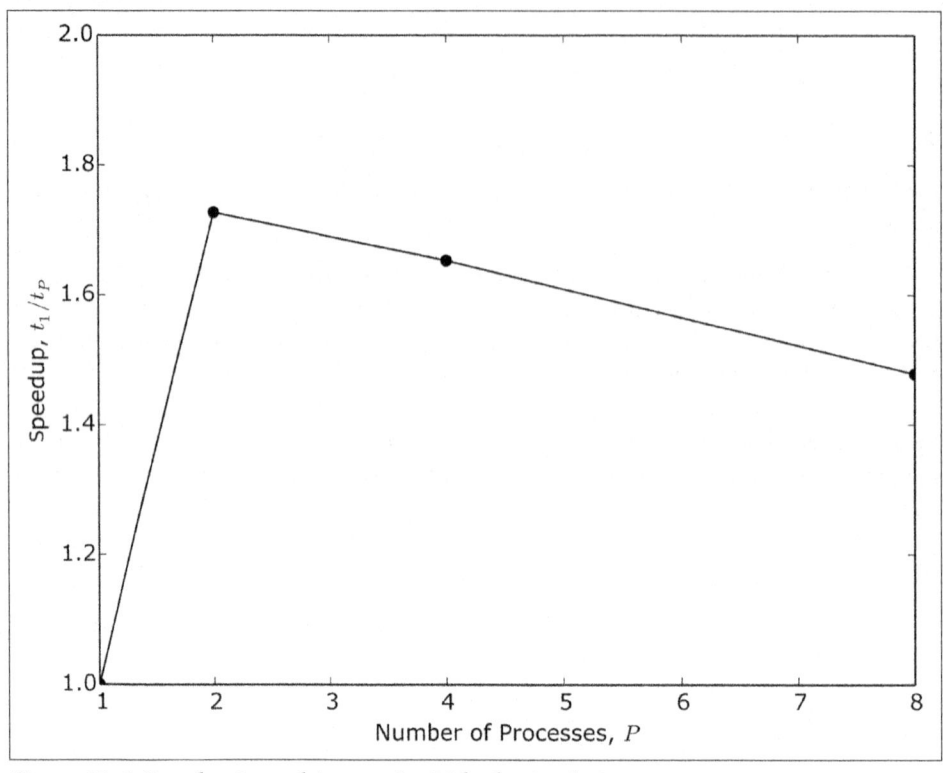

Figure 12-6. Speedup in multiprocessing N-body simulation on two processors

Multiprocessing is a great tool to use when you have a single machine and less than a thousand physical processors. It is great for daily use on a laptop or on a cluster at work. However, multiprocessing-based strategies do not scale up to supercomputers. For that, we will need *MPI*, covered in the next section.

MPI

The gold standard for high-performance parallelism is *MPI*, the *Message-Passing Interface*. As an interface MPI is a specification for how to communicate information between various processes, which may be close to or very far from one another. The MPI-3.0 Standard (*http://www.mpi-forum.org/docs/*) was released in September 2012. There are two primary open source projects that implement MPI: MPICH (*http://www.mpich.org/*) and Open MPI (*http://www.open-mpi.org/*). Since they implement the same standard, these are largely interchangeable. They both take great care to provide the MPI interface completely and correctly.

It is not an understatement to say that supercomputing is built on top of MPI. This is because MPI is an abstraction for parallelism that is independent of the machine.

This allows physicists (and other domain experts) to learn and write MPI code and have it work on any computer. Meanwhile, the architects of the supercomputers can implement a version of MPI that is optimized to the machines that they are building. The architects do not have to worry about who is going to use their version of MPI, or how they will use it.

MPI is a useful abstraction for anyone who buys into its model. It is a successful abstraction because almost everyone at this point *does* buy into it. MPI is huge and very flexible, and we do not have space here to do it justice. It currently scales up to the level of hundreds of thousands to millions of processors. It also works just fine on a handful of processors. If you are serious about parallelism on even medium scales (1,000+ processors), MPI is an unavoidable boon.

As its name states, MPI is all about communication. Mostly this applies to data, but it is also true for algorithms. The basic elements of MPI all deal with how to communicate between processes. For a user, this is primarily what is of interest. As with all good things, there is a Python interface. In fact, there are many of them. The most commonly used one is called `mpi4py`. We will be discussing the `mpi4py` package rather than the officially supported C, C++, or Fortran interfaces.

In MPI terminology, processes are called *ranks* and are given integer identifiers starting at zero. As with the other forms of parallelism we have seen, you may have more ranks than there are physical processors. MPI will do its best to spread the ranks out evenly over the available resources. Often—though not always—rank 0 is considered to be a special "master" process that commands and controls the other ranks.

Having a master rank is a great strategy to use, until it isn't! The point at which this approach breaks down is when the master process is overloaded by the sheer number of ranks. The master itself then becomes the bottleneck for doling out work. Reimagining an algorithm to not have a master process can be tricky.

At the core of MPI are *communicator* objects. These provide metadata about how many processes there are with the `Get_size()` method, which rank you are on with `Get_rank()`, and how the ranks are grouped together. Communicators also provide tools for sending messages from one processor and receiving them on other processes via the `send()` and `recv()` methods. The `mpi4py` package has two primary ways of communicating data. The slower but more general way is that you can send arbitrary Python objects. This requires that the objects are fully *picklable*. Pickling is Python's native storage mechanism. Even though pickles are written in plain text, they are not human readable by any means. To learn more about how pickling works and what it looks like, please refer to the pickling section of the Python documentation (*http://bit.ly/py-pickle*).

NumPy arrays can also be used to communicate with mpi4py. In situations where your data is already in a NumPy arrays, it is most appropriate to let mpi4py use these arrays. However, the communication is then subject to the same constraints as normal NumPy arrays. Instead of going into the details of how to use NumPy and mpi4py together, here we will only use the generic communication mechanisms. This is because they are easier to use, and moving to NumPy-based communication does not add anything to your understanding of parallelism.

The mpi4py package comes with a couple of common communicators already instantiated. The one that is typically used is called COMM_WORLD. This represents all of the processes that MPI was started with and enables basic *point-to-point* communication. Point-to-point communication allows any process to communicate directly with any other process. Here we will be using it to have the rank 0 process communicate back and forth with the other ranks.

As with multiprocessing, the main module must be importable. This is because MPI must be able to launch its own processes. Typically this is done through the command-line utility mpiexec. This takes a -n switch and a number of nodes to run on. For simplicity, we assume one process per node. The program to run—Python, here—is then followed by any arguments it takes. Suppose that we have written our N-body simulation in a file called *n-body-mpi.py*. If we wish to run on four processes, we would start MPI with the following command on the command line:

```
$ mpiexec -n 4 python n-body-mpi.py
```

Now we just need to write the *n-body-mpi.py* file! Implementing an MPI-based solver for the N-body problem is not radically different from the solutions that we have already seen. The remove_i(), initial_cond(), a(), timestep(), and timestep_i() functions are all the same as they were in "Multiprocessing" on page 296.

What changes for MPI is the simulate() function. To be consistent with the other examples in this chapter (and because it is a good idea), we will also implement an MPI-aware process pool. Let's begin by importing MPI and the following helpers:

```
from mpi4py import MPI
from mpi4py.MPI import COMM_WORLD

from types import FunctionType
```

The MPI module is the primary module in mpi4py. Within this module lives the COMM_WORLD communicator that we will use, so it is convenient to import it directly. Finally, types is a Python standard library module that provides base classes for built-in Python types. The FunctionType will be useful in the MPI-aware Pool that is implemented here:

```
class Pool(object):
    """Process pool using MPI."""
```

```
def __init__(self):
    self.f = None  ❶
    self.P = COMM_WORLD.Get_size()  ❷
    self.rank = COMM_WORLD.Get_rank()  ❸

def wait(self):  ❹
    if self.rank == 0:  ❺
        raise RuntimeError("Proc 0 cannot wait!")
    status = MPI.Status()
    while True:
        task = COMM_WORLD.recv(source=0, tag=MPI.ANY_TAG, status=status)  ❻
        if not task:
            break
        if isinstance(task, FunctionType):  ❼
            self.f = task
            continue
        result = self.f(task)  ❽
        COMM_WORLD.isend(result, dest=0, tag=status.tag)

def map(self, f, tasks):  ❾
    N = len(tasks)
    P = self.P
    Pless1 = P - 1
    if self.rank != 0:  ❿
        self.wait()
        return

    if f is not self.f:
        self.f = f
        requests = []
        for p in range(1, self.P):  ⓫
            r = COMM_WORLD.isend(f, dest=p)
            requests.append(r)
        MPI.Request.waitall(requests)

    requests = []
    for i, task in enumerate(tasks):  ⓬
        r = COMM_WORLD.isend(task, dest=(i%Pless1)+1, tag=i)
        requests.append(r)
    MPI.Request.waitall(requests)

    results = []
    for i in range(N):  ⓭
        result = COMM_WORLD.recv(source=(i%Pless1)+1, tag=i)
        results.append(result)
    return results

def __del__(self):
    if self.rank == 0:
        for p in range(1, self.P):  ⓮
            COMM_WORLD.isend(False, dest=p)
```

❶ A reference to the function to execute. The pool starts off with no function.

❷ The total number of processors.

❸ Which processor we are on.

❹ A method for receiving data when the pool has no tasks. Normally, a task is data to give as arguments to the function f(). However, if the task is itself a function, it replaces the current f().

❺ The master process cannot wait.

❻ Receive a new task from the master process.

❼ If the task was a function, put it onto the object and then continue to wait.

❽ If the task was not a function, then it must be a real task. Call the function on this task and send back the result.

❾ A map() method to be used like before.

❿ Make the workers wait while the master sends out tasks.

⓫ Send all of the workers the function.

⓬ Evenly distribute tasks to all of the workers.

⓭ Wait for the results to come back from the workers.

⓮ Shut down all of the workers when the pool is shut down.

The purpose of the Pool class is to provide a map() method that is similar to the map() on the multiprocessing pool. This class implements the rank-0-as-master strategy. The map() method can be used in the same way as for other pools. However, other parts of the MPI pool operate somewhat differently. To start with, there is no need to tell the pool its size. P is set on the command line and then discovered with COMM_WORLD.Get_size() automatically in the pool's constructor.

Furthermore, there will be an instance of Pool on each processor because MPI runs the same executable (python) and script (*n-body-mpi.py*) everywhere. This implies that each pool should be aware of its own rank so that it can determine if it is the master or just another worker. The Pool class has to jointly fulfill both the worker and the master roles.

The `wait()` method here has the same meaning as `Thread.run()` from "Threads" on page 290. It does work when there is work to do and sits idle otherwise. There are three paths that `wait()` can take, depending on the kind of task it receives:

1. If a function was received, it assigns this function to the attribute `f` for later use.
2. If an actual task was received, it calls the `f` attribute with the task as an argument.
3. If the task is `False`, then it stops waiting.

The master process is not allowed to wait and therefore not allowed to do real work. We can take this into account by telling MPI to use `P+1` nodes. This is similar to what we saw with threads. However, with MPI we have to handle the master process explicitly. With Python threading, Python handles the main thread, and thus the master process, for us.

The `map()` method again takes a function and a list of tasks. The tasks are evenly distributed over the workers. The `map()` method is only runnable on the master, while workers are told to wait. If the function that is passed in is different than the current value of the `f` attribute, then the function itself is sent to all of the workers. Sending happens via the "initiate send" (`COMM_WORLD.isend()`) call. We ensure that the function has made it to all of the workers via the call to `MPI.Request.waitall()`. This acts as an acknowledgment between the sender and all of the receivers. Next, the tasks are distributed to their appropriate ranks. Finally, the results are received from the workers.

When the master pool instance is deleted, it will automatically instruct the workers to stop waiting. This allows the workers to be cleaned up correctly as well. Since the `Pool` API here is different enough, a new version of the top-level `simulate()` function must also be written. Only the master process should be allowed to aggregate results together. The new version of `simulate()` is shown here:

```python
def simulate(N, D, S, G, dt):
    x0, v0, m = initial_cond(N, D)
    pool = Pool()
    if COMM_WORLD.Get_rank() == 0:
        for s in range(S):
            x1, v1 = timestep(x0, v0, G, m, dt, pool)
            x0, v0 = x1, v1
    else:
        pool.wait()
```

Lastly, if we want to run a certain case, we need to add a main execution to the bottom of *n-body-mpi.py*. For 128 bodies in 3 dimensions over 300 time steps, we would call `simulate()` as follows:

```python
if __name__ == '__main__':
    simulate(128, 3, 300, 1.0, 1e-3)
```

Given MPI's fine-grained control over communication, how does the N-body problem scale? With twice as many processors, we again expect a 2x speedup. If the number of MPI nodes exceeds the number of processors, however, we would expect a slowdown due to managing the excess overhead. Figure 12-7 shows a sample study on a dual-core laptop.

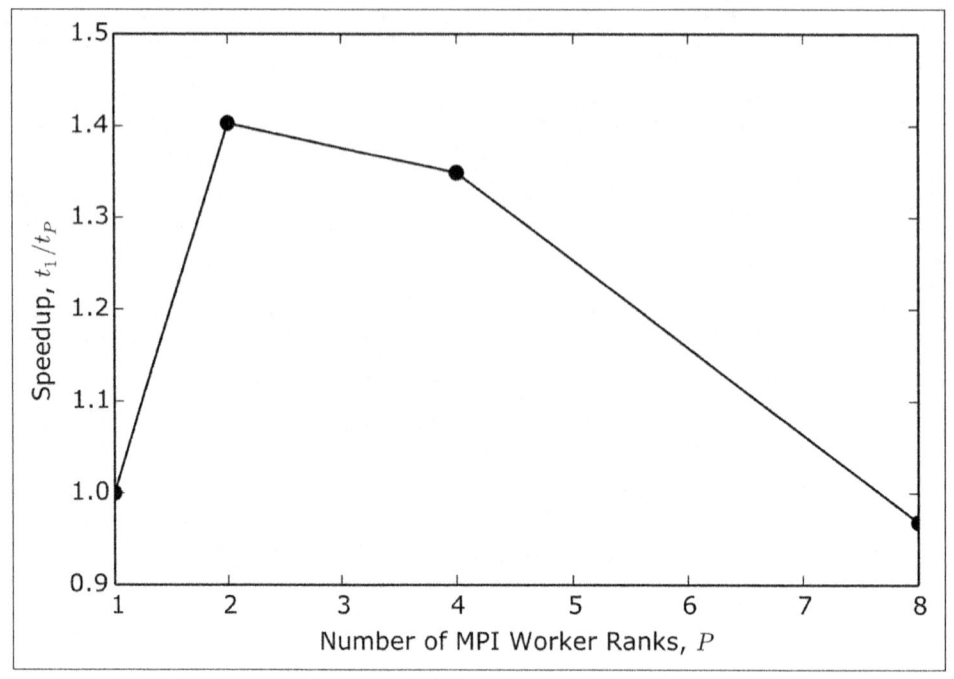

Figure 12-7. Speedup in MPI N-body simulation

While there is a speedup for the P=2 case, it is only about 1.4x, rather than the hoped-for 2x. The downward trend for P>2 is still present, and even steeper than with multiprocessing. Furthermore, the P=1 MPI case is about 5.5x slower than the same simulation with no parallelism. So, for small simulations MPI's overhead may not be worth it.

Still, the situation presented here is a worst-case scenario for MPI: arbitrary Python code with two-way communication on a small machine of unspecified topology. If we had tried to optimize our algorithm at all—by giving MPI more information or by using NumPy arrays to communicate—the speedups would have been much higher.

These results should thus be viewed from the vantage point that even in the worst case, MPI is competitive. MPI truly shines in a supercomputing environment, where everything that you have learned about message passing still applies.

Parallelism Wrap-up

Parallelism is a huge topic, and we have only scratched the surface of what can be accomplished, what mechanisms for parallelism exist, and what libraries implement the various strategies. Having read this chapter, you are well prepared to go forth and learn more about how to implement parallel algorithms in the context of scientific computing. This is very different from the more popular web-based parallelism that permeates our modern lives. The following list presents some excellent parallel systems that you may find interesting or helpful to explore, beyond the essentials covered here:

OpenMP (http://openmp.org/wp/)
  Preprocessor-based, easy to use, low-level parallelism for C, C++, and Fortran

GNU Portable Threads (http://www.gnu.org/software/pth/)
  Cross-platform thread system for C, C++, and Fortran

IPython Parallel (http://bit.ly/IPython_parallel)
  The parallel architecture that IPython uses, based on ZeroMQ (*http://zeromq.org/*)

Twisted (https://twistedmatrix.com/trac/)
  Event-driven parallelism for Python web applications

You should now be familiar with the following ideas:

- There are many ways to measure scale.
- Certain problems are embarrassingly easy to make parallel, while others are very difficult.
- High-performance computing systems are built to handle non-embarrassingly parallel problems.
- High-throughput computing systems are best used for embarrassingly parallel or heterogeneous problems.
- Non-parallel algorithms are faster than parallel code used with one process.
- Stay away from Python threads when number crunching.
- Multitasking is great for problems involving up to around a thousand processes.
- Use MPI when you really need to scale up.

Now that you know how to write software in serial and parallel, it is time to talk about how to get your software to other computers.

CHAPTER 13
Deploying Software

So far, we have mostly been concerned with how to write software that solves physics problems. In this chapter, though, we will discuss how writing software that runs reliably—especially at scale—can prove almost as challenging.

Most software developers want users for their code. For scientific programs this is doubly true, since science runs on a reputation system. Without users, code can hardly be called reproducible. However, even in the event that you do not wish for users, users are still extraordinarily helpful. A new user is a fresh set of eyes that can point out the blind spots of a program better than anyone else. What is broken a new user will break. What is unintuitive a new user will not understand. This process starts at installation.

> New users should realize that if they do not understand something about a project after reading the documentation, then the moral failing lies with the developers and not with themselves. Good users in this situation will then kindly report their struggles back to the developers.

That said, you are your first user. Writing code is a distinctly different process from running the code. Even with tools to mitigate this distinction, such as testing, the difference is never completely gone. Deploying code that you wrote for the first time is an exciting experience. It probably will not work on the first try. As a developer, it is easy to address the feedback that you yourself have as a user. You can iterate through this develop-then-use cycle as many times as necessary until the code really *does* work (for you, at least).

What confounds the deployment problem is that every system is different. Every laptop, desktop, phone, operating system, and environment has its own history. This happens naturally as people use their computers. To make matters worse, oftentimes the system might describe itself incorrectly. This can make installing and running on a new system painful. Deployment often feels like an approximate art rather than an exact science. Trying from the start to ensure that your software will run in as many places as reasonably possible is the only way to preserve your sanity in the long term.

With a lot of attention and care, brand new systems can be forced to be the same. This is incredibly useful to developers: if the software works on one copy of the system, it will likely work on all other copies. Strong guarantees like this often come from *virtualization* in one form or another.

Virtualization is the act of *isolating* the software that you wish to run from the system that is running it. This comes in two forms. A *virtual machine* allows for an entire operating system to be installed and booted within another operating system. A *virtual environment* creates a new, clean space to run code in that is safely separated from other parts of the operating system and environment.

Large-scale deployments of virtualizations can be found on supercomputers and in the cloud. Supercomputers and smaller clusters often have login machines that users connect to in order to launch jobs, compile their code, and perform other tasks. However, the nodes that execute the software typically all run the same well-defined operating system and environment. In the cloud, different virtualizations might exist to fill different needs of an application. In all cases, the virtual environment is well defined.

What these large deployments share is that the users are not allowed to touch the software or environment directly. Users will break things. Such systems are too intricate and expensive to risk by giving even experienced people direct access. Removing users from the mix removes a sure source of error. Virtualization in large computing settings rightly remains a tool for developers. However, developers can prepackage a virtualization and give it to users as a common starting point.

Deployment is a process, a struggle. Users want your software to work for them with whatever they bring to the table. Users running Windows 98 Second Edition do not care if your code works perfectly on a prerelease version of Linux. As in any negotiation, you will end up meeting somewhere in the middle, but it will likely be much closer to the user's side.

Figuring out what works for you as a developer and your users at the same time is one of the great challenges in software engineering. Its importance and difficulty cannot be understated. Adversity breeds success, however, and the rewards for trying are huge for software deployment. This chapter covers various modern tools that are used to help in the deployment process.

Deploying the Software Itself

The first stage of deployment is often to figure out how to *package* the software. This involves creating a file that is *distributable* to a wide audience. Once users have this file, they can run the code after a few special commands to install its contents on their systems.

The internal structure of the package, how the package is distributed, and how it is installed by the user all vary by *package manager*. Package managers are special pieces of software that are responsible for installing other software on a user's computer. Most operating systems now come with at least one package manager preinstalled.

Lacking a package manager, you can always fall back to giving users a download link to the source code and have them follow instructions to install your code manually. This tends to limit your users to other developers.

For our purposes here, package management falls into three broad distribution categories:

1. Source-based
2. Binary
3. Virtualization

Source-based distributions are an automated extension of the "give people a link to the code" idea. The package manager will download the source code and install it onto the user's machine. For dynamic languages this is fine. Installation is fast and errors will crop up at runtime, if relevant. For compiled languages, source-based distribution is somewhat out of vogue. This is because it requires users to have a compiler available on their systems. This is a reasonable assumption on most Linux systems but almost categorically false on Mac and Windows systems, where it is difficult to get a compiler working in the first place.

For this reason, binary package management has proven more successful for both dynamic and compiled languages. With this approach, the developers compile the code into its binary form for every combination of architectures that they wish to support. Architectures are at minimum specified by word size (32 bits, 64 bits) and operating system (Linux, Mac OS X, Windows). The results of compilation are then added to a ZIP file, given to the user, and unzipped by the user into the appropriate location. For users, this is fast and vastly reduces the possibility of error. For developers, the extra work of creating many combinations of packages can be a headache.

Lastly, virtualizations can be packaged up and sent to users. These are similar to binary packages in the sense that the developer expends up-front effort to create a version of the code that should just work for the user. Virtualizations, however, go the extra step of also giving the user the environment that the software was created in.

The user then has to manage the virtualization host to be able to run the software. While this approach is easy for the developer to create and maintain, it is often a little more work for the users. It also sometimes takes away from the users' sense of agency since the code is no longer directly running on their machines.

Which strategy should be pursued depends a lot on the expectations of potential users. The rest of this section, while by no means comprehensive, presents a few popular options for distributing and deploying software to users. Keeping with the theme of the book, we focus on deploying Python software.

pip

Python packaging has had a long and storied history (*http://bit.ly/py-pack-history*). The culmination of this is a tool called `pip`, which is the Python Packaging Authority's (PyPA) recommended way to install Python code. In the past, there have been a broad spectrum of tools for creating and managing packages. Each of these tools had its own proclivities. Almost all of them were based on or compatible with the `distutils` module that lives in the Python standard library. For most scientific software, though, `distutils` is an insufficient choice: it handles compiling code from other languages poorly, but just well enough to lull scientists into a false sense of security. Sincere attempts at fixing `distutils` for the scientific use case have been made, but none to date have been successful. Still, for a purely Python code package, `pip` is a good solution that works well. The endorsement by PyPA gives `pip` a weight that will carry it forward into the future.

`pip` is a command-line, source-based package manager that finds and downloads its packages from the Python Package Index (*https://pypi.python.org/pypi*), or PyPI.[1]

For users, `pip` is easy. Here is an excerpt from the `pip` help:

```
$ pip -h

Usage:
  pip <command> [options]

Commands:
  install                     Install packages.
  uninstall                   Uninstall packages.
  list                        List installed packages.
  show                        Show information about installed packages.
  search                      Search PyPI for packages.
  help                        Show help for commands.

General Options:
```

---

[1] For those current with their Monty Python humor, PyPI is sometimes pronounced *cheeseshop*.

```
-h, --help          Show help.
-V, --version       Show version and exit.
-q, --quiet         Give less output.
```

For example, to install numpy, all the user has to do is execute the following command:

```
$ pip install numpy
```

This will install numpy into the system Python. (On Linux, this command may need to be run via sudo.) Alternatively, to install into user space, simply append the --user switch. Other pip commands follow similarly.

As a developer, it is your job to create a pip-installable package and upload it to PyPI. Luckily, if you have a source code directory structure like that which was presented in "Packages" on page 60, then there are helpers that make this easy. Unfortunately, picking which helper to use can be its own hardship. Historically, the distutils package in the standard library was used to manage package installation. From here, the setuptools package evolved to address certain issues in distutils. From setup tools came the distribute package, which itself gave rise to distutils2. Attempts at getting some version of these back into the standard library in some form have failed. So, in the intervening years, different code packages have used whichever option the developer felt was reasonable at the time. For more on this mess, please see The Hitchhiker's Guide to Packaging (*http://bit.ly/hh-guide-pack*). pip, thankfully, simplifies our lives by recommending that we use setuptools. A strategy that is successful in almost all cases is to use setuptools if it is available and fall back to distu tils when it is not.

> Before you romp off and start deploying packages willy-nilly, it is a good idea to make sure that you are at least mostly following the standard best practices of testing, documentation, and compliance with a style guide (*http://bit.ly/pep-8*). See Chapters 18, 19, and 20 for more details. Oh, and of course, make sure that the code actually works! This is harder than it sounds.

To use distutils or the other helpers to manage your Python package, create a file called *setup.py* at the top level of your directory structure. This should live outside of the modules that you want to install. Going back to the directory structure from "Packages" on page 60, we would place the *setup.py* as follows:

```
setup.py ❶
/compphys
    |- __init__.py
    |- constants.py
    |- physics.py
    /more
```

```
    |- __init__.py
    |- morephysics.py
    |- evenmorephysics.py
    |- yetmorephysics.py
/raw
    |- data.txt
    |- matrix.txt
    |- orphan.py
```

❶ The *setup.py* file is at an equal level with the source code directory.

The sole purpose of *setup.py* is to import and call the `setup()` function with appropriate arguments. This `setup()` function acts as a main function. It provides a command-line interface for installing the software locally and also for making and uploading packages to PyPI. The `setup()` function takes a number of keyword arguments that describe how the package source is laid out on the filesystem. It also describes how the package should be installed on the user's filesystem. The following is an example of a *setup.py* that corresponds to file structure just shown:

```python
import sys
try:
    from setuptools import setup    ❶
    have_setuptools = True
except ImportError:
    from distutils.core import setup    ❷
    have_setuptools = False

setup_kwargs = {    ❸
    'name': 'compphys',
    'version': '0.1',
    'description': 'Effective Computation in Physics',
    'author': 'Anthony Scopatz and Kathryn D. Huff',
    'author_email': 'koolkatz@gmail.com',
    'url': 'http://www.oreilly.com/',
    'classifiers': [
        'License :: OSI Approved',
        'Intended Audience :: Developers',
        'Programming Language :: Python :: 3',
        ],
    'zip_safe': False,
    'packages': ['compphys', 'compphys.more'],
    'package_dir': {
        'compphys': 'compphys',
        'compphys.more': 'compphys/more',
        },
    'data_files': [('compphys/raw', ['*.txt'])],
    }

if __name__ == '__main__':
    setup(**setup_kwargs)    ❹
```

❶ Use `setuptools` if we can.

❷ Use `distutils` otherwise.

❸ Create the package metadata before we call `setup()`.

❹ Call `setup()` like it is a main function.

While most of the keyword arguments here are self-explanatory, a full and complete description of all of the available options can be found in the `distutils` documentation (*http://bit.ly/setup-script*). The two primary commands of the setup script are `build` and `install`. That may be seen from the help:

```
$ python setup.py -h
Common commands: (see '--help-commands' for more)

  setup.py build      will build the package underneath 'build/'
  setup.py install    will install the package

Global options:
  --verbose (-v)      run verbosely (default)
  --quiet (-q)        run quietly (turns verbosity off)
  --dry-run (-n)      don't actually do anything
  --help (-h)         show detailed help message

usage: setup.py [global_opts] cmd1 [cmd1_opts] [cmd2 [cmd2_opts] ...]
   or: setup.py --help [cmd1 cmd2 ...]
   or: setup.py --help-commands
   or: setup.py cmd --help
```

The `build` command builds the software into a *build/* directory. This command will create the directory if it does not already exist. The `install` command installs the contents of the *build* directory onto the system (the computer you are currently logged into). You can use the `--user` flag here to install into a user's home directory instead of installing systemwide. If the `build` command has not been run, the `install` command will run `build` automatically. The following example shows how to install the package from source and install into your user space:

```
$ python setup.py install --user
```

For pure Python code, source-only packages are easily created with the setup script via the `sdist` command. This builds the package and puts it into a ZIP file that `pip` knows how to unzip and install. The easiest way to create a package is by running this command:

```
$ python setup.py sdist
```

At this point, you as a developer now have a ZIP file living on your computer. Thus, `sdist` does not solve the problem of actually getting the software to users. PyPI is the easiest service to use for Python source-based distribution. This is because `setup tools` and `distutils` already plug into PyPI, and PyPI is free. Before you can upload a package, you have to register it with the PyPI server. This ensures that no two packages have *exactly* the same package name. Registration is accomplished through the aptly named `register` command:

```
$ python setup.py register
```

This requires that you have an existing account on PyPI and that you provide some metadata in *setup.py* about the package in its current state, such as the version number. After the package is registered, you may copy it to PyPI with the `upload` command. This command must follow an `sdist` command, as seen here:

```
$ python setup.py sdist upload
```

And that is all there is to it. Your Python package is ready for users to download and install with `pip`. For more information, see the `sdist` command documentation (*http://bit.ly/py-dist*) and the PyPI documentation (*http://bit.ly/py-pi*).

While `pip` is great for pure Python source code, it falls a little flat for multilanguage and compiled code projects. Up next, we will see a package manager that is better suited to the needs of scientific software.

Conda

Conda (*http://conda.pydata.org/*) is a cross-platform binary package manager that aims to solve many of the problems with deploying scientific software. It was developed relatively recently by Continuum (*http://continuum.io/*) in response the deficiencies in using `pip` and `distutils` for scientific programs. Like all good things, Conda is free and open source. Conda (at least initially) took inspiration from Enthought's (*https://www.enthought.com/*) Canopy/EPD Python distribution and package manager, and it has gained considerable popularity lately.

Conda has three properties that jointly distinguish it from most other package managers, and from `pip` in particular:

1. It is general enough to seamlessly handle multilanguage and non-Python code projects.
2. It runs on any operating system, and especially Linux, Mac OS, and Windows.
3. It runs in the user's home space by default and does not try to install into the system.

Many package managers have one or two of these features, but all of them are required to satisfactorily cover the full range of scientific computing use cases. In the past, deploying packages required the developers to target as many Linux distributions as they cared to (*apt*, *portage*, *pacman*, *yum*, etc.), create packages for Mac OS X for *homebrew* and *macports*, and create a custom binary installer for Windows. After all of this effort, the developer had to hope that users had administrative privileges on the machines that they were trying to install the package on. Conda replaces all of that with a single distribution and interface.

The fastest and most reliable way to get Conda is to use the Miniconda distribution (*http://bit.ly/mini-conda*). This is a package that installs Conda and its dependencies (including Python) into a directory of the user's choosing. The default install location is *~/miniconda*. From here, Conda can be used to search for and install all other desired packages. This includes updates to Conda itself.

Using Conda is very similar to using `pip` or other package managers. An abbreviated version of the `conda` help is shown here:

```
$ conda -h
usage: conda [-h] [-V] command ...

conda is a tool for managing environments and packages.

positional arguments:
  command
    help         Displays a list of available conda commands and their help
                 strings.
    list         List linked packages in a conda environment.
    search       Search for packages and display their information.
    create       Create a new conda environment from a list of specified
                 packages.
    install      Install a list of packages into a specified conda
                 environment.
    update       Update conda packages.
    remove       Remove a list of packages from a specified conda environment.
    clean        Remove unused packages and caches
    build        Build a package from a (conda) recipe. (ADVANCED)
```

A user could install `numpy` via `conda` with the following command:

```
$ conda install numpy
```

This grabs `numpy` from the first *channel* that has a `numpy` package that matches the user's platform. A channel is a URL that points to a *channel file*, which in turn contains metadata about what packages are available on the channel and where to find them. A channel can be added to the local Conda configuration with the following command:

```
$ conda config --add channels http://conda.binstar.org/foo
```

Conda comes with a default channel that contains a wide variety of core packages. Additional developer-supplied channels may be found at Binstar (*https://binstar.org/*). Binstar serves the same role as PyPI in that it is a place for developers to upload custom packages and users to download them though Conda.

Conda's package-building infrastructure is more general than that of `distutils`. Rather than requiring a setup script, like `pip`, Conda looks for a directory of a certain structure, known as a *recipe*. The name of the recipe directory is the same as the package name. The recipe may contain the following files:

- *build.sh* (a bash build script for building on Linux, Mac, and other POSIX systems)
- *bld.bat* (a batch script for building on Windows)
- *meta.yaml* (the metadata for the package in YAML format)
- *run_test.py* (an optional file for running tests)
- Optional patches to the source code
- Optional other files that cannot be included in other ways

This system is more general because you can write anything that you want in the build scripts, as long as the resultant software ends up in the correct place. This allows developers to fully customize their packages. However, this freedom can sometimes be daunting. In Conda, it is explicitly the developer's responsibility to ensure that the code builds on all systems where packages are desired. You might not know anything about building code on a Mac platform, but if you want a Conda package for this operating system you have to figure it out. Please consult the Conda build documentation (*http://bit.ly/conda-build*) for more information on writing Conda recipes, and see the conda-recipes GitHub page (*https://github.com/continuumio/conda-recipes*) for many working recipe examples.

Once you have a Conda recipe, it is easy to build a package for the system that you are currently on. Simply pass the path to the recipe to the `build` command. Say we had a `compphys` recipe in the current directory. This could be built with the following command:

```
$ conda build compphys
```

To upload the new package to Binstar, you need to have a Binstar account and the *binstar* command-line utility. You can get an account from the website, and you can obtain the *binstar* command-line utility though Conda itself, as follows:

```
$ conda install binstar
```

This allows you to sign into Binstar from the shell using the `binstar login` command. When you are logged in, any builds that you perform will also prompt you to upload the package to Binstar under your account name. Thus, to both build and upload a package, you just need to run the following commands:

```
$ binstar login
$ conda build compphys
```

Conda is the preferred binary package manager. It solves many of the problems caused by language-based or operating system–based package managers. However, neither source nor binary packages give the user the same execution environment that the developer had when the package was built. To distribute the environment along with the code, we first turn to virtual machines.

Virtual Machines

A virtual machine, or *VM*, is a simulated computer that runs as a process on another computer. The simulated computer is known as the *guest* and the computer that is running the simulation is called the *host*. Everything about modern computer architecture is replicated: the number of processors, memory, disk drives, external storage, graphics processors, and more. This allows the guest VM to run any operating system with nearly any specifications, completely independently of what the host happens to be running as its operating system. As with other recursions, you can nest VMs inside of VMs. You *could* run Ubuntu inside of Vista inside of Windows 7 inside of Red Hat inside of Mac OS X, though your friends might question the sanity of such an endeavor.

Calling a virtual machine a simulation is slightly incorrect. More correctly, VMs are *hypervisors*. While it is true that the hardware interface is simulated, what happens when code is executed on the guest is a little more involved.

Normally when you run software on a computer, the operating system's *kernel* schedules time on a processor for your process to execute. The kernel is the Grand Poobah of the operating system. It is the algorithm that decides what gets run, when it gets run, and with how much memory. For a virtual machine, the hypervisor translates requests for time and space from the guest kernel into corresponding requests made to the host kernel.

To make these requests as speedy as possible, the hypervisor often has hooks into the host's operating system. The guest gets real compute time on the host's processor, but in such a way that it is completely hidden from other parts of the host. Thus, a VM does not simulate the way a processor works; that would be horridly expensive. Rather, it simulates the hardware that the guest *believes* is the computer.

Setting up a virtual machine requires you to specify all of the attributes of the machine that you would like to create: the number of processors, the size of the hard disk, the amount of memory, and so on. However, this process has been greatly simplified in recent years. It now takes just a handful of button clicks to get a new VM up and running.

The effort of moving to a VM buys you reliability and reproducibility. For example, before you touch it at all, a new VM with the latest version of Ubuntu is going to be exactly the same as all other VMs with the latest version of Ubuntu (that have the same virtual hardware specs). Furthermore, you can snapshot a virtual machine to a file, store it, and ship it to your friends and colleagues. This allow them to restart the VM on another machine exactly where you left it. These features are incredibly valuable for reproducing your work, tracking down bugs, or opening Microsoft Office while on a Linux machine.

The virtualization software that we recommend is Oracle's VirtualBox (*https://www.virtualbox.org/*). It is the most popular open source VM software for day-to-day users. VirtualBox may run as a host on Linux, Mac OS X, Windows, and a number of other operating systems. Competitors include VMware (*http://www.vmware.com/*), XenServer (*http://www.xenserver.org/*), and KVM (*http://www.linux-kvm.org/*). Figure 13-1 shows an example virtual machine with a 64-bit Ubuntu host running a 32-bit version of Windows 7 as a guest.

Figure 13-1. VirtualBox with a 64-bit Ubuntu 14.04 host and a 32-bit Windows 7 guest

Virtual machines are incredibly important for large-scale deployment. We will see their primary strength when we get to "Deploying to the Cloud" on page 325. Yet even when you're working in small groups, it is useful to have a working virtual machine with your software. Upload this VM to the Internet, and users can download it and try out your code without actually trying to learn how to install it! The only major disadvantage with this VM-based distribution strategy is that the virtual machines can become quite large. The snapshot of a VM can easily range from 1 to 10 GB, depending on the guest operating system and the size of the code base.

The size and startup times of VMs can be crippling for small, automated tasks. This has led to another valuable approach to packaging and deployment.

Docker

Motivated by the fact that virtual machines and hypervisors can be large and take a long time to start up, a new technology called *containers* has recently rocketed to the top of many people's lists of deployment strategies. A container may be thought of as *operating system–level virtualization*. Rather than the guest having to go through a separate hypervisor, the operating system itself provides an interface for the guest to request and access resources, safely shielded from other parts of the operating system. This makes containers much lighter-weight and faster than traditional virtual machines. It also adds restrictions, since the guest has to know about the host's kernel in order to access it.

Containers are typified by the Docker (*https://www.docker.com/*) project. Initially released in March 2013, Docker saw a stable v1.0 in June 2014. The velocity of its rise and adoption is remarkable.

So why now? And why so fast? Without diving into too many of the details, Linux Containers (*https://linuxcontainers.org/*) (LXC) have been around since Linux v2.6.24, which was released in January 2008. However, they had some pretty large security holes. It was not until Linux v3.8, released in February 2013, that these holes were fixed sufficiently to be viable for large-scale deployment. The Docker project was started, and the rest is history.

The main limitation of containers is that the guest operating system must be the same as the host operating system. Furthermore, LXC is a Linux-only technology right now. Microsoft and Docker have recently announced a collaboration, so Windows containers are on their way, but the Mac OS X platform has yet to start to catch up. Critics of LXC will sometimes point out that other operating systems, such as FreeBSD and Solaris, had container support long before Linux. For various historical reasons, though, none of these container technologies gained the popularity that LXC currently enjoys.

> Since Docker is currently limited to Linux (and soon, hopefully, Windows), feel free to skip the rest of this section if that is not your platform of choice. What follows is a tutorial on how to use Docker. You have now learned what you need to about containers and their importance as a tool for collaboration; you can come back to this section when you have a personal need for Docker itself.

Using Docker is nearly synonymous with using Docker Hub (*https://hub.docker.com/*), the online hosting service for Docker. Tight integration between the Docker command-line interface and Docker the Internet service is part of what makes it so popular. If you do not already have an account on Docker Hub, you should go create one now. It is easy and free.

Assuming you have Docker installed and usable (on Ubuntu, run the command `sudo apt-get install -y docker docker.io`), you can run a simple "hello world" container with the following command:

```
$ sudo docker run ubuntu:14.04 echo "Hello, World!"
```

This executes `docker` with the `run` command, downloading from Docker Hub an Ubuntu 14.04 *image*, as specified by the `ubuntu:14.04`. The remaining arguments are any bash commands that you wish to run from inside the container. Here, we simply run echo. If you have not run `docker` before, the output of this command will look like the following:

```
$ sudo docker run ubuntu:14.04 echo "Hello, World!"  ❶
[sudo] password for scopatz:  ❷
Unable to find image 'ubuntu:14.04' locally  ❸
Pulling repository ubuntu
c4ff7513909d: Download complete
511136ea3c5a: Download complete
1c9383292a8f: Download complete
9942dd43ff21: Download complete
d92c3c92fa73: Download complete
0ea0d582fd90: Download complete
cc58e55aa5a5: Download complete
Hello, World!  ❹
```

❶ Note that you need root privileges to run `docker`.

❷ That's a secret!

❸ Docker will intelligently stash containers for future use.

❹ The output of our echo command.

This shows that the Ubuntu image, which was only around 225 MB, could not be found locally, so Docker automatically downloaded it from Docker Hub for us. Docker then executed the echo command. Compared to downloading and setting up a whole virtual machine, using Docker is easy. (Of course, this is a "hello world" example, so it *should* be easy!) Naturally, there are other tweaks you can make to this process, such as specifying private resources other than Docker Hub for finding containers. Note that the image that was downloaded was cached for later use. Rerunning the same command will not require downloading the image again. The second time around, we should only see the output of echo:

```
$ sudo docker run ubuntu:14.04 echo "Hello, World!"
Hello, World!
```

A list of all Docker images that are on the local system can be printed out with the images command, as follows:

```
$ sudo docker images
REPOSITORY       TAG       IMAGE ID        CREATED         VIRTUAL SIZE
ubuntu           14.04     c4ff7513909d    2 weeks ago     225.4 MB
```

To avoid the business of downloading images when you want to run them, the pull command allows you to download them ahead of time. Say we wanted to run the latest version of the WordPress blog. We could grab the corresponding image by passing in wordpress:latest:

```
$ sudo docker pull wordpress:latest
```

Of course, you have to check the Docker Hub website to see what repositories (ubuntu, wordpress, etc.) and what tags (14.04, latest, etc.) are available before you can pull down an image.

You may also delete local Docker images from your system with the "remove image," or rmi, command. Suppose that we decided that we were not that into writing blogs anymore and wanted to get rid of WordPress. This could be performed with the command:

```
$ sudo docker rmi wordpress
```

Now, say that we wanted to add numpy to the ubuntu container so that it would be readily available for future use. This kind of container customization is exactly what Docker was built for, and it does it quite well. The first step is to launch the ubuntu container in interactive mode. We do so by using the run command along with the -t option to give us a terminal and the -i option to make it interactive. We will probably want to run bash so that we can be effective once inside of the container, too. When we run the following command we are dropped into a new interactive terminal within the container:

```
$ sudo docker run -t -i ubuntu:14.04 /bin/bash
root@ae37c22b3c49:/#
```

While inside the container's shell, Docker will automatically record anything that we do. Let's install Ubuntu's package manager, install numpy, and then leave. These steps are shown here:

```
$ sudo docker run -t -i ubuntu:14.04 /bin/bash
root@ae37c22b3c49:/# apt-get update   ❶
...
```

```
root@ae37c22b3c49:/# apt-get install -y python-numpy
...
root@ae37c22b3c49:/# exit
```

❶ Note that while we are inside of the container, we have root privileges.

Back on our host machine, to truly save our work we have to *commit* the changes. This creates a new image based on the original one, any modifications we may have made, and metadata about the change that we supply. The docker `commit` command takes the identifier that we saw in the container (here, `ae37c22b3c49`), a message string via the `-m` option, an author name via the `-a` option, and a repository name for the new image (here, `ubuntu-numpy`). When following along at home, be sure to substitute your own Docker Hub username for `<user>`. Putting this all together, we can commit our changes with the command:

```
$ sudo docker commit -m "with numpy" -a "<name>" ae37c22b3c49 <user>/ubuntu-numpy
73188d24344022203bee5ef5d6cb31ccaa8b5f38085ae69fcf9502828220f81d
```

Our new container now shows up in the images list and is available for future use. Running the `images` command from before now produces the following output on my computer:

```
$ sudo docker images
REPOSITORY              TAG      IMAGE ID       CREATED             VIRTUAL SIZE
scopatz/ubuntu-numpy    latest   73188d243440   About a minute ago  225.4 MB
ubuntu                  14.04    c4ff7513909d   2 weeks ago         225.4 MB
```

Running docker with `<user>/ubuntu-numpy` will save us time, because `numpy` is preloaded. We could also have built this same container using a `Dockerfile`. Docker files are more effort to set up, though also more reproducible. For most regular Docker tasks, the interactive shell is good enough. Please see the Docker documentation (*https://docs.docker.com/userguide/*) for more details.

At this point, the `ubuntu-numpy` image still lives only on our computer. However, we can upload it to Docker Hub for ourselves and others to freely use. This is done with the push command. This command will ask you to log into Docker Hub if you have not done so already. As you can see here, push requires that you specify the image that you want to upload:

```
$ sudo docker push scopatz/ubuntu-numpy
The push refers to a repository [scopatz/ubuntu-numpy] (len: 1)
Sending image list

Please login prior to push:
Username: scopatz
Password:
Email: scopatz@gmail.com
Login Succeeded
The push refers to a repository [scopatz/ubuntu-numpy] (len: 1)
```

```
Sending image list
Pushing repository scopatz/ubuntu-numpy (1 tags)
511136ea3c5a: Image already pushed, skipping
1c9383292a8f: Image already pushed, skipping
9942dd43ff21: Image already pushed, skipping
d92c3c92fa73: Image already pushed, skipping
0ea0d582fd90: Image already pushed, skipping
cc58e55aa5a5: Image already pushed, skipping
c4ff7513909d: Image already pushed, skipping
73188d243440: Image successfully pushed
Pushing tag for rev [73188d243440] on {https://cdn-registry-1.docker.io/v1/
repositories/scopatz/ubuntu-numpy/tags/latest}
```

Docker is a sophisticated tool that puts power into its users' hands. We have only scratched the surface of what it can do here and not discussed how it works internally. However, those details are not needed to use Docker to deploy physics code and you can already see that it is an efficient and masterful way of creating, customizing, and sharing software. Rightfully so, Docker is quickly replacing other, more traditional methods of software deployment.

Now that you know how to deploy software through a variety of mechanisms, let's go on to *where* you might deploy it.

Deploying to the Cloud

Lately, it seems impossible to avoid hearing about *the cloud*: cloud systems, cloud architectures, cloud business solutions, and the like. As a group of remote computers that combine to provide a wide range of services to a local user or machine, the cloud could easily be dismissed as just another phrase for the Internet itself. And to users, there does not seem to be much distinction.

While there is no formal agreement on what the cloud *is*, a reasonable definition is that it is the deployment of and interaction between three reliability strategies: Software-as-a-Service (*SaaS*), Platform-as-a-Service (*PaaS*), and Infrastructure-as-a-Service (*IaaS*). These technologies are enabled by virtual machines and containers, which have already been covered in this chapter. You can envision cloud computing as the stack shown in Figure 13-2.

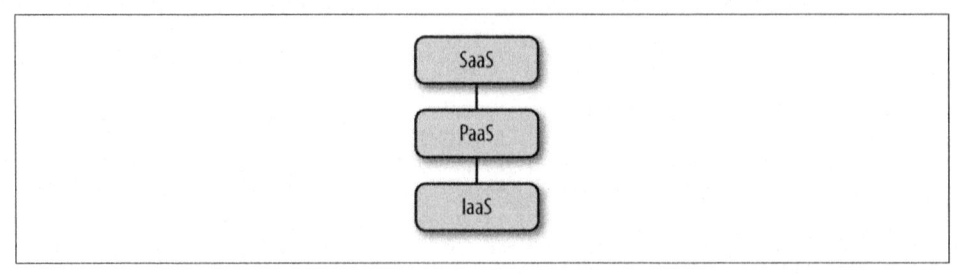

Figure 13-2. Cloud computing stack

The cloud is not a revolution in terms of the kinds of technology that are deployed (websites, email, databases, etc.). Rather, it is an evolution of who does the deploying, at what scale, and with what level of redundancy. For example, if 10 years ago you wanted an email server for your group, it often meant you would go over to the old server in your closet, install the email server software, and then assign yourself and your friends any email addresses you liked. This worked just fine until your hard drive spun itself into oblivion, vermin ate through your ethernet cable, or your Internet service provider suddenly decided to start blocking port 25.[1] Alternatively, suppose that you were running a website from your home and it suddenly became extraordinarily popular.[2] Your brand new DSL connection would probably not have been able to handle the load. This would invariably lead to the site being down until it was no longer popular. This used to happen so frequently that the phenomenon earned the name *the Slashdot effect*, after the news website (*http://slashdot.org/*) whose references caused many pages to go down temporarily.

The cloud solves these problems by offloading services to a larger, more stable, and better connected third party, such as Google or Amazon or Rackspace. It also allows you to scale your services up or down as needed. Cloud service providers make it very easy to provision new machines, bandwidth, or storage space as you need them. They also make it easy to remove such resources once you are finished with them.

Starting at the bottom of the stack, Infrastructure-as-a-Service is where you rent some or all of the physical parts of a large-scale computer system: hard drives, servers, networking devices, electricity, an amazing Internet connection, and the roof above all of this expensive machinery. You do not get an operating system, but you are able to configure the kind of system you want temporarily. This is great to have if you want to have a single machine to experiment with, 10 to do a trial run, 100 to do a full release with, and then finally scale back to 50 when you realize you purchased too many machines. IaaS maximizes flexibility and minimizes risk.

---

[1] Author's note: all of these have happened to me.

[2] This has not.

In the middle of the cloud stack live the Platform-as-a-Service vendors. In the PaaS model, a developer will write an application and deploy it. The PaaS typically offers a common way to set up and write applications, making it easy to do so once you have adopted its platform. PaaS applications are often run on IaaS machinery. Examples of PaaS include Google App Engine (GAE) and Heroku.

At the top level of the stack are user-facing Software-as-a-Service tools. These are how most people interact with the cloud on a daily basis. Almost always, these tools are websites that are either available publicly or to a limited set of people. The "software" here refers to the fact that a website is code-developed as a service-oriented application, distinct from the hardware that runs the code. The classic example of SaaS is Gmail, though in truth anything that involves an active user doing more than looking at websites could be considered SaaS. Content generation sites such as WordPress blogs represent this well.

Cloud service providers (Google, Amazon, Rackspace) tend to supply their customers with IaaS, PaaS, and SaaS in whatever mix they believe best suits customers' needs. The lines between these three categories are often blurred. A company that sets out to sell software might also end up selling hard disk space because it finds that its users want more storage. A business that just wants to sell time on processors for virtual machines will typically also end up adding a web interface to manage those VMs.

The power of the cloud comes from the realization that you can be more efficient if you can offload at least part of the minutiae of managing your services to someone else who specializes at it. How you deploy to a particular platform depends on your needs and the supplier. Every company has its own guides and documentation. Once you think you know what you want, it can help to look for guides and to shop around online before committing to a particular vendor.

In the physical sciences, the cloud is most often utilized when you or your group have outgrown the resources that you currently have available. Rather than buying and setting up expensive new machinery, you can rent the resources you need, when you need them, for as long as you need them. Since the price point is so much lower than buying your own in-house capabilities, the cloud effectively brings large-scale computing services to everyone.

That said, the cloud is centered around a high-throughput computing model. The next section covers deploying high-performance computing applications, which tend to be more common in computational physics.

Deploying to Supercomputers

Supercomputers are notoriously finicky devices. They are large computer engineering *experiments*. Unlike the cloud, a supercomputer was not built to serve you or your needs. If you reach the point where you need to have or you do have access to a supercomputer for your work, it will be clear that supercomputers are shared resources. You will almost never need to access the whole machine. If you do, that just means that you'll have to wait your turn in line even longer.

Deployment in a supercomputing environment is embodied by the following three features:

- Uniform processors and memory across nodes
- A parallel filesystem
- A scheduler

As was touched on in Chapter 12, having the same processor and the same amount of memory available everywhere in a supercomputer means that the code that you write for one part of the machine is executable on all other parts. This is great. However, you do not have direct access to the supercomputer's compute nodes. First, you must sign in to the *login node* to access the machine at all. From the login node, you may then submit jobs to be executed by the scheduler. Typically, supercomputing environments require you to compile custom versions of your code in order for that code to be able to run. Furthermore, every machine is its own special work of art. The customizations that you make for one supercomputer do not necessarily apply to another machine. These customizations often break when going from one generation of machines to the next.

Supercomputers, as a shared resource, also have a parallel filesystem. This is because many people and many processes will attempt to access files simultaneously. As a user, this will look and feel much like a normal filesystem that you have on your laptop or desktop or flash drive. However, it will perform much differently. On a parallel filesystem, every file you access incurs a certain overhead. For a large number of files this overhead becomes unbearable, because each file has to be checked individually. For example, on a laptop, executing the `ls` command in a directory with a hundred thousand files might take 1–3 seconds. On a parallel filesystem, this same operation could take half an hour or longer. Limiting the number of files that you have to access is the key to being successful. File metadata is just that much slower on these systems. There are usually tricks to make commands faster, but these are not necessarily well known. For example, using `ls --color=never` will sometimes bring the listing runtime back down to sane levels, but not if you use the `-l` option or any other possible slowdowns.

Lastly, to actually run code on a supercomputer you need to go through the *scheduler*. Common schedulers include TORQUE (*http://bit.ly/torque-rm*) and Grid Engine (*http://gridscheduler.sourceforge.net/*). This program is responsible for keeping track of how much total time each user has allocated and how much has been used, and for determining which *jobs* get to be run and which ones remain in the *queue* until the next job has finished. The point of the scheduler is to keep the usage of the machine fair. As with most "fair" systems, it can leave its constituents frustrated and annoyed. Almost all schedulers have a time limit. If a job exceeds the time limit, the scheduler will abruptly kill the job. Time limits typically range from three days to a week. Smaller and shorter jobs will typically move through the queue more quickly than larger and longer jobs. However, since you are on a supercomputer you likely need the greater resources. There is a balance that has to be struck.

Since every supercomputer is special, you will need to consult the machine-specific documentation for yours to deploy to it. If this does not work, please consult your friendly and overworked system administrator to help figure out where you went wrong. Note that kindness serves you well in these scenarios. When nothing is working and everything has gone wrong, five minutes of social engineering can easily save a month's worth of computer engineering. Being nice helps.

Deployment Wrap-up

We have now seen a wide range of strategies for deploying software. You should be familiar with these main points:

- Deployment is ultimately about your users.
- You are your first user.
- Source-based package managers are good for high-level languages, like Python.
- Binary package managers are good for compiled languages because the users do not need to have compilers themselves.
- Virtual machines package up the entire environment, including the operating system.
- Containers are much lighter-weight than virtual machines, but have the same advantages.
- Cloud computing is useful when reliability is needed.
- Supercomputers can be frustrating to interact with because they are shared resources, while simultaneously being computer engineering experiments. However, the benefits outweigh the costs when your problem demands such a device.

Now that you know how to write and deploy software, it is time to learn the best practices of computational science.

PART III
Getting It Right

CHAPTER 14
Building Pipelines and Software

The most effective analysis and software tools are reused and shared often. Physics software, libraries, and analysis pipelines are no exception. Indeed, this book is intentionally titled *Effective Computation in Physics*. Effective computational physicists streamline repetitive work and liberate themselves from the mundane. They are also able to build upon the work of others by installing and using external libraries. This chapter will cover both strategies, since they are intricately related by an emphasis on automation and by a versatile tool called *make*.

By the end of this chapter, you should be able to get started with the following tasks:

- Automating complex workflows and analysis pipelines
- Configuring and installing external libraries
- Creating build systems for your own software

By automating the tedious steps, you are much more likely to encourage colleagues to extend your work and to, therefore, positively impact the future of your field. This chapter will help you to assist your peers in building, installing, and reproducing your analysis pipeline by automating the various steps of building that pipeline and your software:

1. Configuration
2. Compilation
3. Linking
4. Installation

The first step, configuration, detects platform-dependent variables and user-specified options. It then uses those to customize later steps in the process. For example, an analysis pipeline might run a Python program on a data file to create a particular plot. However, the location of the data file may vary from analysis to analysis or from computer to computer. Therefore, the configuration step occurs before the program is executed. The configuration step seeks out the proper data path by querying the environment and the user to configure the execution accordingly.

The next two steps, compilation and linking, are only necessary you're when building software written in a compiled language. The compilation step relies on a compiler to convert source code into a machine-readable binary format. The linking step attaches that binary-formatted library or executable to other libraries on which it may depend. These two steps prepare the software for installation.

In Chapter 13, the installation step was addressed in a Python context. This step is when compiled libraries, executables, data, or other files are placed in an accessible place in the filesystem. This is all followed by the execution step, when the user actually runs the analysis pipeline or software.

This chapter will address each of these five steps in the context of the *make* utility and its *makefiles*. These are ubiquitous in the computational physical sciences and are integral to automating nearly any computational process.

make

make is a command-line utility that determines which elements of a process pipeline need to be executed, and then executes them. It can be used to automate any process with a tree of dependencies—that is, any process that builds files based on others.

Our case study in this chapter will be Maria Goeppert Mayer, who won the 1963 Nobel Prize in Physics for having theorized the nuclear shell model. She was a remarkably effective scientist. Thus, we can be fairly certain that if she were still doing science these days, she would definitely use *make* to automate her work. Any time a new experiment generated new data or her theoretical model was tweaked, she might want to update the figures in her draft theory paper.

make can automate this. Let us imagine that Professor Mayer is working on a paper describing a new theory and that one day, she receives additional data to support her theory. She would like to update her paper accordingly.

> ### LaTeX and Its Dependencies
>
> In the following examples, Prof. Mayer will be writing her papers in LaTeX. This program will be covered in great depth in Chapter 20, but some basic information about it will be helpful for following along in this section. In particular, LaTeX is a program that can be used to combine plain-text files and image files to create documents. Fundamentally, an author creates some number of *.tex* files, image files, and others. The LaTeX program converts those into a *.dvi* file, which is much like a *.pdf*.

To achieve this, first and foremost, she needs to add the new data to the list of data files that she analyzes. Next, the data analysis program must be rerun using all of the old data *plus* the new data. The results of the data analysis affect one of the figures in her paper ("Figure 4: Photon-Photon Interactions"), so just adding the data and rerunning the analysis won't be enough; she also needs to rerun the plotting program that generates the image for this figure. Of course, when any changes to the figures or *.tex* files are made, the actual paper must be rebuilt using LaTeX (see Chapter 20).

This should sound familiar. It is the vast rabbit hole of tasks down which many physicists lose uncountably many research hours. These things that need to be done, the tasks, are the nodes of a complex dependency tree. The file dependency tree for Prof. Mayer's project might resemble that in Figure 14-1. When a new data file is introduced, like *1948-12-21.h5*, many of the files that depend on it must be regenerated.

A simple bash script like the ones discussed in Chapter 1 could be written to execute every command on the tree, regenerating all of the figures and the paper any time it is called. However, since the new data only affects one figure, not all of the figures need be rebuilt. Such a bash script would spend a lot of time regenerating figures unnecessarily: replotting the rest of the figures would be a waste of time, since they are already up to date.

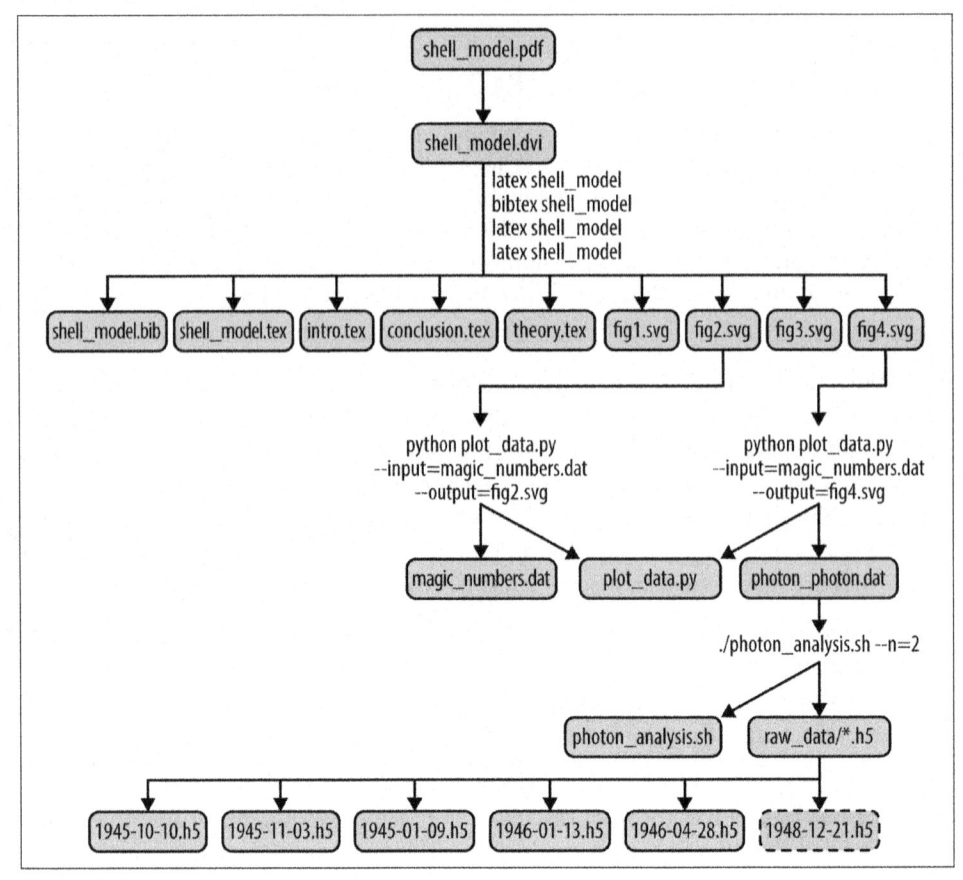

Figure 14-1. Mayer dependency tree

The *make* utility is superior. It can be used to automate every step of this situation more efficiently, because it keeps track of *how things depend on one another* and detects *which pieces are not up to date*. Given the file dependency tree and a description of the processes that compile each file based on the others, *make* can execute the necessary processes in the appropriate order.

> Because it detects which files in the dependency tree have changed, *make executes only the necessary processes, and no more.* This saves time, especially when some actions take a long time to execute but are not always necessary.

Can you tell what processes would need to be reexecuted if a new file in the *raw_data* directory (*1948-12-21.h5*) were introduced? Try drawing a path up the branches of the tree to the top. Which commands do you pass on the way?

Some platforms, like Windows and Mac OS X, do not have *make* enabled in the Terminal by default. Try opening a terminal and typing `which make`. If the *make* utility is available, this command will output the path to the executable. If it is not available, an error will be emitted: "make: command not found." In the latter case, return to the Preface for instructions on how to enable *make* on your platform.

When a new data file is added, *make* can determine what analysis files, figures, and documents are affected. It can then execute the processes to update them. In this way, it can automatically rerun Prof. Mayer's data analysis, regenerate appropriate plot, and rebuild the paper accordingly.

It sounds glorious—too good be true, really. Prof. Mayer would like to try running *make*.

Running make

make can be run on the command line with the following syntax:

```
make [ -f makefile ] [ options ] ... [ targets ] ...
```

It looks like the `make` command can be run without any arguments. So, in some directory where Prof. Mayer holds these interdependent files, she can try typing this magical `make` command:

```
~/shell_model $ make
```

```
make: *** No targets specified and no makefile found.  Stop.
```

Uh oh, it looks like it may not be magic after all. But what is a makefile?

Makefiles

A makefile is just a plain-text file that obeys certain formatting rules. Its purpose is to supply the *make* utility with a full definition of the dependency tree describing the relationships between files and tasks. The makefile also describes the steps for updating each file based on the others (i.e., which commands must be executed to update one of the nodes).

If the `make` command is run with no arguments, then the *make* utility seeks a file called *Makefile* in the current directory to fulfill this purpose. The error response occurs because Prof. Mayer has not yet created a *Makefile* in the directory where she holds her analysis.

If the makefile has any name other than *Makefile*, its name must be provided explicitly. The -f flag indicates the location of that file to the *make* utility. Makefiles with names other than *Makefile* are typically only necessary if more than one makefile must exist in a single directory. By convention, makefiles not called *Makefile* end in the *.mk* extension.

This section will discuss how to write makefiles by hand. Such makefiles can be used to automate simple analysis and software pipelines. Prof. Mayer will create one to update the plots in her paper based on new data.

Targets

First and foremost, the makefile defines *targets*. Targets are the nodes of the dependency tree. They are typically the files that are being updated. The makefile is made up mostly of a series of target-prerequisite-action maps defined in the following syntax:

```
target : prerequisites ❶
        action ❷
```

❶ A colon separates the target name from the list of prerequisites.

❷ Note that the action must be preceded by a single tab character.

The analyzed *.dat* files depend on the raw *.h5* files in the *raw_data* directory. They also depend on the bash scripts that churn through the *.h5* files to convert them into useful *.dat* files. Therefore, the *photon_photon.dat* target depends on two prerequisites, the set of *./raw_data/*.h5* files and the *photon_analysis.sh* shell script.

Let us imagine the shell script is quite clever, having been written by Prof. Mayer herself. It has been written to generically model various interactions and accepts arguments at runtime that modify its behavior. One of the arguments it accepts is the number of photons involved in the interaction. Since the *photon_photon.dat* file describes the two-photon interaction, the shell script can be modified with the special flag -n=2 indicating the number of photons. The following definition in a makefile sets up the target with its prerequisites and passes in this argument:

```
# Building the Shell Model Paper
photon_photon.dat : photon_analysis.sh ./raw_data/*.h5 ❶
        ./photon_analysis.sh -n=2 > photon_photon.dat ❷
```

❶ The target file to be created or updated is the *photon_photon.dat* file. The prerequisites (the files on which it depends) are the shell script and the *.h5* files.

❷ This command is the action that must be taken to update the target using the prerequisites.

In this example, the first line is a comment describing the file. That's just good practice and does not affect the behavior of the makefile. The second line describes the target and the prerequisites, and the third line describes the action that must be taken to update *photon_photon.dat* in the event that *make* detects any changes to either of its prerequisites.

If this source code is saved in a file called *Makefile*, then it will be found when make is executed.

Exercise: Create a Makefile

1. In the *make* directory of the files associated with this book, create an empty file called *Makefile*.
2. Add the *photon_photon.dat* target as described.
3. Save the file.

Now that the makefile defines a target, it can be used to update that target. To build or update a target file using *make*, you must call it with the name of the target defined in the makefile. In this case, if make photon_photon.dat is called, then *make* will:

1. Check the status of the prerequisites and *photon_photon.dat*.
2. If their timestamps do not match, it will execute the action.
3. However, if their timestamps do match, *nothing will happen*, because everything is up to date already.

The makefile is built up of many such target-prerequisite-action maps. The full dependency tree can accordingly be built from a set of these directives. The next node Prof. Mayer might define, for example, is the one that rebuilds Figure 4 any time the *photon_photon.dat* file is changed. That figure is generated by the *plot_response.py* Python script, so any changes to that script should also trigger a rebuild of *fig4.svg*. The makefile grows accordingly as each target definition is added. The new version might look like this:

```
# Building the Shell Model Paper
photon_photon.dat : photon_analysis.sh ./raw_data/*.h5
        ./photon_analysis.sh -n=2 > photon_photon.dat

fig4.svg : photon_photon.dat plot_response.py   ❶ ❷
        python plot_dat.py --input=photon_photon.dat --output=fig4.svg  ❸
```

❶ A new target, *fig4.svg*, is defined.

❷ The *fig4.svg* file depends on *photon_photon.dat* as a prerequisite (as well as a Python script, *plot_response.py*).

❸ The action to build *fig4.svg* executes the Python script with specific options.

Since the figure relies on *photon_photon.dat* as a prerequisite, it also, in turn, relies on prerequisites of *photon_photon.dat*. In this way, the dependency tree is made. So, when `make fig4.svg` is called, *make* ensures that all the prerequisites of its prerequisites are up to date.

> **Exercise: Add Additional Targets**
>
> 1. Open the *Makefile* created in the previous exercise.
> 2. Add the *fig4.svg* target as above.
> 3. Can you tell, from Figure 14-1, how to add other targets? Try adding some.
> 4. Save the file.

The final paper depends on all of the figures and the *.tex* files. So, any time a figure or the *.tex* files change, the LaTeX commands must be reissued. The LaTeX program will be covered in much greater detail in Chapter 20. At that time, you may combine your knowledge of *make* with your knowledge of LaTeX to determine what targets should be included in a makefile for generating a LaTeX-based document.

Special Targets

The first target in a file is usually run by default. That target is the one that is built when `make` is called with no arguments. Often, the desired default behavior is to update everything. An "all" target is a common convention for this. Note that the target name does not have to be identical to the filename. It can be any word that is convenient. The "all" target simply needs to depend on all other top-level targets.

In the case of Prof. Mayer's paper, the `all` target might be defined using the wildcard character (*):

```
# Building the Shell Model Paper
all: figure*.svg *.dat *.tex *.pdf     ❶

photon_photon.dat : photon_analysis.sh ./raw_data/*.h5
        ./photon_analysis.sh -n=2 > photon_photon.dat

fig4.svg : photon_photon.dat
        python plot_response.py --input=photon_photon.dat --output=fig4.svg
```

...

❶ Note how the `all` target does not define an action. It just collects prerequisites.

The `all` target tells *make* to do exactly what is needed. That is, when this target is called (with `make` or `make all`), *make* ensures that all prerequisites are up to date, but performs no final action.

> **Exercise: Create a Special Target**
>
> Another common special target is `clean`. This target is typically used to delete generated files in order to trigger a fresh reupdate of everything.
>
> 1. Open the *Makefile* you have been working with.
> 2. Create a "clean" target.
> 3. What are the appropriate prerequisites? Are there any?
> 4. What is the appropriate command to delete the auxiliary files created by LaTeX?

Now that she knows how to create a makefile, Prof. Mayer can use it to manage dependencies for the entire process of building her paper from the raw data. This is a common use for makefiles and facilitates many parts of analysis, visualization, and publication. Another common use for makefiles is configuring, compiling, building, linking, and installing software libraries. The next section will cover many aspects of this kind of makefile.

Building and Installing Software

Python is called a compiled language because it does not need to be compiled. That is, Python is *precompiled*. However, that compilation step is not handled so nicely by all programming languages. C, C++, Fortran, Java, and many others require multiple stages of building before they are ready to run. We said in the introduction to this chapter that these stages were:

1. Configuration
2. Compilation
3. Linking
4. Installation

From a user's perspective, this maps onto the following set of commands for installing software from source :

```
~ $ .configure [options]  ❶
~ $ make [options]  ❷
~ $ make test  ❸
~ $ [sudo] make install  ❹
```

❶ The configuration step may be called with a different command (i.e., ccmake or scons). This step creates a makefile based on user options and system characteristics.

❷ The build step compiles the source code into binary format and incorporates file path *links* to the libraries on which it depends.

❸ Before installing, it is wise to execute the `test` target, if available, to ensure that the library has built successfully on your platform.

❹ The installation step will copy the build files into an appropriate location on your computer. Often, this may be a location specified by the user in the configuration step. If the install directory requires super-user permissions, it may be necessary to prepend this command with `sudo`, which changes your role during this action to the super-user role.

For installation to succeed, each of these steps requires commands, flags, and customization specific to the computer platform, the user, and the environment. That is, the "action" defined by the makefile may involve commands that should be executed differently on different platforms or for different users.

For example, a compilation step can only use the compiler available on the computer. Compilation is done with a command of the form:

```
compiler [options] <source files> <include files> [-l linked libraries]
```

For C++ programs, one user may use *g++* while another uses *clang* and a third uses *gcc*. The appropriate command will be different for each user. The makefile, therefore, must be *configured* to detect which compiler exists on the machine and to adjust the "action" accordingly. That is, in the action for the compilation step, the compiler command and its arguments are not known *a priori*. Configuration, Compilation, Linking, and Installation depend on the computer environment, user preferences, and many other factors.

For this reason, when you are building and installing software libraries, makefiles can become very complex. However, at their core, their operation is no different than for simple analysis pipeline applications like the one in the previous section. As the dependency tree grows, more targets are added, and the actions become more complex or system-dependent, more advanced makefile syntax and platform-specific configuration becomes necessary. Automation is the only solution that scales.

Configuration of the Makefile

It would be tedious and error-prone to write a custom makefile appropriate for each conceivable platform-dependent combination of variables. To avoid this tedium, the most effective researchers and software developers choose to utilize tools that *automate* that configuration. These tools:

- Detect platform and architecture characteristics
- Detect environment variables
- Detect available commands, compilers, and libraries
- Accept user input
- Produce a customized makefile

In this way, configuration tools (a.k.a. "build systems") address all aspects of the project that may be variable in the build phase. Additionally, they enable the developer to supply sensible default values for each parameter, which can be coupled with methods to override those defaults when necessary.

Why Not Write Your Own Installation Makefile?

Writing your own makefile from scratch can be time-consuming and error-prone. Furthermore, as a software project is adopted by a diversity of users and scales to include dependencies on external libraries, generating an appropriate array of makefiles for each use case becomes untenable. So, the makefile should be generated by a sophisticated build system, which will enable it to be much more flexible across platforms than would otherwise be possible.

Some common build system automation tools in scientific computing include:

- CMake (*http://www.cmake.org/*)
- Autotools (Automake (*http://bit.ly/gnu-automake*) + Autoconf (*http://bit.ly/gnu-autoconf*))
- SCons (*http://www.scons.org/*)

Rather than demonstrating the syntax of each of these tools, the following sections will touch on shared concepts among them and the configurations with which they assist.

First among these, most build systems enable customization based on the computer system platform and architecture.

Platform configuration

Users have various computer platforms with similarly various architectures. Most software must be built differently on each. Even the very simplest things can vary across platforms. For example, libraries have different filename extensions on each platform (perhaps *libSuperPhysics.dll* on Windows, *libSuperPhysics.so* on Linux, and *libSuperPhysics.dyld* on Unix). Thus, to define the makefile targets, prerequisites, and actions, the configuration system must detect the platform. The operating system may be any of the following, and more:

- Linux
- Unix
- Windows
- Mobile
- Embedded

Additionally, different computer architectures store numbers differently. For example, on 32-bit machines, the processors store integers in 32-bit-sized memory blocks. However, on a 64-bit machine, an integer is stored with higher precision (64 bits). Differences like this require that the configuration system detect how the current architecture stores numbers. These specifications often must be included in the compilation command.

Beyond the platform and architecture customizations that must be made, the system environment, what libraries are installed, the locations of those libraries, and other user options also affect the build.

System and user configuration

Most importantly, different computers are controlled by different users. Thus, build systems must accommodate users who make different choices with regard to issues such as:

- What compiler to use
- What versions of libraries to install
- Where to install those libraries
- What directories to include in their PATH and similar environment variables
- What optional parts of the project to build

- What compiler flags to use (debugging build, optimized build, etc.)

The aspects of various systems that cause the most trouble when you're installing a new library are the environment variables (such as PATH) and their relationship to the locations of installed libraries. In particular, when this relationship is not precise and accurate, the build system can struggle to find and link dependencies.

Dependency configuration

When one piece of software depends on the functionality of another piece of software, the second is called a *dependency*. For example, if the SuperPhysics library relies on the EssentialPhysics library and the ExtraPhysics library, then they are its dependencies. Before attempting to install the SuperPhysics library, you must install EssentialPhysics and ExtraPhysics.

The build can fail in either of these cases:

- The build system cannot locate a dependency library.
- The available library is not the correct version.

The build system seeks the libraries listed in the PATH, LD_LIBRARY_PATH, and similar environment variables. Thus, the most common problems in building software arise when too many or not enough dependency libraries appear in the directories targetted by the environment.

When too many versions of the ExtraPhysics library are found, for example, the wrong version of the library might be linked and an error may occur. At the other extreme, if no EssentialPhysics library is found, the build will certainly fail. To fix these problems, be sure all dependencies are appropriately installed.

Once all dependencies, environment variables, user options, and other configurations are complete, a makefile or installation script is generated by the build system. The first action it conducts is the compilation step.

Compilation

Now that the makefile is configured, it can be used to compile the source code. The commands in the makefile for a software build will be mostly compiler commands. Without getting into too much detail, compilers are programs that turn source code into a machine-readable binary format.

The build system, by convention, likely generated a makefile with a default target designed to compile all of the source code into a local directory. So, with a simple make command, the compiled files are generated and typically saved (by the makefile) in a temporary directory as a test before actual installation. Additionally, once compiled, the build can usually be tested with `make test`.

If the tests pass, the build system can also assist with the next step: installation.

Installation

As mentioned in Chapter 13, the key to attracting users to your project is making it installable.

On Windows, this means creating a *Setup.exe* file. With Python, it means implementing a *setup.py* or other distribution utility. For other source code on Unix systems, this means generating a makefile with an `install` target so that `make install` can be called.

Why not just write a simple script to perform the installation?

The user may eventually want to upgrade or even uninstall your program, fantastic as it may be. By tradition, the installation program is usually created by the application developer, but the uninstall program is usually the responsibility of the operating system. On Windows, this is handled by the Add/Remove Programs tool. On Unix, this is the responsibility of the package manager. This means the installation program needs special platform-dependent capabilities, which are usually taken care of by the build system.

For example, on Linux, `make install` is not used when creating packages. Instead, `make DESTDIR=<a_fake_root_dir> install` installs the package to a fake root directory. Then, a package is created from the fake root directory, and uninstallation is possible because a manifest is generated from the result.

The build system will have created this target automatically. If the installation location chosen in the configuration step is a restricted directory, then you must execute the `make install` command with `sudo` in order to act as the superuser:

    sudo make install

At that point, the software should be successfully installed.

Building Software and Pipelines Wrap-up

At the end of this chapter, you should now feel comfortable with automating pipelines and building software using makefiles. You should also now be familiar with the steps involved in building a non-Python software library from source:

1. Configuration
2. Compilation

3. Linking
4. Installation

Additionally, with the knowledge of how these steps and the makefile relate to the platform, environment, user, and dependencies, you should feel prepared to understand a wide array of installation issues more fully.

If this chapter has succeeded in its purpose, you may be interested in researching automated build systems (e.g., CMake, autotools, and SCons) more fully in order to implement a build system for your own software project.

CHAPTER 15
Local Version Control

In science, reproducibility is paramount. A fundamental principle of science, reproducibility is the requirement that experimental results from independent laboratories should be commensurate. In scientific computation simulations, data munging and analysis pipelines are experimental analogs. To ensure that results are repeatable, it must be possible to unwind code and analysis to previous versions, and to replicate plots. The most essential requirement is that all previous versions of the code, data, and provenance metadata must be robustly and retrievably archived. The best practice in scientific computing is called version control.

Rather than inventing a system of indexed directories holding full versions of your code from each day in the lab, the best practice in software development is to use a version control system that automates archiving and retrieval of text documents such as source code.

This chapter will explain:

- What version control is
- How to use it for managing files on your computer
- And how to use it for managing files in a collaboration

First up, this chapter will discuss what version control is and how it fits into the reproducible workflow of an effective researcher in the physical sciences.

What Is Version Control?

Very briefly, version control is a way to:

- Back up changing files

- Store and access an annotated history
- And manage merging of changes between different change sets

There are many tools to automate version control. Wikipedia provides both a nice vocabulary list and a fairly complete table (*http://bit.ly/revcontrol*) of some popular version control systems and their equivalent commands.

Fundamentally, version control provides capabilities similar to those that a laboratory notebook historically has provided in the workflow of the experimental scientist. In this way, it can be considered a sort of laboratory notebook for scientists who use computation.

The Lab Notebook of Computational Physics

The Wu Experiment, one of the foundational experiments in nuclear physics, demonstrated a violation of the Law of Conservation of Parity. Dr. Chien Shiung Wu, a Chinese physicist at Columbia University, took great pains to construct a reproducible experiment. Toward this goal, she even moved her experimental setup to the National Bureau of Standards Headquarters in Maryland while her colleagues reproduced her work back at her home institution. This work led to a Nobel Prize for her theorist colleagues (and the inaugural Wolf Prize for Madame Wu).

A modern Dr. Wu (we will call her Dr. Nu) might simulate this physics experiment with software. Dr. Nu is a persistent researcher, so she works on this important software project day and night until it accurately represents the theory of her colleagues. When she finishes one day before dinnertime, she plots her results and is relieved to be ready to submit them to a journal in the morning.

An unconscionable number of months later, she receives the journal's review of her work. It is a glowing review, but asks that the results be presented (in all plots, equations, and analysis) with the inconvenient positive-current convention for charged particle currents.

Since so many months have passed since the article was first submitted, the code has changed significantly and plots are now rendered differently in preparation for a new journal submission. For an alarming percentage of physicists, this would be a minor disaster involving weeks of sorting through the files to recall the changes that have happened in the last year (Merali 2010). It might even be impossible to roll back the code to its previous state.

However, Dr. Nu breathes a sigh of relief. She has been using version control. With version control, she can execute a single command to examine the record of her actions over the last several months. Her version control system has kept, essentially, a laboratory notebook of her software development history. When satisfied with her understanding of the logs, she can execute another simple command to revert the code to the state it was in the night she made the journal-ready plots. Before afternoon tea, Dr. Nu makes the simple change of sign convention in the plot, reruns her plotting script, and submits the revisions. Once tea is over, she can bring the repository back up to date and get back to work, as if nothing had changed.

What would happen if you received a review asking for a convention change to results you completed a year ago? If that's a scenario that is simply too terrible to imagine, you are not alone. This chapter will explain how Dr. Nu reached this enviable position, so that you can do the same.

We will start by explaining the types of version control available to a scientist. Then, the rest of the chapter will explain version control concepts in the context of one version control system in particular.

Version Control Tool Types

Version control systems come in two fundamentally different categories. More modern version control systems are "distributed" rather than "centralized." Centralized systems designate a single (central) definitive location for a repository's source, while distributed version control systems treat all (distributed) locations as equals. Some common version control systems in each category include:

- Centralized
    - Concurrent Versions System (*cvs*)
    - Subversion (*svn*)
    - Perforce (*p4*)
- Distributed
    - Decentralized CVS (*dcvs*)
    - mercurial (*hg*)
    - bazaar (*bzr*)
    - Git (*git*)

Recently, distributed version control systems have become more popular. They are better suited to collaborative projects, since their capabilities for managing and merging together changes from multiple developers are more powerful and user-friendly.

Choosing the appropriate option from among these should depend largely on the expertise of your colleagues and their collaboration style. Due to its popularity, flexibility, and collaborative nature, we will demonstrate version control concepts using the Git tool. Git, written by Linux creator Linus Torvalds, is an example of a distributed version control system. It has a somewhat steep learning curve, so the sooner we can get started, the better.

Getting Started with Git

Git needs to be installed and configured before it can be used to control the versions of a set of files. When Dr. Nu was first getting started with her simulation, she knew she should keep versions of everything, just like in a laboratory notebook. So, she decided to install Git on the computer where she writes her code and conducts her analysis.

Installing Git

The first step for using Git is installing it. Dr. Nu wasn't sure if she already had Git installed on her computer, so to check she used the which command we met in Chapter 1.

> To determine whether Git is already installed and to find help using it, try executing which git in the terminal. Does it respond with a path to the program, or does it return nothing?

If which git returns no executable path, then Git is not installed. To install it, she'll need to:

1. Go to the Git website (*http://git-scm.com*).
2. Follow the instructions for her platform.

On some platforms, the default version of Git is quite old (developers call that *stable*). Unfortunately, it may not have all the features of an up-to-date version. So, even if Git is installed, you may consider updating it anyway.

Once Git has been installed (or updated), it can be used. But how?

Getting Help (git --help)

The first thing Dr. Nu likes to know about any tool is how to get help. From the command line, she types:

```
~ $ man git
```

The manual entry for the Git version control system appears before her, rendered in less. She may scroll through it using the arrow keys, or she can search for keywords by typing / followed by the search term. Dr. Nu is interested in help, so she types /help and then hits Enter.

By doing this, Dr. Nu finds that the syntax for getting help with Git is git --help.

Manual (man) pages are rendered in less. To exit the man page, therefore, type the letter q and hit Enter.

To try this help syntax, Dr. Nu exits the man page and tests what happens when she types:

```
~ $ git --help
```

Excellent! Git returns, to the terminal, a list of commands it is able to help with, as well as their descriptions:

```
usage: git [--version] [--exec-path[=<path>]] [--html-path]
           [-p|--paginate|--no-pager] [--no-replace-objects]
           [--bare] [--git-dir=<path>] [--work-tree=<path>]
           [-c name=value] [--help]
           <command> [<args>]

The most commonly used git commands are:
   add        Add file contents to the index
   bisect     Find by binary search the change that introduced a bug
   branch     List, create, or delete branches
   checkout   Checkout a branch or paths to the working tree
   clone      Clone a repository into a new directory
   commit     Record changes to the repository
   diff       Show changes between commits, commit and working tree, etc
   fetch      Download objects and refs from another repository
   grep       Print lines matching a pattern
   init       Create an empty git repository or reinitialize an existing one
   log        Show commit logs
   merge      Join two or more development histories together
   mv         Move or rename a file, a directory, or a symlink
   pull       Fetch from and merge with another repository or a local branch
   push       Update remote refs along with associated objects
   rebase     Forward-port local commits to the updated upstream head
   reset      Reset current HEAD to the specified state
   rm         Remove files from the working tree and from the index
   show       Show various types of objects
   status     Show the working tree status
   tag        Create, list, delete or verify a tag object signed with GPG

See 'git help <command>' for more information on a specific command.  ❶
```

❶ The help command even has a metacomment about how to get more specific information about a particular command. Based on this, how would you learn more about the git init command?

That's a lot of commands, but it isn't everything Git can do. This chapter will cover many of these and a few more. The first command we will show will complete the setup process: `git config`.

Control the Behavior of Git (git config)

To complete the setup process for Git, a configuration step is necessary. Knowing your name, email address, favorite text editor, and other data will help Git to behave optimally and to correctly provide attribution for the work that you do with your files.

Dr. Nu knows that version control provides an exceptional attribution service to scientists collaborating on code. When a change is made to code under version control, it must be attributed to the author. To ensure that she is appropriately attributed for her excellent work and held accountable for any code bugs, Dr. Nu configures her instance of Git thus:

```
~ $ git config --global user.name "Nouveau Nu"     ❶
~ $ git config --global user.email "nu@university.edu"  ❷
~ $ git config --global core.editor "nano"         ❸
```

❶ Changes made by Dr. Nu will be attributed to the name she provides to Git.

❷ Her university email address will also be stored with each logged change.

❸ Git can behave more optimally when it is aware of the author's preferred text editor. Later, we will see how.

Not only do scientists love to see their names in print next to a piece of excellent work, but authorship metadata is essential to provide provenance and answer the question, "Where did this work come from?" Indeed, attribution is central to the scientific process, since accountability is one of the fundamental motivators of scientific excellence.

> **Exercise: Configure Git on Your Computer**
>
> 1. Use the preceding example as a model to inform Git of *your* name, email address, and favorite text editor.
> 2. List your new configuration settings with `git config --list`.

Now that Git is set up on her system, Dr. Nu is able to use it to manage the versions of the files stored on her computer.

Local Version Control with Git

All researchers in scientific computing have at least one computer full of data, text files, images, plots, scripts, and other software. Those files often constitute the bulk of the day-to-day efforts of that researcher. Code and data are often created, manipulated, dealt with, and stored primarily on a single computer. Controlling versions of those files involves:

- Creating a repository where those changes are stored
- Adding files to that repository so that they can be tracked
- Taking snapshots of incremental versions, so that they are logged
- Undoing changes
- Redoing them
- Trying new ideas in separate sandboxes

We will talk about all of these tasks in this section. So first, we must learn how to create a repository.

Creating a Local Repository (git init)

Dr. Nu would like to write code that simulates Dr. Chien-Shiung Wu's landmark experiment. Writing this code may take years and involve the effort of many graduate students, and the code will undergo many iterations. To keep track of numerous versions of her work without saving numerous copies, Dr. Nu can make a local *repository* for it on her computer.

> A repository is where the tracked files live and are edited. For each version of those files, Git records the *change set*, or "diff"—the line-by-line differences between the new version and the one before it.

To begin keeping a record of files within a directory, Dr. Nu must enter that directory and execute the command `git init`. This creates an empty repository:

```
~ $ mkdir parity_code  ❶
~ $ cd parity_code  ❷
~/parity_code $ git init  ❸
Initialized empty Git repository in /filespace/people/n/nu/parity_code/.git/  ❹
```

❶ First, she creates the directory where she will do the work.

❷ She navigates to that directory.

❸ She initializes a repository within that directory.

❹ Git responds positively. An empty repository has been created here.

Because she is a scientist, Dr. Nu is curious about what happened. She can browse the directory's hidden files to see what happened here:

```
~/parity_code $ ls ❶
~/parity_code $ ls -A ❷
.git ❸
~/parity_code $ cd .git && ls- A ❹
HEAD        config      description hooks      info       objects     refs
```

❶ A simple listing of the directory contents results in nothing. The directory appears to be empty. Where is the repository?

❷ Curious, Dr. Nu lists all of the contents of the repository.

❸ A hidden directory, *.git*, is visible.

❹ Navigating into that directory and listing all of its contents reveals the mechanism by which the repository operates.

With ordinary use of Git, none of those hidden files will ever need to be altered. However, it's important to note that the infrastructure for the repository is contained within this hidden subdirectory (*.git*) at the top level of your repository.

> A whole repository directory can be moved from one location to another on a filesystem as long as the *.git* directory inside moves along with it.

This means that moving the entire repository directory to another location is irrelevant to the behavior of the repository. However, moving files or directories outside of the repository will move them outside of the space governed by the hidden infrastructure.

Exercise: Create a Local Repository

1. From the Terminal, create a new directory like Dr. Nu's and use `git init` to make it an empty local repository.
2. Browse the files in the hidden directory and find out what you can learn in one minute.

Now that a repository has been initialized, work can begin in this directory. As work progresses, files must first be added to the directory.

Staging Files (git add)

Now, Dr. Nu has created a repository directory to start working in. So, she gets to work creating a "readme" file. This is an important part of the documentation of any software; it indicates basic information about the project, as well as where to find more details.

First, she can create an empty file with the `touch` command:

```
~/parity_code $ touch readme.rst
```

Now the file exists in the directory. However, it is not yet being tracked by the repository. For the Git repository to know which files Dr. Nu would like to keep track of, she must add them to the list of files that the repository knows to watch. This is called "staging" the files. It is analogous to arranging people on a stage so that they are ready for a photo to be taken. In this case, we are staging the file so that it can be included in the upcoming snapshots.

Thus, to make Git aware of the file, she adds it to the repository with the `git add` command:

```
~/parity_code $ git add readme.rst
```

> **Exercise: Add a File to a Local Repository**
>
> 1. Create a readme file within your repository directory.
> 2. With `git add`, inform Git that you would like to keep track of future changes in this file.

Now that something has been added to the repository, the state of the repository has changed. Often, it is important to be able to check the state. For this, we will need the `git status` command.

Checking the Status of Your Local Copy (git status)

The files you've created on your machine are your local "working" copy. The repository, as we have already said, stores versions of the files that it is made aware of. To find out the status of those files, a `status` command is available:

```
~/parity_code $ git status   ❶
On branch master   ❷

Initial commit   ❸
```

Local Version Control with Git | 357

```
Changes to be committed:
  (use "git rm --cached <file>..." to unstage)  ❹

        new file:   readme.rst  ❺
```

❶ Check the status of the repository in the current directory.

❷ Git has something called "branches." We are on the "master" branch by default. We will talk more about branches later in the chapter.

❸ Git knows that we have not yet used the `commit` command in this repository.

❹ Git gives us a hint for what to do if we did not really intend to add the readme file. We can "unstage" it with `git rm`!

❺ Git reports that there is a new file "to be committed." We will discuss committing in the next section.

This result indicates the current difference between the repository records (which, so far, are empty) and the *parity_code* directory contents. In this case, the difference is the existence of this new *readme.rst* file. Git suggests that these changes are "to be committed." This means that now that the file has been added to the watch list, the scene is set and the repository is ready for a snapshot to be taken. We save snapshots with the `git commit` command.

Saving a Snapshot (git commit)

In order to save a snapshot of the current state of the repository, we use the `commit` command. This command:

1. Saves the snapshot, officially called a "revision"
2. Gives that snapshot a unique ID number (a revision hash)
3. Names you as the author of the changes
4. Allows you, the author, to add a message

The `git commit` command conducts the first three of these tasks automatically. However, the fourth requires input from the author. When executing the `git commit` command, the author must provide a "commit message" describing the changes represented by this commit and indicating their purpose. Informative commit messages will serve you well someday, so make a habit of never committing changes without at least a full sentence description.

> Log messages have a lot of power when used well. To this end, some open source projects even suggest information to include in all commit messages. This helps the developers to more systematically review the history of the repository. For example, the Pandas project uses various three-letter keys to indicate the type of commit, such as (ENH)ancement, BUG, and (DOC)umentation.

In the same way that it is wise to often save a document that you are working on, so too is it wise to save numerous revisions of your code. More frequent commits increase the granularity of your "undo" button.

> Commit often. Good commits are *atomic*, the smallest change that remains meaningful. They should not represent more work than you are willing to lose.

To commit her work, Dr. Nu simply types `git commit` into the command line. Git responds by opening up an instance of the *nano* text editor, where she can add text to a document recording the change. She does so by adding a message: "This is my first commit. I have added a readme file." When she saves the file, the commit is complete.

> **Exercise: Commit Your Changes**
>
> 1. Use `git commit` to save the staged changes of the file you've added to your repository.
> 2. Git will send you to your preferred text editor. There, create a message, then save and exit.
> 3. Admire your work with the `git status` command. You should see something like:
>
>    ```
>    $ git status
>    # On branch master
>    nothing to commit (working directory clean)
>    ```
>
>   If instead you receive a warning indicating that Git is not configured, you will need to return to "Control the Behavior of Git (git config)" on page 354 and configure it.

Now, as she makes changes to the files in her repository, Dr. Nu can continue to commit snapshots as often as she likes. This will create a detailed history of her efforts.

Local Version Control with Git | 359

Exercise: Stage and Commit New Changes

1. Edit your readme file. It should say something like:

   ```
   Welcome

   This is my readme file for a project on parity
   violations in the standard model.
   ```

1. Stage it for the snapshot (`git add`).
2. Commit the snapshot (`git commit`).
3. Add a meaningful commit message.

So far, we've learned that the workflow should be :

1. Make changes.
2. `git add` the files you want to stage for a commit.
3. `git commit` those files.
4. Fill out the log message.
5. Repeat.

Since that is a lot of steps, note that command-line flags can cut this way down. Some useful flags for `git commit` include:

```
-m: add a commit message from the command line
-a: automatically stage tracked files that have been modified or deleted
-F: add a commit message from a file
--status: include the output of git status in the commit message
--amend: fix the commit message at the repository tip
```

These can be used in combination to reduce the add/commit/message process to one command: `git commit -am "<useful log message goes here>"`.

Exercise: Commit and Add a Message in One Step

1. Edit your readme file to tell us whose it is (e.g., "This is Dr. Nu's readme...").
2. Add the file, commit the changes, and append your log message with one command.

Whichever method you use to write commit messages, be sure to make those messages useful. That is, write commit messages as if they were the scientific notation

360 | Chapter 15: Local Version Control

that they are. Like section headings on the pages of your lab notebook, these messages should each be one or two sentences explaining the change represented by a commit.

> The frequency of commits is the resolution of your undo button. Committing frequently makes merging contributions and reversing changes less painful.

Now we have successfully taken a snapshot representing an incremental change in the repository and provided a message to go along with it. The fundamental innovation in version control is that a record of that work has now been kept. To view that record, we use the log command.

git log: Viewing the History

A log of the commit messages is kept by the repository and can be reviewed with the log command:

```
~/parity_code $ git log  ❶
commit cf2631a412f30138f66d75c2aec555bb00387af5  ❷
Author: Nouveau Nu <nu@university.edu>  ❸
Date:   Fri Jun 21 18:21:35 2013 -0500  ❹

    I have added details to the readme to describe the parity violation project.  ❺

commit e853a4ff6d450df7ce3279098cd300a45ca895c1
Author: Nouveau Nu <nu@university.edu>
Date:   Fri Jun 21 18:19:38 2013 -0500

    This is my first commit. I have added a readme file.
```

❶ The log command prints the logged metadata for each commit.

❷ Each commit possesses a unique (hashed) identification number that can be used to refer to that commit.

❸ The metadata from the configuration step is preserved along with the commit.

❹ Git automatically records the date and time at which the commit occurred.

❺ Finally, the log message for each commit is printed along with that commit.

As more changes are made to the files, more snapshots can be committed, and the log will reflect each of those commits. After making a number of commits, Dr. Nu can review the summary of her work in the form of these commit messages.

When she wants to review her work in more detail than the commit messages can offer, she may want to review the actual *changes* that were made between certain commits. Such differences between file versions are called "diffs," and the tools that display them are diff tools.

Viewing the Differences (git diff)

Let's recall the behavior of the `diff` command on the command line. Choosing two files that are similar, the command:

```
diff <file1> <file2>
```

will output the lines that differ between the two files. This information can be saved as what's known as a *patch*, but we won't go deeply into that just now. Suffice it to say that there are many diff tools. Git, however, comes with its own diff system.

The only difference between the command-line diff tool and Git's diff tool is that the Git tool is aware of all of the revisions in your repository, allowing each revision of each file to be treated as a full file.

Thus, `git diff` will output the changes in your working directory that are not yet staged for a commit. If Dr. Nu adds a definition of parity to her readme file, but does not yet commit it, those changes are *staged*. When she asks Git for the diff, the following occurs:

```
~/parity_code $ git diff ❶
diff --git a/readme.rst b/readme.rst ❷
index 28025a7..a5be27f 100644 ❸
--- a/readme.rst
+++ b/readme.rst
@@ -2,3 +2,5 @@ Welcome

 This is my readme file for a project on parity violations in the standard
 model.
+
+In the context of quantum physics, parity is a type of symmetric relation.
```

❶ Dr. Nu executes the `git diff` command.

❷ `diff` reports that the differences *not staged for commit* exist only in the readme file versions *a* and *b*.

❸ And indicates that these versions are those from the commits ending in `28025a7` and `a5be27f`, respectively.

To see how this works, make a change in your *readme.rst* file, but don't yet commit it. Then, try `git diff`.

A summarized version of this output can be seen with the `--stat` flag:

```
~/parity_code $ git diff --stat
readme.rst | 2 ++
1 file changed, 2 insertions(+)
```
❶

❶ For each line where one or more characters has been added, an insertion is counted. If characters are deleted, this is called a deletion.

To see only what is staged for commit, you can try:

```
$ git diff --cached
```

What is the difference shown in the cached diff? What does this mean about what files are staged?

Sometimes what you have staged is not what you actually want to commit. In the same way, sometimes after reviewing a change that she has already committed, Dr. Nu thinks better of it and would prefer to roll a file back to an earlier version. In both of those instances, the `git reset` command can be used.

Unstaging or Reverting a File (git reset)

If, after reviewing the log, Dr. Nu decides that she prefers a past version of some file to the previous revision, she can use `git reset`. This command can be used either to unstage a staged file or to roll back a file or files to a previous revision.

If you added a file to the staging area that you didn't mean to add, you can use `reset` to "unstage" it (i.e., take it out of the staged set of commits):

```
git reset <filename>
```

In this case, `reset` acts like the opposite of `add`. However, `reset` has another use as well. If you want to return the repository to a previous version, you can use `reset` for that too. Just use the commit number:

```
git reset [<mode>] [<commit>]
```

`reset` has some useful mode flags:

`--soft`
    Leaves the contents of your files and repository index alone, but resets repository head

`--mixed`
    Resets the index and repository head, but not the contents of your files

`--hard`
    Returns the contents of all files and the repository index to the commit specified

Using `reset`, you can therefore undo changes that have already been committed. For changes that have not yet been committed, you can use `git checkout`. This unstages modifications:

```
git checkout -- <filename>
```

Note that `git checkout` has other purposes, which we'll see soon.

> **Exercise: Discard Modifications**
>
> 1. Create five files in your directory, with one line of content in each file.
> 2. Commit the files to the repository.
> 3. Change two of the five files and commit them.
> 4. Undo the changes in step 3.
> 5. Print out the last entry in the log.

Using `reset` or `checkout`, however, does not delete the commits permanently. The record of those commits is still stored in the repository, and they can be accessed with their commit revision hash numbers via `git checkout`. A more permanent option is `git revert`.

Discard Revisions (git revert)

Much like `git reset --hard`, but with more permanence, `git revert` is a helpful tool when you *really* want to erase history—for example, if you've accidentally committed something with private or proprietary information. The syntax for `git revert` is:

```
git revert <commit>
```

While she was working on her readme file, Dr. Nu decided to add contact information for herself. In doing so, she committed the following change to the readme:

```
diff --git a/readme.rst b/readme.rst
index a5be27f..0a07497 100644
--- a/readme.rst
+++ b/readme.rst
@@ -1,5 +1,8 @@
 Welcome

+To contact Dr. Nouveau Nu, please send an email to nu@university.edu or call
+her cell phone at 837-5309.
+
 This is my readme file for a project on parity violations in the standard
 model.
```

A few seconds after committing this change, she regretted making her cell phone number available. She could edit the readme to remove the number. However, even after she commits that change, her number can still be accessed with `git checkout`. To remove the record entirely, she must use the `revert` command. First, she needs to know what commit number to revert, so she uses the `log` command:

```
~/parity_code $ git log
commit fc06a890ecba5d16390a6fb4514cb5ba45546952  ❶
Author: Nouveau Nu <nu@university.edu>
Date:   Wed Dec 10 14:00:26 2014 -0800

    Added my email address and phone number
...
```

❶ Here, Dr. Nu finds the hash number to use.

She can use the whole hash number or as few as the first eight characters to uniquely identify the commit she would like to revert:

```
~/parity_code $ git revert fc06a890
[master 2a5b0e1] Revert "Added my email address and phone number"
 1 file changed, 3 deletions(-)
```

Now it has been completely removed from the history; she can breathe a sigh of relief and move on. Now Dr. Nu would like to start programming in seriousness. However, she is concerned about something. Since science involves trying things out and making mistakes, will she have to spend a lot of her time rewinding changes like this to remove them from the master branch? What if she wants to try two new things at once? In the next section, we will see that the answer to both of these questions is "Using branches will make everything easier."

Listing, Creating, and Deleting Branches (git branch)

Branches are parallel instances of a repository that can be edited and version controlled in parallel. They are useful for pursuing various implementations experimentally or maintaining a stable core while developing separate sections of a code base.

Without an argument, the `git branch` command lists the branches that exist in your repository:

```
~/parity_code $ git branch
* master
```

The "master" branch is created when the repository is initialized. This is the default branch and is conventionally used to store a clean master version of the source code.

With an argument, the `branch` command creates a new branch with the given name. Dr. Nu would like to start a branch to hold some experimental code—that is, some code that she is just trying out:

```
~/parity_code $ git branch experimetal  ❶
~/parity_code $ git branch  ❷
  experimetal
* master  ❸
```

❶ She creates a branch called "experimetal."

❷ To check that this worked, she lists the branches using the `branch` command.

❸ The asterisk indicates which branch she is *currently* in. We'll demonstrate how to change branches shortly.

Whoops—Dr. Nu forgot to type the *n* in experimental. Simple typos like this happen all the time in programming, but they are nothing to fear. Very few typos will break everything. In this case, deleting the branch and trying again is very simple. To delete a branch, she can use the `-d` flag:

```
~/parity_code $ git branch -d experimetal  ❶
Deleted branch experimetal (was 2a5b0e1).  ❷
~/parity_code $ git branch  ❸
* master  ❹
~/parity_code $ git branch experimental
~/parity_code $ git branch
  experimental
* master
```

❶ She deletes the misspelled branch.

❷ Git responds that, yes, it has been deleted (and provides the hash number for the `HEAD`, or most recent commit, of that branch).

❸ Just to double-check, she can list the branches again.

❹ And *voilà*, it's gone. She can try again—this time without the typo.

At this point, Dr. Nu has created the "experimental" branch. However, she is still currently working in the master branch. To tell Git that she would like to work in the experimental branch, she must switch over to it with the `checkout` command.

Switching Between Branches (git checkout)

The `git checkout` command allows context switching between branches as well as abandoning local changes and viewing previous commits.

To switch between branches, Dr. Nu can "check out" that branch:

```
~/parity_code $ git checkout experimental
Switched to branch 'experimental'  ❶
~/parity_code $ git branch
```

```
* experimental
  master
```

❶ Git is actually very good at keeping the user well informed. It can be very reassuring.

How can you tell when you've switched between branches? When we used the `branch` command before, there was an asterisk next to the master branch. Now it's next to the experimental branch—the asterisk indicates which branch you're currently in.

Now, Dr. Nu can safely work on code in the experimental branch. When she makes commits, they are saved in the history of the experimental branch, but are not saved in the history of the master branch. If the idea is a dead end, she can delete the experimental branch without polluting the history of the master branch. If the idea is good, however, she can decide that the commit history made in the experimental branch *should* be incorporated into the master branch. For this, she will use a command called `merge`.

Merging Branches (git merge)

At some point, the experimental branch may be ready to become part of the master. The method for combining the changes in two parallel branches is the `merge` command. To merge the changes from the experimental branch into the master, Dr. Nu executes the `merge` command from the master branch:

```
~/parity_code $ git checkout master
~/parity_code $ git merge experimental
```

Now, the logs in the master branch should include all commits from each branch. Give it a try yourself with the following long exercise.

Exercise: Create Two New Branches

1. Create a new repository and commit an empty readme file:

   ```
   ~ $ mkdir geography
   ~ $ cd geography
   ~/geography $ git init
   ~/geography $ touch readme.rst
   ~/geography $ git add readme.rst
   ~/geography $ git commit -am "first commit"
   ```

2. Create two new branches and list them:

   ```
   ~/geography $ git branch us
   ~/geography $ git branch texas
   ```

3. Add files describing each entity. In the "us" branch, include at least a file called *president*. For "texas," of course, you'll need a file called *governor*. You'll probably also want one called *flower*:

   ```
   ~/geography $ git checkout us
   Switched to branch 'us'
   ~/geography $ touch president
   ~/geography $ git add president
   ~/geography $ git commit -am "Added president to the us branch."
   ~/geography $ git checkout texas
   Switched to branch 'texas'
   ~/geography $ touch flower
   ~/geography $ git add flower
   ~/geography $ git commit -am "Added bluebonnets to the
                   texas branch."
   ```

4. Merge the two branches into the master branch:

   ```
   ~/geography $ git checkout master
   Switched to branch 'master'
   ~/geography $ git merge texas
   Updating d09dfb9..8ce09f1
   Fast-forward
    flower | 0
    1 file changed, 0 insertions(+), 0 deletions(-)
    create mode 100644 flower
   ~/geography $ git merge us
   Merge made by the 'recursive' strategy.
    president | 0
    1 file changed, 0 insertions(+), 0 deletions(-)
    create mode 100644 president
   ```

The ability to automatically merge commits is powerful and quite superior to having multiple versions of directories cluttering your filesystem. That said, the `merge` command is only capable of combining changes that do not conflict with one another. In the next section, you'll get a taste of this problem.

Dealing with Conflicts

Both Texas and the United States have a national anthem. However, we notice that the national anthem isn't there, so we add a file called *national_anthem* to the "us" branch:

```
~/geography $ git checkout us
~/geography $ echo "Star-Spangled Banner" > national_anthem
~/geography $ git add national_anthem
~/geography $ git commit -am "Added star spangled banner to the us branch."
```

Next, of course, we put on our Wranglers and Stetsons and do the same for the "Texas" branch, which does not yet have a national anthem file.

```
~/geography/$ git checkout texas
~/geography/$ echo "Texas, Our Texas" > national_anthem
~/geography/$ git add national_anthem
~/geography/$ git commit -am "Added Texas, Our Texas to the texas branch."
```

If we merge them into one another or into the master branch, what happens?

What happens is a conflict. This is a common issue when two different people are working independently on different branches of a respository and try to merge them. Since that is the context in which conflicts are most commonly encountered, the explanation of how to deal with conflicts will be addressed in the next chapter. For now. abort the merge with `git merge --abort`.

Version Conrol Wrap-Up

In this chapter, we have shown how to use git for recording versions of files, rewinding changes, and merging independent changes. These are the first steps toward reproducible scientific computation. Having read this chapter, you are now prepared to go forth and version control the work you do day to day. In fact, take a moment now to place those analysis scripts you are working on under version control. Since Git is an expansive and complex tool, you may find a need for additional resources. We can recommend, in particular:

- Pro Git Book (*http://git-scm.com/book/en/v2*)
- Software Carpentry's Git Lessons (*http://bit.ly/sc-git-lesson*)
- The Software Carpentry quick reference (*http://bit.ly/quick-git-ref*)

Now that you are comfortable with managing versions of files and source code locally, you can move forward and program reproducibly. Next, you will need to know how to harness the power of Git for collaboration. The next chapter will cover the use of Git in combination with the immense power of the Internet.

CHAPTER 16
Remote Version Control

Now that you have learned how to version files locally with Git, you are ready to revolutionize the way you collaborate on software, papers, data, and everything else. This chapter will cover the immense power of Git when it is combined with the broad reach of the Internet.

Chapter 15 described tasks related to the local working copy of your repository. However, the changes you make in this local copy aren't backed up online automatically. Until you send those changes to the Internet, the changes you make are local changes. This chapter will discuss syncing your local working copy with remote copies online and on other computers on your network. In particular, this chapter will explain how to use Git and the Internet for:

- Backing up your code online
- Forking remote repositories to enable collaboration
- Managing files in a collaboration
- Merging simultaneous changes
- Downloading open source code to keep track of updates

First among these, this chapter will cover backing up code online.

Repository Hosting (github.com)

Repositories can be stored and accessed through repository hosting servers online. Many people store their source code repositories on common repository hosting services such as:

- Launchpad (*https://launchpad.net*)

- Bitbucket (*https://bitbucket.org*)
- Google Code (*https://code.google.com*)
- SourceForge (*http://sourceforge.net*)
- GitHub (*https://github.com*)

This chapter will use GitHub as an example. It provides tools for browsing, collaborating on, and documenting code. These include:

- Landing page support
- Wiki support
- Network graphs and time histories of commits
- Code browser with syntax highlighting
- Issue (ticket) tracking
- User downloads
- Varying permissions for various groups of users
- Commit-triggered mailing lists
- Other service hooks (e.g., Twitter)

These services allow anyone with a repository to back up their work online and optionally share it with others. They can choose for it to be either open source or private. Your home institution may have a repository hosting system of its own. To find out, ask your system administrator.

> Setting up a repository on GitHub requires a GitHub username and password. Please take a moment to create a free GitHub account (*https://github.com/signup/free*).
>
> Additionally, you may find it helpful to set up SSH keys for automatic authentication (*http://bit.ly/genkeys*).

Dr. Nu can use GitHub as a way to back up her parity work, share it with her graduate students, and demonstrate its fidelity to paper reviewers. Since her *parity_code* simulation software from the previous chapter already exists, she can upload it to GitHub in four simple steps:

1. Create a user account on GitHub.
2. Create a space for her repository on GitHub.
3. Point to that remote from the local copy.
4. Push her repository to that location.

The first two of these steps occur within the GitHub interface online.

Creating a Repository on GitHub

Setting up a user account creates a space for a user like Dr. Nu to collect all of the repositories she uses. Creating a repository names location within that space for a certain piece of software.

When she creates a username (*NouveauNu*) on GitHub, a location on its servers is reserved for her at *github.com/NouveauNu*. If she navigates in her browser to that location, she can click a big green button that says "New Repository." She can supply the repository name *parity_code*, and GitHub will respond by creating an empty space at *github.com/NouveauNu/parity_code*.

This location is called a "remote" location because it is distant from the local working copy. Now that the repository location has been created, Git can be used to send commits from the local copy to the remote. For this, Dr. Nu needs to alert the local copy to the existence of the remote.

Declaring a Remote (git remote)

Remote repositories are just like local repositories, except they are stored online. To synchronize changes between the local repository and a remote repository, the location of the remote must be registered with the local repository.

The `git remote` command allows the user to register remote repository URLs under shorter aliases. In particular, this command can be used to add, name, rename, list, and delete remote repository aliases. The original remote repository, with which a local copy is meant to synchronize, is called "origin" by convention. In our example, this repository is the one where Dr. Nu holds the master copy of *parity_code*. So, from Dr. Nu's local working copy of the *parity_code* repository, she creates an alias to the remote thusly:

```
$ git remote add origin https://github.com/NouveauNu/parity_code.git ❶
```

❶ `git remote` command declares that Git should register something about a remote repository. `add` declares that a remote repository alias should be added. Dr. Nu chooses to use the conventional alias for this repository, *origin*. She then associates the alias with the online location of the repository. This URL can be copied from the GitHub page holding this repository.

Once she has executed this command, her local repository is now synced with the one online. She is then capable of sending and receiving commits from the remote repository. She can see a list of remotes registered with the `remote` command and its "verbose" flag:

```
~/parity_code $ git remote -v ❶
origin   https://github.com/NouveauNu/parity_code.git (fetch)
origin   https://github.com/NouveauNu/parity_code.git (push) ❷
```

❶ The -v flag is common and means "verbose." In this case, it means "verbosely list the remotes."

❷ The *origin* alias is associated with the URL Dr. Nu provided. The meanings of fetch and push will be covered very shortly.

She can now use this remote alias to "push" a full copy of the current status of the *parity_code* repository onto the Internet.

Sending Commits to Remote Repositories (git push)

The git push command pushes commits in a local working copy to a remote repository. The syntax is:

```
git push [options] <remote> <branch>
```

To push a copy of her *parity_code* repository up to the Internet, Dr. Nu can therefore execute the command:

```
~/parity_code (master) $ git push origin master ❶
Username for 'https://github.com': nouveaunu ❷
Password for 'https://nouveaunu@github.com': ❸
Counting objects: 22, done.
Delta compression using up to 4 threads.
Compressing objects: 100% (15/15), done.
Writing objects: 100% (22/22), 2.19 KiB | 0 bytes/s, done.
Total 22 (delta 3), reused 0 (delta 0)
To https://github.com/NouveauNu/parity_code
 * [new branch]      master -> master ❹
```

❶ Dr. Nu pushes the current (master) branch up to the *origin* remote.

❷ GitHub requires a username.

❸ And a password.

❹ The master branch has been pushed online.

This sends the full history of the local master branch up to the "origin" remote on GitHub, as seen in Figure 16-1. For security, the GitHub servers ask Dr. Nu for her username and password before the push is accepted. Only users with the appropriate permissions can push changes to this repository. In this case, Dr. Nu is the only user with permission to do so.

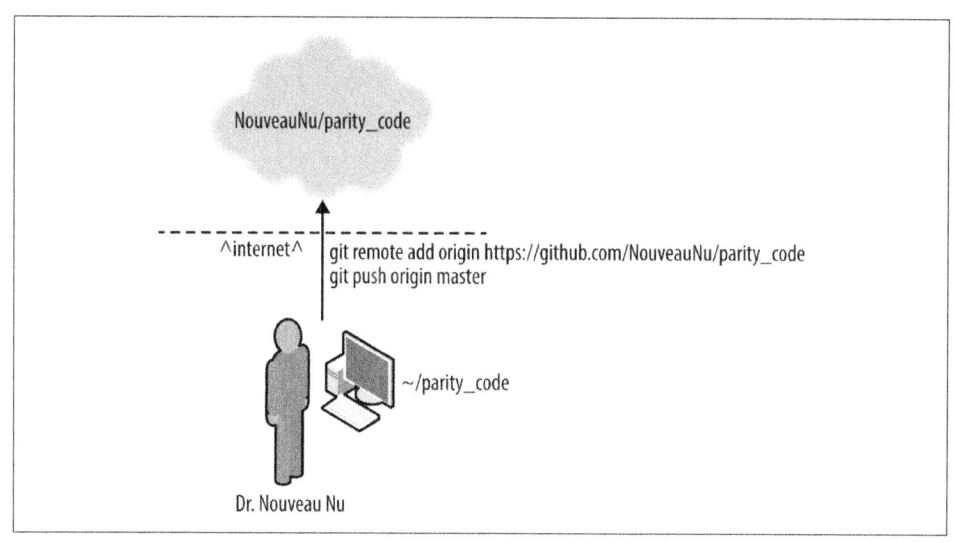

Figure 16-1. Pushing to a remote

To access the files that are now online, Dr. Nu can navigate in a browser to the location of that repository online. Indeed, so can her collaborators. This is where the magic begins. Since Dr. Nu has collaborators at other universities who rely on her software to do their analysis, GitHub can be very helpful for sharing that software with them. In particular, Fran Faraway, a postdoc on another continent, can now keep up to date without any emails or phone calls to Dr. Nu. Now that Dr. Nu's code is online, Fran can use Git to download it from that location on GitHub using the `clone` command.

Downloading a Repository (git clone)

Like Dr. Nu's parity code, many useful open source scientific software libraries are kept in repositories online. With the help of Git, scientists relying on these scientific libraries can acquire up-to-date source code for their use and modification. The best way to download such a repository is to *clone* it, as illustrated in Figure 16-2.

When a repository is cloned, a local copy is created on the local computer. It will behave as a fully fledged local repository where local branches can be created, edits can be made, and changes can be committed. To clone the *parity_code* repository, Fran can use the syntax:

```
~/useful_software $ git clone https://github.com/NouveauNu/parity_code.git
```

This command downloads the online repository into a directory called *parity_code*.

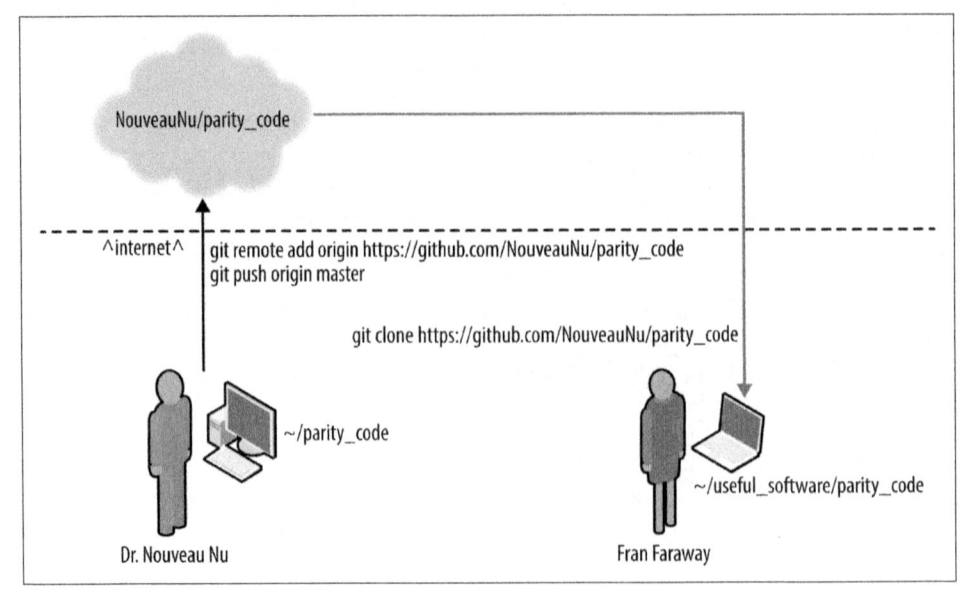

Figure 16-2. Cloning a repository

Exercise: Clone a Repository from GitHub

1. Pick any repository you like. There are many cool projects hosted on GitHub. Take a few minutes here to browse GitHub and pick a piece of code of interest to you.

2. Clone it. If you didn't find anything cool, we can suggest cloning the AstroPy libraries:

    ```
    ~ $ git clone https://github.com/astropy/astropy.git
    Cloning into astropy...
    remote: Counting objects: 24, done.
    remote: Compressing objects: 100% (21/21), done.
    remote: Total 24 (delta 7), reused 17 (delta 1)
    Receiving objects: 100% (24/24), 74.36 KiB, done.
    Resolving deltas: 100% (7/7), done.
    ```

3. You should see many files download themselves onto your machine. These files will have been placed in a directory with the name of the repository. Let's make sure it worked. Change directories, and list the contents:

    ```
    ~/ $ cd astropy
    ~/ $ ls
    ```

Now that she has cloned Dr. Nu's repository, Fran has a full copy of its history. Fran can even edit the source code for her own purposes and push that to her own repository online without affecting Dr. Nu whatsoever. An example of this process is shown in Figure 16-3.

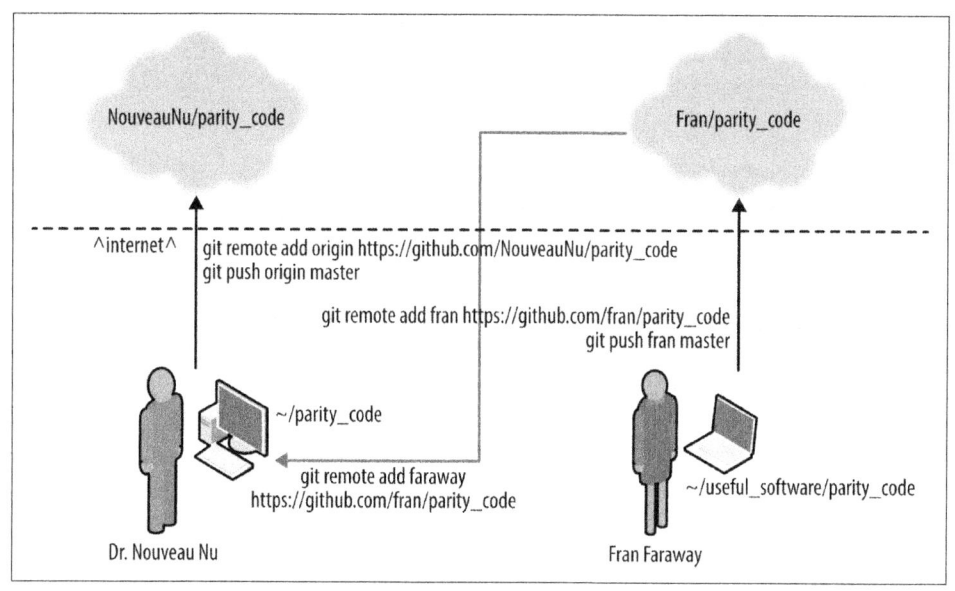

Figure 16-3. Creating new remotes

In Figure 16-3, Fran has pushed her own changes up to her own GitHub repository. Sometimes, when it is initialized in a particular way, this is called a "fork" of the original repository. Forks are a GitHub notion, rather than a Git notion. Basically, they are mutually aware remote repositories. To create them, simply locate the Fork button at the top-righthand corner of any repository on GitHub. That creates a new repository in your user space from which you can clone your work. A project managed in this way between the Curies might have the structure demonstrated in Figure 16-4.

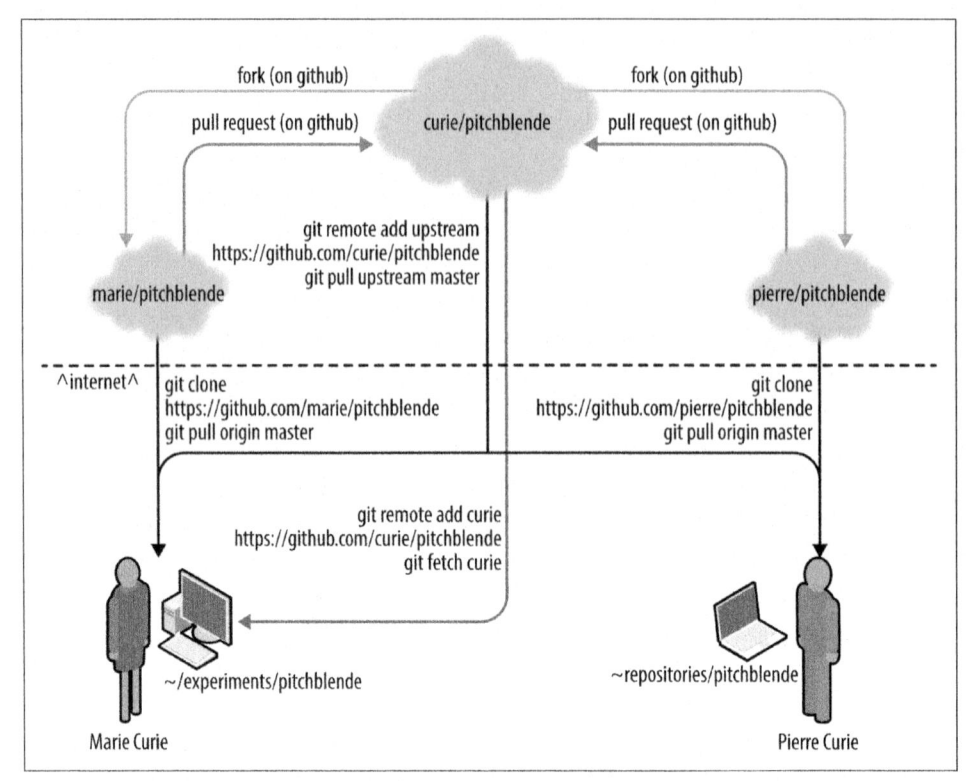

Figure 16-4. Forks in the Curie Pitchblende collaboration

This distributed system of remotes scales nicely for larger collaborations.

Since Dr. Nu's code is under active development, Fran must update her own local repository when Dr. Nu makes improvements to the code. By default, Fran's cloned repository is configured to be easily updated: the *origin* remote alias is registered by default and points to the cloned URL. To download and incorporate new changes from this remote repository (*origin*), Fran will require the `git fetch` and `git merge` commands.

Exercise: Fork the GitHub Repository

While you may already have a copy of this repository, GitHub doesn't know about it until you've made a fork. You'll need to tell GitHub you want to have an official fork of this repository.

1. Go to *github.com/nouveaunu/parity_code* in your Internet browser, and click on the Fork button.

2. Clone it. From your terminal:

   ```
   $ git clone https://github.com/<you>/parity_code.git ❶
   $ cd parity_code
   ```

   ❶ In the place of <you>, put your actual GitHub username.

3. Now, create an alias for the remote repository:

   ```
   $ git remote add nu \
           https://github.com/nouveaunu/parity_code.git ❶
   $ git remote -v ❷
   origin   https://github.com/YOU/parity_code (fetch) ❸
   origin   https://github.com/YOU/parity_code (push)
   nu   https://github.com/nouveaunu/parity_code (fetch)
   nu   https://github.com/nouveaunu/parity_code (push)
   ```

   ❶ Create a remote alias called *nu* that points at the original repository.

   ❷ List the remotes to see the effect.

   ❸ The *origin* remote is set by default during the cloning step.

Fetching the Contents of a Remote (git fetch)

Since the cloned repository has a remote that points to Dr. Nu's online repository, Git is able to fetch information from that remote. Namely, the `git fetch` command can retrieve new commits from the online repository. In this case, if Fran wants to retrieve changes made to the original repository, she can `git fetch` updates with the command:

```
~/useful_software/parity_code $ git fetch origin
```

The `fetch` command merely pulls down information about recent changes from the original master (*origin*) repository. By itself, the `fetch` command does not change Fran's local working copy. To actually merge these changes into her local working copy, she needs to use the `git merge` command

Merging the Contents of a Remote (git merge)

To incorporate upstream changes from the original master repository (in this case, *NouveauNu/parity_code*) into her local working copy, Fran must both fetch and merge. If Fran has made many local changes and commits, the process of merging may result in conflicts, so she must pay close attention to any error messages. This is where version control is very powerful, but can also be complex.

> **Exercise: Fetch and Merge the Contents of a GitHub Repository**
>
> 1. In the repository you cloned, fetch the recent remote repository history:
>
>    ```
>    $ git fetch origin
>    ```
> 2. Merge the *origin* master branch into your master branch:
>
>    ```
>    $ git merge origin/master
>    ```
> 3. Find out what happened by browsing the directory.

This process of fetching and merging should be undertaken any time a repository needs to be brought up to date with a remote. For brevity, both of these steps can be achieved at once with the command `git pull`.

Pull = Fetch and Merge (git pull)

The `git pull` command is equivalent to executing `git fetch` followed by `git merge`. Though it is not recommended for cases in which there are many branches to consider, the `pull` command is shorter and simpler than fetching and merging as it automates the branch matching. Specifically, to perform the same task as we did in the previous exercise, the `pull` command would be:

```
$ git pull origin master
Already up-to-date.
```

When there have been remote changes, the pull will apply those changes to your local branch. It may require conflict resolution if there are conflicts with your local changes.

When Dr. Nu makes changes to her local repository and pushes them online, Fran must update her local copy. She should do this especially if she intends to contribute back to the upstream repository and particularly before making or committing any changes. This will ensure Fran is working with the most up-to-date version of the repository:

```
~/useful_software/parity_code $ git pull
Already up-to-date.
```

The "Already up-to-date" response indicates that no new changes need to be added. That is, there have not been any commits to the original repository (*origin*) since the most recent update.

Conflicts

If Dr. Nu and Fran Faraway make changes in different files or on different lines of the same file, Git can merge these changes automatically. However, if for some reason they make different changes on the same line of a certain file, it is not possible for the `merge` (or `pull`) command to proceed automatically. A conflict error message will appear when Fran tries to merge in the changes.

This is the trickiest part of version control, so let's take it very carefully.

In the *parity_code* repository, you'll find a file called *readme.rst*. This is a standard documentation file that appears rendered on the landing page for the repository in GitHub. To see the rendered version, visit your fork on GitHub. The first line of this file is "Welcome."

For illustration, let's imagine that both Dr. Nu and Fran Faraway suddenly decide they would like to welcome visitors in the tongue of their home nation. Since these two collaborators are from two different places, there will certainly be disagreements about what to say instead of "Welcome." This will cause a conflict.

First, since Dr. Nu is French, she alters, commits, and pushes the file such that it reads:

```
Bonjour

This is my readme file for a project on parity violations in the standard
model.

In the context of quantum physics, parity is a type of symmetric relation.
```

Fran, however, is from Texas, so she commits her own version of "Welcome."

```
Howdy

This is my readme file for a project on parity violations in the standard
model.

In the context of quantum physics, parity is a type of symmetric relation.
```

Before pushing her change to her own remote, Fran updates her repository to include any changes made by Dr. Nu. The result is a conflict:

```
~/useful_software/parity_code $ git merge origin
Auto-merging readme.rst
CONFLICT (content): Merge conflict in readme.rst
Automatic merge failed; fix conflicts and then commit the result.
```

Since the two branches have been edited on the same line, Git does not have an algorithm to merge the changes correctly.

Resolving Conflicts

Now what?

Git has paused the merge. Fran can see this with the `git status` command:

```
$ git status
# On branch master
# Unmerged paths:
#   (use "git add/rm <file>..." as appropriate to mark resolution)
#
#       unmerged:    readme.rst
#
no changes added to commit (use "git add" and/or "git commit -a")
```

The only thing that has changed is the *readme.rst* file. Opening it, Fran sees something like this:

```
<<<<<<< HEAD  ❶
Howdy  ❷
=======  ❸
Bonjour  ❹
>>>>>>> master  ❺

This is my readme file for a project on parity violations in the standard
model.

In the context of quantum physics, parity is a type of symmetric relation.
```

❶ Git has added this line to mark where the conflict begins. It should be deleted before a resolution commit.

❷ The change that Fran committed.

❸ Git has added this line to mark the separation between the two conflicting versions. It should be deleted before a resolution commit.

❹ The change that Dr. Nu committed.

❺ Git has added this line to mark where the conflict ends. It should be deleted before a resolution commit.

The intent is for Fran to edit the file intelligently and commit the result. Any changes that Fran commits at this point will be accepted as a resolution to this conflict.

Fran knows now that Dr. Nu wanted the "Welcome" to say "Bonjour." However, Fran also wants it to say "Howdy," so she should come up with a compromise. First, she should delete the marker lines. Since she wants to be inclusive, she then decides to change the line to include both greetings. Decisions such as this one must be made by a human, and are why conflict resolution is not handled more automatically by the version control system:

```
Howdy and Bonjour

This is my readme file for a project on parity violations in the standard
model.

In the context of quantum physics, parity is a type of symmetric relation.
```

This results in the status:

```
$ git status
# On branch master
# Unmerged paths:
#   (use "git add/rm <file>..." as appropriate to mark resolution)
#
# both modified:      readme.rst
#
...
no changes added to commit (use "git add" and/or "git commit -a")
```

Now, to alert Git that she has made appropriate alterations, Fran follows the instructions it gave her in the status message (namely, `git add` and `git commit` those changes):

```
$ git commit -am "Compromises merge conflict to Howdy and Bonjour"  ❶
$ git push origin master  ❷
Counting objects: 10, done.
Delta compression using up to 2 threads.
Compressing objects: 100% (6/6), done.
Writing objects: 100% (6/6), 762 bytes, done.
Total 6 (delta 2), reused 0 (delta 0)
To https://github.com/username/repositoryname.git
```

❶ Explain your solution to the merge conflict in the log message.

❷ Push the results online.

And that is it. Now the repository contains the history of both conflicting commits as well as a new commit that merges them intelligently. The final result is the version Fran has just committed.

Remote Version Control Wrap-up

In this chapter, we have shown how to use Git along with GitHub for downloading, uploading, and collaborating on code. Combined with remote repository hosting sites, the skills learned in Chapter 15 allow the scientist to manage files and change sets, merge simultaneous work among collaborators, and publish that work on the Internet.

Having read this chapter, you are now prepared to open your code to your collaborators. Go forth and version control the work you do day to day. In fact, take a moment now to place those analysis scripts you are working on under version control. Since it's an expansive and complex tool, you may find a need for additional resources on Git. We can recommend, in particular:

- *Pro Git* book (*http://git-scm.com/book/en/v2*) (Apress), a free and open source ebook
- Software Carpentry's Version Control with Git (*http://bit.ly/sc-git-lesson*)
- Software Carpentry's Git Reference (*http://bit.ly/quick-git-ref*)

Now that you are comfortable with pulling, pushing, fetching, merging, and dealing with conflicts, you should be able to collaborate with your colleagues on code and papers more smoothly and reproducibly. Next, you will need to know how to find and fix bugs so that your collaborative, reproducible software is reproducibly *correct*.

CHAPTER 17
Debugging

> The scientific method's central motivation is the ubiquity of error—the awareness that mistakes and self-delusion can creep in absolutely anywhere and that the scientist's effort is primarily expended in recognizing and rooting out error.
>
> —Donoho 2009

In the very early days of computing, Admiral Grace Hopper and her team on the Mark II computer encountered errors in the performance of the computer. Ultimately, a moth was discovered in one of the relays. Admiral Hopper reportedly remarked that they were "debugging" the system. Though the term had been used before in engineering, this event popularized the terms *bug* and *debugging* for the causes and solutions, respectively, of errors in computer code and performance.

Bugs are errors in code, and they are ubiquitous reminders of our humanity. That is, computers, by their very nature, do exactly what we tell them to do. Therefore, bugs are typically imperfections in syntax and logic introduced by humans. However careful we are, bugs will be introduced while we are developing code. They begin to be introduced as soon as we start writing a piece of code. For this reason, we must be vigilant, and we must be prepared to fix them when they arise. This chapter will prepare you to recognize, diagnose, and fix bugs using various tools and methods for "debugging" your code. It will do so by introducing:

- When, how, and by whom bugs are encountered
- Methods of diagnosing bugs
- Interactive debugging, for diagnosing bugs quickly and systematically
- Profiling tools to quickly identify memory management issues
- Linting tools to catch style inconsistencies and typos

Of these, the most time will be spent on using the *pdb* interactive debugger in Python, since it is essential for debugging issues whose source is not obvious from tests or simple print statements. First, however, we will discuss the instigating event itself: encountering a bug.

Encountering a Bug

A bug may take the form of incorrect syntax, imperfect logic, an infinite loop, poor memory management, failure to initialize a variable, user error, or myriad other human mistakes. It may materialize as:

- An unexpected error while compiling the code
- An unexpected error message while running the code
- An unhandled exception from a linked library
- An incorrect result
- An indefinite pause or hang-up
- A full computer crash
- A segmentation fault
- Silent failure

Developers may encounter bugs that they or their colleagues have introduced in their source code. Users, also, test the limits of a program when they deploy it for their own uses. Irrespective of when a bug is found or who finds it, bugs must be fixed. Faced with any of them, whether it be a build error, a runtime exception, or a segmentation fault, the user or developer should first try to track down and fix the bug.

Bugs can also be encountered at any time. In the next chapter, we will explain how testing should do most of the verification work in a piece of scientific software. That is, most bugs in a well-developed piece of software should be found by the tests before they make it to the trusted version of the code.

> A well-written test suite should do the heavy lifting in finding and diagnosing bugs. Interactive debugging lifts the remainder.

However, tests rarely cover all edge and corner cases, so bugs can slip through the cracks. The longer a bug exists undetected in a piece of trusted software, the more dire the situation:

1. If a bug is found in testing, it can be fixed before the software is ever used.

2. If a bug is found before there are users, it can be fixed before it affects anyone running the code.
3. If a bug is found when the code is run, it can be fixed before analysis is done on the results.
4. If a bug is found when the results of the code are analyzed, it can be fixed before the results are published in a journal article.
5. If a bug is found after the results are published, the paper has to be retracted.

Many papers are retracted every year due to bugs in code. Chapter 18 will show you how to improve your software tests to avoid getting past stage 1 in the preceding list. However, bugs occasionally slip through the cracks, so you must be ready to encounter and fix them at any stage of your project.

While bugs can be found by anyone, they are usually only diagnosable and fixable by people who know what the code is meant to do. Without knowing what the code is meant to do, it is nearly impossible to know when a result is incorrect, when a long pause is suspicious, or when silent termination indicates failure.

Now that you know how a bug is encountered, you are ready to learn how to diagnose its cause. In order to walk before we run, we will first introduce a simplistic way to diagnose bugs in code: print statements.

Print Statements

Print statements are every developer's first debugger. Because of this, we'll start here —but know that they are not the best practice for effective computing and that we will be covering better methods later in the chapter. Printing is typically a check that asks one or both of these questions:

- Is the bug happening before a certain line?
- What is the status of some variable at that point?

In a simple, buggy program, a print statement can answer the first question if it is inserted at a place where the program is suspected of misbehavior.

> In Python 3, print(x) is a function call, and thus an expression. In Python 2, print x was a statement and used a slightly different syntax. Still, the term *print statement* is used across many programming languages, even when printing to the screen is not technically a statement on its own (like in Python 3, C, and C++).

In the following example, something about the code is causing it to "hang." That is, it simply seems to run forever, as if stalled:

```
def mean(nums):
    bot = len(nums)
    it = 0
    top = 0
    while it < len(nums):
        top += nums[it]
    return float(top) / float(bot)

if __name__ == "__main__":
    a_list = [1, 2, 3, 4, 5, 6, 10, "one hundred"]
    mean(a_list)
```

It is likely that you can determine the cause of this problem by visual inspection. However, in the case that you cannot, a print statement can be inserted where the code is suspected to be hanging:

```
def mean(nums):
    bot = len(nums)
    it = 0
    top = 0
    print("Still Running at line 5")   ❶
    while it < len(nums):
        top += nums[it]
        print(top)   ❷
    return float(top) / float(bot)

if __name__ == "__main__":
    a_list = [1, 2, 3, 4, 5, 6, 10, "one hundred"]
    mean(a_list)
```

❶ This `print()` is added to determine where the error is happening.

❷ This one is added to determine what is happening to the variables during the loop.

Once a print statement is inserted at the suspected point of misbehavior, the program can be executed, and the print statement either appears before the exception, or does not appear at all. In the case shown here, many things are wrong with the code. However, the most fatal is the infinite `while` loop. Since the print statement appears before the code enters the infinite loop, the troublemaking line must be after the print statement. In this case, the first print statement is printed, so it is clear that the error occurs after line 5. Additionally, the second print statement results in "1" being printed infinitely. Can you tell what is wrong in the code using this information? The infinite loop can certainly be fixed as in the code shown here:

```
def mean(nums):
    top = sum(nums)   ❶
    bot = len(nums)
    return float(top) / float(bot)
```

```
if __name__ == "__main__":
    a_list = [1, 2, 3, 4, 5, 6, 10, "one hundred"]
    mean(a_list)
```

❶ Rather than looping needlessly, the `sum()` function can be applied to the list.

Print statements like this can provide very helpful information for pinpointing an error, but this strategy does not scale. In a large code base, it usually takes more than a few tries or a few more print statements to determine the exact line at which the error occurred. Additionally, the number of potentially problem-causing variables increases as the size of the code base increases. A more scalable solution is needed: interactive debugging.

Interactive Debugging

Rather than littering one's code base with print statements, interactive debuggers allow the user to pause during execution and *jump into the code* at a certain line of execution. Interactive debuggers, as their name suggests, allow the developer to query the state of the code in an interactive way. They allow the developer to move forward through the code execution to determine the source of the error.

Interactive debugging tools generally enable the user to:

- Query the values of variables
- Alter the values of variables
- Call functions
- Do minor calculations
- Step line by line through the call stack

All of this can help the developer to determine the cause of unexpected behavior, and these features make interactive debuggers an excellent exploratory tool if used systematically and intentionally. That is, without a strong notion of the expected behavior or a consistent plan of action, an interactive debugger only enables a developer to attempt random changes to the code and to query variables at random, just hoping for a change in behavior. This strategy is inefficient and error-prone. There are ways to ensure that you are being systematic:

- Before investigating a line of code, ask yourself how the error could be caused by that part of the code.
- Before querying the value of a variable, determine what you expect the correct value to be.
- Before changing the value of a variable, consider what the effect of that change should be.

- Before stepping forward through the execution, make an educated guess about what will indicate an error or success.
- Keep track of the things you try and the changes you make. Use version control to track changes to the files and use a pen and paper to track leads followed.

Now that you have some rules to live by, we can get started debugging interactively. The next sections of this chapter will cover an interactive debugger in the Python standard library, pdb.

Debugging in Python (pdb)

For Python code, interactive debugging can be achieved with the Python Debugger (*pdb*). It provides an interactive prompt where code execution can be paused with a *trace* and subsequent *breakpoints*. Then, the state of the code and variables can be queried, stepped through line by line, restarted, and modified. This section will describe all of these in the context of the still-failing mean code from the previous example.

That is, even though we have fixed the infinite loop in the previous example, another, different error arises.

Running

```
$ python a_list_mean.py
```

returns

```
Traceback (most recent call last):
  File "a_list_mean.py", line 9, in <module>
    mean(a_list)
  File "a_list_mean.py", line 2, in mean
    top = sum(nums)
TypeError: unsupported operand type(s) for +: 'int' and 'str'  ❶
```

❶ There is still some kind of error. It looks like it has to do with the types of the values in the list. Maybe we can use the debugger to check if changing the non-int list values to a number will resolve the error.

To diagnose this error with *pdb*, we must first import the pdb module into the script:

```
import pdb  ❶

def mean(nums):
    top = sum(nums)
    bot = len(nums)
    return float(top) / float(bot)

if __name__ == "__main__":
```

```
a_list = [1, 2, 3, 4, 5, 6, 10, "one hundred"]
mean(a_list)
```

❶ Import pdb into the file containing the suspiciously buggy code.

We must make one more edit to the file in order to begin. To tell the debugger where in the source code we would like to "jump into" the execution, we must set a trace.

Setting the Trace

Rather than inserting a new print statement on a new line every time new information is uncovered, you can set a trace point at the line where you would like to enter the program interactively in the debugger. You do so by inserting the following line into the source code:

```
pdb.set_trace()
```

This trace *pauses* the execution of the program at the line where it appears. When the program is paused, *pdb* provides an interface through which the user can type *pdb* commands that control execution. Using these commands, the user can print the state of any variable that is in scope at that point, step further forward through the execution one line at a time, or change the state of those variables.

> **Exercise: Set a Trace**
>
> 1. Create a file containing the buggy mean code.
> 2. Import pdb in that file.
> 3. Decide where you would like to set a trace and add a line there that reads pdb.set_trace().
> 4. Save the file. If you try running it, what happens?

In the mean script, an appropriate trace point to set might be at the very beginning of execution. It is a short program, and starting at the beginning will cover all the bases:

```
import pdb

def mean(nums):
    top = sum(nums)
    bot = len(nums)
    return float(top) / float(bot)

if __name__ == "__main__":
    pdb.set_trace()  ❶
    a_list = [1, 2, 3, 4, 5, 6, 10, "one hundred"]
    mean(a_list)
```

Debugging in Python (pdb) | 391

❶ The trace point is set at the beginning of execution.

Now, when the script is run, the Python debugger starts up and drops us into the code execution at that line.

python a_list_mean.py returns

```
> /filespace/users/h/hopper/bugs/a_list_mean.py(10)<module>()
-> a_list = [1, 2, 3, 4, 5, 6, 10, "one hundred"]
(pdb) ❶
```

❶ The *pdb* prompt looks like (pdb). This is where you enter debugging commands.

Since the location of the trace was set before anything happens at all in the program, the only object in scope is the definition of the mean() function. The next line initializes the a_list object. If we were to step forward through the execution, we would expect to see that happen. The interactive debugger enables us to do just that.

Stepping Forward

In any interactive debugger, once a trace point is reached, we can explore further by stepping slowly forward through the lines of the program. This is equivalent to adding a print statement at each line of the program execution, but takes much less time and is far more elegant.

The first time using a tool, you should find out how to get help. In *pdb*, typing help provides a table of available commands. Can you guess what some of them do?

```
Documented commands (type help <topic>):
========================================
EOF    bt         cont       enable   jump   pp     run      unt
a      c          continue   exit     l      q      s        until
alias  cl         d          h        list   quit   step     up
args   clear      debug      help     n      r      tbreak   w
b      commands   disable    ignore   next   restart  u      whatis
break  condition  down       j        p      return   unalias where

Miscellaneous help topics:
==========================
exec   pdb

Undocumented commands:
======================
retval   rv
```

To move forward through the code, for example, we would use the command step. Note also the s command listed above step. This is a shorthand for the step function. Either s or step can be used to move forward through execution one step.

Exercise: Step Through the Execution

1. Run your script from the last exercise.
2. Determine the expected effects of stepping through the execution by one line.
3. Type s. What just happened?

After the step, the program state is paused again. Any variables in scope at that line are available to be queried. Now that we have stepped forward one line, the `a_list` object should be initialized. To determine whether that is truly the case when the code is run, and whether `a_list` has been assigned the `list` that we expect, we can use *pdb* to print the value of the `a_list` variable that is suspicious.

Querying Variables

Since valid Python is valid in the *pdb* interpreter, simply typing the name of the variable will cause *pdb* to print its value (alternatively, the `print` function could be used):

Code	Returns
(Pdb) s	> /filespace/users/h/hopper/bugs/ a_list_mean.py(10)<module>() -> mean(a_list)
(Pdb) a_list	[1, 2, 3, 4, 5, 6, 10, 'one hundred']

Now, while it is clear that the variable is being set to the value we expect, it is suspect. If you recall, the error we received involved a type mismatch during the summation step. The string value *one hundred* may not be a valid input for the summation function. If we can change the value of that element to an `int`, it may be a more valid input for the summation. To test this with the debugger, we will need to execute a command that resets the value of the last element of `a_list`. Then, if we continue the execution of the code, we should see the summation function succeed.

Now, how do we change the last element of `a_list` while we are in *pdb*?

Setting the State

Since we have a guess about what the variable *should* be at this point, we can make that happen in the interactive debugger with simple interactive Python. Just as we could in the Python interpreter, we can set the value of the last element to `100` with `a_list[-1]=100`:

Code	Returns
(Pdb) a_list[-1] = 100 (Pdb) a_list	[1, 2, 3, 4, 5, 6, 10, 100]

Excellent. That was easy! Now that the program should be in a state that will not crash the summation function, we should check that the summation function works. How do we execute functions within the debugger?

Running Functions and Methods

In addition to variables, all functions and methods that are in scope at the breakpoint are also available to be run within the debugging environment. So, just as we would in the Python interpreter, we can execute sum(a_list):

Code	Returns
(Pdb) sum(a_list)	131

It turns out that our initial hunch was correct. Changing the string version *one hundred* to the integer version (100) allowed the summation function not to choke. Now we would like to tell *pdb* to continue the execution in order to see whether our change allows the program to run through to its finish without error. How do we continue the execution?

Continuing the Execution

Rather than stepping through the rest of the code one line at a time, we can continue the execution through to the end with the continue command. The shorthand for this command is c. If the execution succeeds, we, the developers, will know that changing the code in the Python script will solve our problem.

> **Exercise: Continue the Execution to Success**
>
> 1. Run the script from the previous exercise.
> 2. Step forward one line.
> 3. Change *one hundred* to 100 in a_list.
> 4. Continue execution with c. What happened? Was the mean of the list printed correctly? Why?

Now that the final element of the list is no longer a string (it has been set to the integer 100), the execution should succeed when the continue command is entered. The continue command, as you can see, proceeds with the execution until the program ends. The actual file can now be edited to capture this bug fix. The script that calculates the mean should now be similar to the following:

```
def mean(nums):
    top = sum(nums)
    bot = len(nums)
    return float(top) / float(bot)

if __name__ == "__main__":
    a_list = [1, 2, 3, 4, 5, 6, 10, 100]
    result = mean(a_list)
    print result
```

Sometimes, however, you may not be interested in running the execution all the way to the end. There may be some other place in the execution where the state of the variable should be checked. For this reason, the continue command stops if a *breakpoint* is reached. What is a breakpoint?

Breakpoints

If there is only one suspicious point in the execution, then setting the trace at that point or shortly before it is sufficient. However, sometimes a variable should be checked at many points in the execution—perhaps every time a loop is executed, every time a certain function is entered, or right before as well as right after the variable should change values. In this case, breakpoints are set.

In *pdb*, we can set a breakpoint using the break or shorthand b syntax. We set it at a certain line in the code by using the line number of that place in the code or the name of the function to flag:

```
b(reak) ([file:]lineno | function)[, condition]
```

With breakpoints, new lines can be investigated as soon as they become suspicious. Just set the breakpoint and call the continue function. The execution will continue until *pdb* encounters the line at which you have set the breakpoint. It will then pause execution at that point.

However, for this to work, you have to know where to put the breakpoint. In order to know that, the developer often has to know the code execution path that led to the error or crash. That list is called the *backtrace*, and it can be accessed from the *pdb* debugger quite easily with the bt command, which outputs the stack of commands that led up to the current state of the program. Sometimes also called the call stack, execution stack, or traceback, it answers the question "How did we get here?"

With that, you should have enough information to begin debugging your code. However, the job is not done. Even when your code is no longer exhibiting actual errors, there may still be issues that slow it down or are otherwise nonoptimal. To increase the speed of your code, it is helpful to know which parts are the slowest. The next section focuses on just how to find that out.

Profiling

Tools called *profilers* are used to sketch a profile of the time spent in each part of the execution stack. Profiling goes hand in hand with the debugging process. When there are suspected memory errors, profiling is the same as debugging. When there are simply memory inefficiencies, profiling can be used for optimization.

For example, certain for loops may be the source of slowdowns in a piece of software. Since we can often reduce for loops by vectorizing them, it is tempting to guess that the best solution is to rewrite all for loops in this more complex manner. However, that is a lower-level programming task that takes programmer effort. So, instead of vectorizing all for loops, it is best to find out which ones are the slowest, and focus on those.

In Python, cProfile is a common way to *profile* a piece of code. For our *fixed_mean.py* file, in which the bugs have been fixed, *cProfile* can be executed on the command line, as follows:

```
$ python -m cProfile -o output.prof fixed_mean.py  ❶ ❷
```

❶ Give the output file a name. It typically ends in the *prof* extension.

❷ Provide the name of the Python code file to be examined.

That creates a profile file in a binary format, which must be read by an interpreter of such files. The next section will discuss such an interpreter.

Viewing the Profile with pstats

One fast option is to use the *pstats* module. In an interactive Python session, the print_stats() function within the *pstats* package provides a breakdown of the time spent in each major function:

```
In [1]: import pstats

In [2]: p = pstats.Stats('output.prof')

In [3]: p.print_stats()
Mon Dec  8 19:43:12 2014    output.prof

         5 function calls in 0.000 seconds  ❶

   Random listing order was used

   ncalls  tottime  percall  cumtime  percall filename:lineno(function)  ❷
```

```
          1    0.000    0.000    0.000    0.000 fixed_mean.py:1(<module>)
          1    0.000    0.000    0.000    0.000 {sum}
          1    0.000    0.000    0.000    0.000 fixed_mean.py:1(mean)
          1    0.000    0.000    0.000    0.000 {method 'disable' of ...
          1    0.000    0.000    0.000    0.000 {len}
```

❶ A summary of the run. `print_stats` doesn't have very fine resolution.

❷ The `print_stats` function prints the number of calls to each function, the total time spent in each function, the time spent each time that function was called, the cumulative time elapsed in the program, and the place in the file where the call occurs.

This view is more helpful for programs that take longer to run. The many zeros in this example indicate that the time per function was never higher than 0.0009 seconds. Since the *fixed_mean.py* script runs so quickly, `pstats` does not, by default, print with fine enough resolution to capture the variable time spent in each function. By using various configuration options, we can make `pstats` print with finer resolution. That exercise is left up to the reader. A more effective way to view this information is with a graphical interface. We will move along to the next section to learn more about that option.

Viewing the Profile Graphically

Many more beautiful and detailed ways to view this output exist. One is a program called RunSnakeRun (*http://bit.ly/runsnake*).

RunSnakeRun is a common graphical interpreter for profiler output from `cProfile` and the *kernprof* tool (which we'll meet in the next section). With the simple command `runsnake <file.prof>` on the command line, RunSnakeRun opens a GUI for browsing the profile output. The results from our simple mean function are shown in Figure 17-1. In RunSnakeRun, the total amount of colored area is the amount of time spent in the program. Within that box, any calls to functions are shown by the amount of time spent in them, hierarchically.

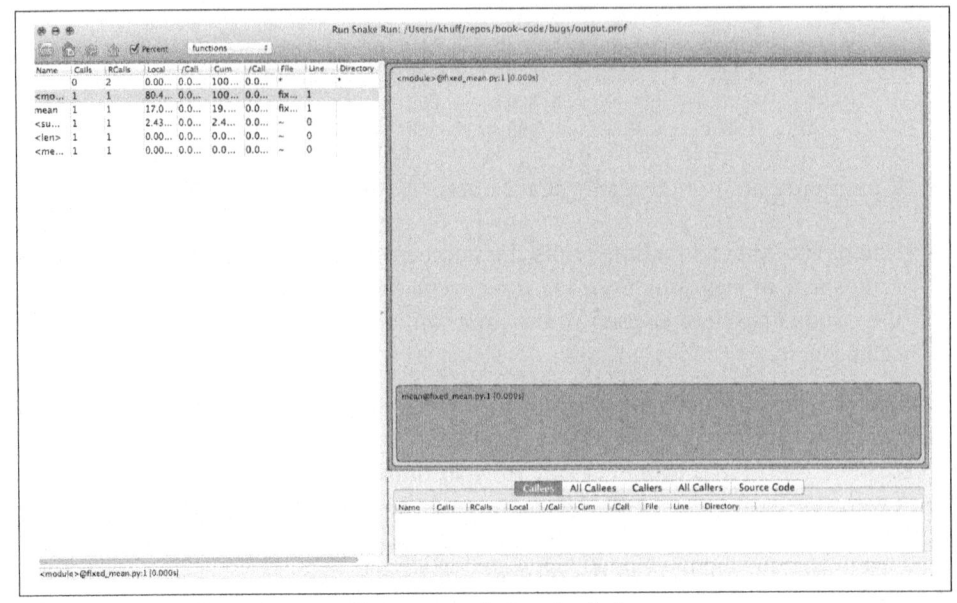

Figure 17-1. Profiling the mean function with RunSnakeRun

However, that example is not very exciting. For more complicated programs, the results can be quite interesting, as seen in Figure 17-2.

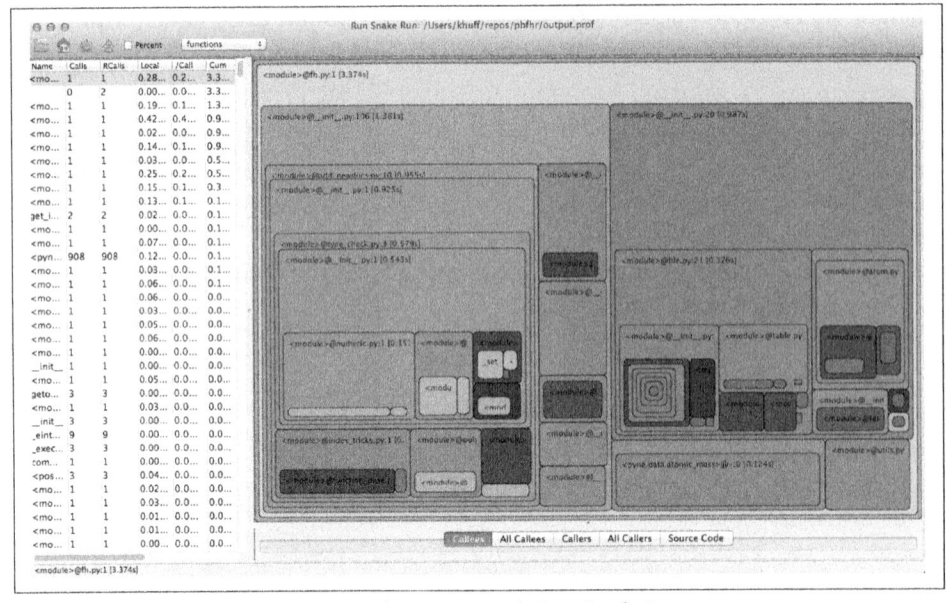

Figure 17-2. Profiling a more complex script with RunSnakeRun

At the top is a percent button. That button will show a breakdown of the percentage of time spent in each part of the code. This interactive graphic demonstrates the behavior of each section of the code so that you can quickly see where time is being wasted.

Another option, inspired by RunSnakeRun, is an in-browser viewer called SnakeViz. To use SnakeViz, first make sure it is installed by running `which snakeviz`. If it is not present, try installing it with `pip` (`pip install snakeviz`) using your package manager or downloading it from its website. Next, in the command line, type:

```
$ snakeviz output.prof
```

The *SnakeViz* program will cause a web browser to open and will provide an interactive infographic of the data in *output.prof*. The results for our simple code are shown in Figure 17-3.

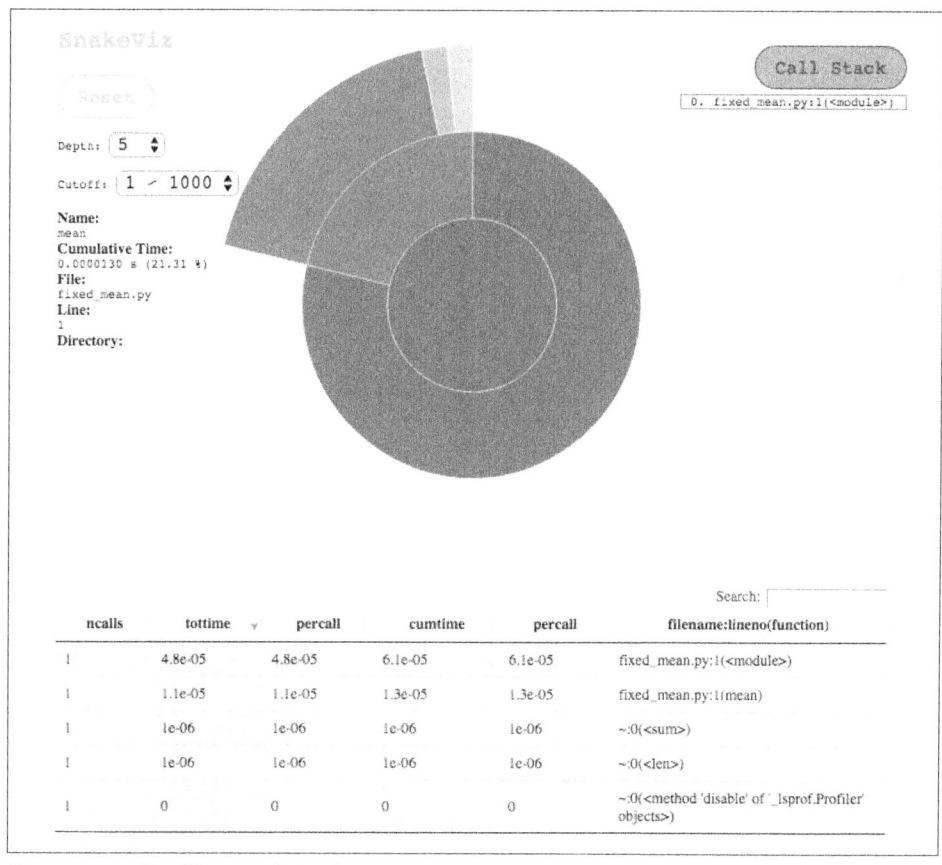

ncalls	tottime	percall	cumtime	percall	filename:lineno(function)
1	4.8e-05	4.8e-05	6.1e-05	6.1e-05	fixed_mean.py:1(<module>)
1	1.1e-05	1.1e-05	1.3e-05	1.3e-05	fixed_mean.py:1(mean)
1	1e-06	1e-06	1e-06	1e-06	~:0(<sum>)
1	1e-06	1e-06	1e-06	1e-06	~:0(<len>)
1	0	0	0	0	~:0(<method 'disable' of '_lsprof.Profiler' objects>)

Figure 17-3. Profiling with SnakeViz

With SnakeViz, the execution of the code can be browsed on a function-by-function basis. The time spent in each function is rendered in radial blocks. The central circle represents the top of the call stack—that is, the function from which all other functions are called. In our case, that is the main body of the module in the final four lines of the file.

The next radial annulus describes the time spent in each function called by the main function, and so on. When the mouse hovers over some section of the graph, more information is shown. To learn more about SnakeViz and how to interperet its contents, see its website (*https://jiffyclub.github.io/snakeviz/*).

Combined with `cProfile`, these graphical interfaces for profiling are an efficient way to pinpoint functions with efficiency issues. Sometimes, though, it is even more helpful to know how much time you spend on each line. For this, consider *kernprof*.

Line Profiling with Kernprof

For showing the specific lines at fault for slowdowns, you can use a line profiler called *kernprof*. To use *kernprof*, you must alter the file itself with a decorator (`@profile`) above each function definition of interest. The mean code becomes:

With that decorator in place, *kernprof* can then be run verbosely in line-by-line mode thus:

```
kernprof -v -l fixed_mean.py
```

When *kernprof* is run in that way, the profile of time spent is printed to the terminal in much greater detail than with the previous tools:

```
16.375  ❶
Wrote profile results to fixed_mean.py.lprof
Timer unit: 1e-06 s  ❷

Total time: 7e-06 s
File: fixed_mean.py
Function: mean at line 1  ❸

Line #      Hits         Time  Per Hit   % Time  Line Contents  ❹
==============================================================
     1                                           @profile
     2                                           def mean(nums):
     3         1            2      2.0     28.6      top = sum(nums)
     4         1            0      0.0      0.0      bot = len(nums)
     5         1            5      5.0     71.4      return float(top)/float(bot)
```

❶ Since the code is run from start to finish, the code output is printed.

❷ *kernprof* intelligently guesses the magnitude of time resolution to print.

❸ The only profiled lines are those within the function that we decorated.

❹ Each line has its own row in this table.

When you're inspecting these results, the fifth column is the most important. It indicates the percentage of time spent on each line in the mean function. The results here indicate that most of the time is spent calculating and returning the quotient. Perhaps some speedup can be achieved. Can you think of any simplifications to the code? Try making a change to determine whether it has an effect on the speed of execution.

Now that our code no longer exhibits errors and can be optimized for speed, the only remaining debugging task is clean up. A tool used to cleanup code is called a *linter*.

Linting

Linting removes "lint" from source code. It's a type of cleanup that is neither debugging nor testing nor profiling, but can be helpful at each of these stages of the programming process. Linting catches unnecessary imports, unused variables, potential typos, inconsistent style, and other similar issues.

Linting in Python can be achieved with the *pyflakes* tool. Get it? Errors are more than just lint, they're flakes!

As an example of how to use a linter, recall the *elementary.py* file from Chapter 6. To lint a Python program, execute the `pyflakes` command on it:

```
$ pyflakes elementary.py
```

pyflakes responds with a note indicating that a package has been imported but remains unused throughout the code execution:

```
elementary.py:2: 'numpy' imported but unused
```

This information is more than just cosmetic. Since importing packages takes time and occupies computer memory, reducing unused imports can speed up your code.

That said, most linting tools do focus on cosmetic issues. Style-related linting tools such as *flake8*, *pep8*, or *autopep8* can be used to check for errors, variable name misspelling, and PEP8 compatibility. For more on the PEP8 style standard in Python, see Chapter 19. To use the *pep8* tool, simply call it from the command line:

```
$ pep8 elementary.py
```

It will analyze the Python code that you have provided and will respond with a line-by-line listing of stylistic incompatibilities with the PEP8 standard:

```
elementary.py:4:1: E302 expected 2 blank lines, found 1
elementary.py:5:3: E111 indentation is not a multiple of four
elementary.py:7:31: E228 missing whitespace around modulo operator
```

This indicates that the *elementary.py* file has a few insufficiencies related to the PEP8 Style Guide (*http://bit.ly/pep-8*). The combined information of both tools can be retrieved with the much more strict *pylint* tool on the command line:

```
$ pylint -rn elementary.py
```

The -rn flag simply tells *pylint* not to print its full report. The report provided by *pylint* by default is quite lengthy indeed and could easily occupy half of the pages in this chapter:

```
No config file found, using default configuration
************* Module elementary
W:  5, 0: Bad indentation. Found 2 spaces, expected 4 (bad-indentation)
W:  6, 0: Bad indentation. Found 4 spaces, expected 8 (bad-indentation)
W:  7, 0: Bad indentation. Found 4 spaces, expected 8 (bad-indentation)
W:  8, 0: Bad indentation. Found 4 spaces, expected 8 (bad-indentation)
W:  9, 0: Bad indentation. Found 4 spaces, expected 8 (bad-indentation)
C:  1, 0: Missing module docstring (missing-docstring)
C:  6, 4: Invalid attribute name "s" (invalid-name)
C:  7, 4: Invalid attribute name "isFermion" (invalid-name)
C:  8, 4: Invalid attribute name "isBoson" (invalid-name)
C:  4, 0: Missing class docstring (missing-docstring)
W:  5, 2: __init__ method from base class 'Particle' is not called...
```

Once the incorrect indentation, invalid names, and missing docstrings are fixed, your code will be ready for prime time.

Debugging Wrap-up

Having read this chapter, you should feel ready to use an interactive debugger to more efficiently and systematically:

- Understand bugs
- Track down their cause
- Prototype solutions
- Check for success

Additionally, this chapter should have prepared you to use profilers and linters to optimize and clean your code once you've fixed the bugs. Now that you are prepared to deal with bugs and inefficiencies that arise in your code, your focus can turn to keeping them from appearing in the first place. In the next chapter, we will show you how to avoid bugs with comprehensive, systematic testing.

CHAPTER 18
Testing

Before relying on a new experimental device, a good physicist will establish its accuracy. A new detector will always have its responses to known input signals tested. The results of this calibration are compared against the *expected* responses. If the device is trustworthy, then the responses received will fall within acceptable bounds of what was expected. To make this a fair test, the accuracy bounds are set prior to the test. The same goes for testing in computational science and software development.

Code is assumed guilty until proven innocent. This applies to software written by other people, but even more so to software written by yourself. The mechanism that builds trust that software is performing correctly is called *testing*.

Testing is the process by which the expected results of code are compared against the observed results of actually having run that code. Tests are typically provided along with the code that they are testing. The collection of all of the tests for a given piece of code is known as the *test suite*. You can think of the test suite as a bunch of precanned experiments that anyone can run. If all of the tests pass, then the code is at least partially trustworthy. If any of the tests fail, then the code is known to be incorrect with respect to whichever case failed.

Now, you may have noticed that the test code itself is part of the software package. Since the tests are just as likely to have bugs as the code they are testing, it is tempting to start writing tests that test the tests. However, this quickly runs into an incompleteness problem. There is no set of tests that is the set of all possible tests. Suppose you write a test suite for some code. Now your test suite is untested, so you add a test for the test suite. Now your test suite tester is untested, so you write a test for that, and so on. It is possible to escape this infinite-work trap using recursion, as discussed in Chapter 5, but it probably is not worth your time.

Even one level of testing—just testing main code and not the tests themselves—is incredibly beneficial. Almost all of the scientific value comes from this first pass. This is because the first level is where the physics is put directly in question. A sufficiently rigorous test suite will find all of the physical and computational errors without having to worry about the philosophical and mathematical ramifications of whether a test is itself sufficiently tested.

Testing is so central to both the scientific method and modern software development that many computational scientists consider it a moral failing for a scientific program not to include tests. They also know to not trust a code when the tests do not pass. Neither should you. For software that you do not write, it is always a good idea to run the test suite when you first start working with the code. The documentation will typically include instructions on how to run the tests, since they can be different from project to project.

In this chapter, we will be discussing testing in the context of Python. Specifically, we will be using the nose *testing framework*. This is a package of tools that make writing and running tests easy. Though other test frameworks exist in Python (`pytest`, `unittest`), nose has become the standard testing tool for scientific Python. It helps that it is also easier to use and understand than some of the others.

We start this chapter by asking a series of questions that illuminate good testing practices that everyone should follow.

Why Do We Test?

Testing is a great practice that aids all of software development. However, not practicing good habits alone is not a moral failing. Testing is considered a core principle of *scientific software* because its impact is at the heart of knowledge generation.

In most other programming endeavors, if code is fundamentally wrong, even if it goes uncorrected for years at a time, the impact of this error can be relatively small. Perhaps a website goes down, or a game crashes, or a day's worth of writing is lost when the computer crashes. Scientific code, on the other hand, controls planes, weapons systems, satellites, agriculture, and (most importantly) physics simulations and experiments. If the software that governs a computational or physical experiment is wrong, then any decisions that are made based on its results will be completely untrustworthy.

This is not to say that physicists have a monopoly on software testing. Arguably, testing is just as important in arenas such as finance, government, and health care. Gross failures in these areas, however, tend to affect lives and livelihoods rather than knowledge itself.

We would like to think that scientists are rigorous enough to realize the importance of testing, but mistakes of negligence happen all too frequently. Everyone who has been involved with scientific software for any length of time has a horror story or two. The truth of the matter is that most scientists are poorly equipped to truly test their code. The average blog or image-sharing website is better tested than most scientific software.

This chapter is here to help remedy the poor testing situation by explaining the motivation behind testing and giving you the tools you need to do better.

When Should We Test?

Always.

Testing should be a seamless part of the scientific software development process. Tests should be created along with the code that is to be tested. At the very minimum, at least one test should be written immediately following the initial implementation of a function or a class. At the beginning of a new project, tests can be used to help guide the overall architecture of the project. This is analogous to experiment design in the experimental science world. The act of writing tests can help clarify how the software should be performing. Taking this idea to the extreme, you could start to write the tests *before* you even write the software that will be tested. We will discuss this practice in greater detail in "Test-Driven Development" on page 419.

In *Working Effectively with Legacy Code* (Prentice Hall), Michael Feathers defines legacy code as "any code without tests." This definition draws on the fact that after its initial creation, tests provide a powerful guide to other developers (and to your forgetful self, a few months in the future) about how each function in a piece of code is meant to be used. Without runnable tests to provide examples of code use, even brand new programs are unsustainable.

Where Should We Write Tests?

While writing code, you can add exceptions and assertions to sound an alarm as runtime problems come up. These kinds of tests, however, are embedded in the software itself. It is better to separate the code from the tests as much as possible. External tests require a bit more sophistication, but are better suited to checking the implementation against its expected behavior. These external tests are what is referred to as the *test suite*. The runtime exceptions and assertions do not count as part of the test suite.

Many projects choose to have a top-level directory named after the project or called *src/*. Similarly, many projects also have a top-level *tests/* directory where the test suite lives. This often mirrors the directory structure of the source directory. Mirroring makes it obvious where the test lives for any corresponding piece of source code.

Alternatively, some projects choose to place the tests right next to the source code that they are testing. Say you had a module called *physics.py*. In this schema, the tests would live in a *test_physics.py* file to keep them somewhat separate. This strategy is not recommended, though you will sometimes encounter it in the wild.

As with everything in software, the most important aspect of where to put the tests is to *be consistent*. Choose one approach and follow that for all of the tests that you write in a given project. If you are working on a more established project, be sure to conform to whatever pattern was set before you started.

What and How to Test?

Consider again the analogy to a detector in a physical experiment. The behavior of the detector must be characterized and calibrated across the valid range of interest. However, it is often unnecessary to characterize the response to every possible valid input. Most detectors rely on the physical quantity that they measure being either continuous or discrete. Testing only a few key signals, typically at the upper and lower edges of its range and some points in between, is enough to determine if and how well the machine is working. This "test what is important" mindset applies equally to scientific software development. Software tests should cover behavior from the common to the extreme, but not every single value within those bounds.

Let's see how this mindset applies to an actual physics problem. Given two previous observations in the sky and the time between them, Kepler's Laws provide a closed-form equation for the future location of a celestial body. This can be implemented via a function named `kepler_loc()`. The following is a stub interface representing this function that lacks the actual function body:

```
def kepler_loc(p1, p2, dt, t):
    ...
    return p3
```

As a basic test of this function, we can take three points on the planet Jupiter's actual measured path and use the latest of these as the expected result. We will then compare this to the result that we observe as the output of the `kepler_loc()` function.

> Tests compare expected outputs versus observed outputs for known inputs. They do not inspect the body of the function directly. In fact, the body of a function does not even have to exist for a valid test to be written.

To start testing, we will raise an exception as a way of signaling that the test failed if the expected value is not equal to the observed value. Frequently, tests are written as functions that have the same name as the code that they are testing with the word test either before or after it. The following example is pseudocode for testing that the measured positions of Jupiter, given by the function jupiter(), can be predicted with the kepler_loc() function:

```
def test_kepler_loc():        ❶
    p1 = jupiter(two_days_ago)
    p2 = jupiter(yesterday)   ❷
    exp = jupiter(today)      ❸
    obs = kepler_loc(p1, p2, 1, 1)  ❹
    if exp != obs:            ❺
        raise ValueError("Jupiter is not where it should be!")
```

❶ The test_kepler_loc() function tests kepler_loc().

❷ Get the inputs to kepler_loc().

❸ Obtain the expected result from experimental data.

❹ Obtain the observed result by calling kepler_loc().

❺ Test that the expected result is the same as the observed result. If it is not, signal that the test failed by raising an exception.

Now, calling the test_kepler_loc() function will determine whether kepler_loc() is behaving as intended. If a ValueError is raised, then we know something is wrong. The test_kepler_loc() function follows a very common testing pattern:

1. Name the test after the code that is being tested.
2. Load or set the expected results.
3. Compute the observed result by actually running the code.
4. Compare the expected and observed results to ensure that they are equivalent.

This pattern can be boiled down to the following pseudocode:

```
def test_func():
    exp = get_expected()
    obs = func(*args, **kwargs)
    assert exp == obs
```

It is critical to understand that tests should usually check for equivalence (==) and not equality (is). It is more important that the expected and observed results are effectively the same than that they are actually the same exact object in memory. For the floating-point data that is common in physics, it is often more pertinent for the expected and observed results to be *approximately* equal than it is for them to have precisely the same value. Floats are an approximation, and this needs to be accounted for when you're testing.

Testing equivalence via exceptions is rather like hammering a nail with a grenade. The nail will probably go in (the test will run), but the grenade will take everything else (i.e., the Python interpreter) along with it. A slightly more subtle way to accomplish the same task would be to use assertions. From Table 2-1, recall that an `assert` statement in Python ensures that the expression following it evaluates to `True`. If the assertion is true, then Python continues on its merry way. If the assertion is false, then an `AssertionError` is raised. We could rewrite `test_keppler_loc()` as follows:

```
def test_keppler_loc():
    p1 = jupiter(two_days_ago)
    p2 = jupiter(yesterday)
    exp = jupiter(today)
    obs = keppler_loc(p1, p2, 1, 1)
    assert exp == obs  ❶
```

❶ Now with an assertion instead of an exception.

The assertion approach still lacks subtlety, though all that we know when the test fails is that it failed. We do not see the values of the expected and observed results to help us determine where the fault lies. To get this kind of extra information in the event of a failure, we need to supply a *custom* assertion. Rich and descriptive assertions are exactly what a test framework like nose provides.

nose has a variety of helpful and specific assertion functions that display extra debugging information when they fail. These are all accessible through the `nose.tools` module. The simplest one is named `assert_equal()`. It takes two arguments, the expected and observed results, and checks them for equivalence (`==`). We can further rewrite `test_kepler_loc()` as seen here:

```
from nose.tools import assert_equal  ❶

def test_kepler_loc():
    p1 = jupiter(two_days_ago)
    p2 = jupiter(yesterday)
    exp = jupiter(today)
    obs = keppler_loc(p1, p2, 1, 1)
    assert_equal(exp, obs)  ❷
```

❶ To obtain functionality from nose, first we have to import it.

❷ Python's assertion can be replaced with nose's.

Using the test framework is the best way to write tests. Executing each of your tests by hand, however, becomes tiresome when you have more than a handful in your test suite. The next section goes over how to manage all of the tests you have written.

Running Tests

The major boon a testing framework provides is a utility to find and run the tests automatically. With nose, this is a command-line tool called *nosetests*. When *nosetests* is run, it will search all the directories whose names start or end with the word *test*, find all of the Python modules in these directories whose names start or end with *test*, import them, and run all of the functions and classes whose names start or end with *test*. In fact, nose looks for any names that match the regular expression `(?:^|[\\b_\|.-])[Tt]est`. This automatic registration of test code saves tons of human time and allows us to focus on what is important: writing more tests.

When you run *nosetests*, it will print a dot (.) on the screen for every test that passes, an F for every test that fails, and an E for every test where there was an unexpected error. In rarer situations you may also see an S indicating a skipped test (because the test is not applicable on your system) or a K for a known failure (because the developers could not fix it promptly). After the dots, *nosetests* will print summary information. Given just the one `test_kepler_loc()` test from the previous section, *nosetests* would produce results like the following:

```
$ nosetests
.
Ran 1 test in 0.224s

OK
```

As we write more code, we would write more tests, and *nosetests* would produce more dots. Each passing test is a small, satisfying reward for having written quality scientific software. Now that you know how to write tests, let's go into what can go wrong.

Edge Cases

What we saw in "What and How to Test?" on page 406 is called an *interior* test. The precise points that we tested did not matter. Any two initial points in an orbit could have been used to predict further positions. Though this is not as true for cyclic problems, more linear scenarios tend to have a clear beginning, middle, and end. The output is defined on a valid range.

The situation where the test examines either the beginning or the end of a range, but not the middle, is called an *edge case*. In a simple, one-dimensional problem, the two edge cases should always be tested along with at least one internal point. This ensures that you have good *coverage* over the range of values.

Anecdotally, it is important to test edges cases because this is where errors tend to arise. Qualitatively different behavior happens at boundaries. As such, they tend to have special code dedicated to them in the implementation. Consider the following simple Fibonacci function:

```
def fib(n):
    if n == 0 or n == 1:
        return 1
    else:
        return fib(n - 1) + fib(n - 2)
```

This function has two edge cases: zero and one. For these values of n, the fib() function does something special that does not apply to any other values. Such cases should be tested explicitly. A minimally sufficient test suite for this function would be:

```
from nose.tools import assert_equal

from mod import fib

def test_fib0():  ❶
    # test edge 0
    obs = fib(0)
    assert_equal(1, obs)

def test_fib1():  ❷
    # test edge 1
    obs = fib(1)
    assert_equal(1, obs)

def test_fib6():  ❸
    # test regular point
    obs = fib(6)
    assert_equal(13, obs)
```

❶ Test the edge case for zero.

❷ Test the edge case for one.

❸ Test an internal point.

Different functions will have different edge cases. Often, you need not test for cases that are outside the valid range, unless you want to test that the function fails. In the fib() function negative and noninteger values are not valid inputs. You do not need to have tests for these classes of numbers, though it would not hurt. Edge cases are not where the story ends, though, as we will see next.

Corner Cases

When two or more edge cases are combined, it is called a *corner case*. If a function is parametrized by two independent variables, a test that is at the extreme of both variables is in a corner. As a demonstration, consider the case of the function (sin(x) / x) * (sin(y) / y), presented here:

```
import numpy as np

def sinc2d(x, y):
    if x == 0.0 and y == 0.0:
        return 1.0
    elif x == 0.0:
        return np.sin(y) / y
    elif y == 0.0:
        return np.sin(x) / x
    else:
        return (np.sin(x) / x) * (np.sin(y) / y)
```

The function sin(x)/x is called the sinc() function. We know that at the point where x = 0, then sinc(x) == 1.0. In the code just shown, sinc2d() is a two-dimensional version of this function. When both x and y are zero, it is a corner case because it requires a special value for both variables. If either x or y (but not both) is zero, these are edge cases. If neither is zero, this is a regular internal point.

A minimal test suite for this function would include a separate test for the corner case, each of the edge cases, and an internal point. For example:

```
import numpy as np
from nose.tools import assert_equal

from mod import sinc2d

def test_internal():  ❶
    exp = (2.0 / np.pi) * (-2.0 / (3.0 * np.pi))
    obs = sinc2d(np.pi / 2.0, 3.0 * np.pi / 2.0)
    assert_equal(exp, obs)

def test_edge_x():  ❷
    exp = (-2.0 / (3.0 * np.pi))
    obs = sinc2d(0.0, 3.0 * np.pi / 2.0)
    assert_equal(exp, obs)

def test_edge_y():  ❸
    exp = (2.0 / np.pi)
    obs = sinc2d(np.pi / 2.0, 0.0)
    assert_equal(exp, obs)

def test_corner():  ❹
    exp = 1.0
    obs = sinc2d(0.0, 0.0)
    assert_equal(exp, obs)
```

❶ Test an internal point.

❷ Test an edge case for x and internal for y.

❸ Test an edge case for y and internal for x.

❹ Test the corner case.

Corner cases can be even trickier to find and debug than edge cases because of their increased complexity. This complexity, however, makes them even more important to explicitly test.

Whether internal, edge, or corner cases, we have started to build up a classification system for the tests themselves. In the following sections, we will build this system up even more based on the role that the tests have in the software architecture.

Unit Tests

All of the tests that we have seen so far have been unit tests. They are so called because they exercise the functionality of the code by interrogating individual functions and methods. Functions and methods can often be considered the atomic units of software because they are indivisible from the outside.

However, what is considered to be the *smallest code unit* is subjective. The body of a function can be long or short, and shorter functions are arguably more unit-like than long ones. Thus, what reasonably constitutes a code unit typically varies from project to project and language to language. A good rule of thumb is that if the code cannot be made any simpler logically (you cannot split apart the addition operator) or practically (a function is self-contained and well defined), then it is a unit. The purpose behind unit tests is to encourage both the code and the tests to be as small, well-defined, and modular as possible. There is no one right answer for what this means, though. In Python, unit tests typically take the form of test functions that are automatically called by the test framework.

Additionally, unit tests may have *test fixtures*. A fixture is anything that may be added to the test that creates or removes the environment required by the test to successfully run. They are not part of expected result, the observed result, or the assertion. Test fixtures are completely optional.

A fixture that is executed before the test to prepare the environment is called a *setup* function. One that is executed to mop up side effects after a test is run is called a *teardown* function. nose has a decorator that you can use to automatically run fixtures no matter whether the test succeeded, failed, or had an error. (For a refresher on decorators, see "Decorators" on page 112.)

Consider the following example that could arise when communicating with third-party programs. You have a function f() that will write a file named *yes.txt* to disk with the value 42 but only if a file *no.txt* does not exist. To truly test that the function works, you would want to ensure that neither *yes.txt* nor *no.txt* existed before you ran your test. After the test, you would want to clean up after yourself before the next test comes along. You could write the test, setup, and teardown functions as follows:

```
import os

from nose.tools import assert_equal, with_setup

from mod import f

def f_setup():  ❶
    files = os.listdir('.')
    if 'no.txt' in files:
        os.remove('no.txt')
    if 'yes.txt' in files:
        os.remove('yes.txt')

def f_teardown():  ❷
    files = os.listdir('.')
    if 'yes.txt' in files:
        os.remove('yes.txt')

def test_f():
    f_setup()  ❸
    exp = 42
    f()
    with open('yes.txt', 'r') as fhandle:
        obs = int(fhandle.read())
    assert_equal(exp, obd)
    f_teardown()  ❹
```

❶ The f_setup() function tests ensure that neither the *yes.txt* nor the *no.txt* file exists.

❷ The f_teardown() function removes the *yes.txt* file, if it was created.

❸ The first action of test_f() is to make sure the filesystem is clean.

❹ The last action of test_f() is to clean up after itself.

This implementation of test fixtures is usually fine. However, it does not guarantee that the f_setup() and f_teardown() functions will be called. This is because an unexpected error anywhere in the body of f() or test_f() will cause the test to abort before the teardown function is reached. To make sure that both of the fixtures will be executed, you must use nose's with_setup() decorator. This decorator may be applied to any test and takes a setup and a teardown function as possible arguments. We can rewrite test_f() to be wrapped by with_setup(), as follows:

```
@with_setup(setup=f_setup, teardown=f_teardown)
def test_f():
    exp = 42
    f()
    with open('yes.txt', 'r') as fhandle:
        obs = int(fhandle.read())
    assert_equal(exp, obd)
```

Note that if you have functions in your test module that are simply named `setup()` and `teardown()`, each of these is called automatically when the entire test module is loaded in and finished.

> Simple tests are the easiest to write. For this reason, functions should be small enough that they are easy to test. For more information on writing code that facilitates tests, we recommend Robert C. Martin's book *Clean Code* (Prentice Hall).

Having introduced the concept of unit tests, we can now go up a level in complexity.

Integration Tests

You can think of a software project like a clock. Functions and classes are the gears and cogs that make up the system. On their own, they can be of the highest quality. Unit tests verify that each gear is well made. However, the clock still needs to be put together. The gears need to fit with one another.

Integration tests are the class of tests that verify that multiple moving pieces of the code work well together. They ensure that the clock can tell time correctly. They look at the system as a whole or at subsystems. Integration tests typically function at a higher level conceptually than unit tests. Thus, programming integration tests also happens at a higher level.

Because they deal with gluing code together, there are typically fewer integration tests in a test suite than there are unit tests. However, integration tests are no less important. Integration tests are essential for having adequate testing. They encompass all of the cases that you cannot hit through plain unit testing.

Sometimes, especially in probabilistic or stochastic codes, the precise behavior of an integration test cannot be determined beforehand. That is OK. In these situations it is acceptable for integration tests to verify average or aggregate behavior rather than exact values. Sometimes you can mitigate nondeterminism by saving seed values to a random number generator, but this is not always going to be possible. It is better to have an imperfect integration test than no integration test at all.

As a simple example, consider the three functions a(), b(), and c(). The a() function adds one to a number, b() multiplies a number by two, and c() composes them. These functions are defined as follows:

```
def a(x):
    return x + 1

def b(x):
    return 2 * x

def c(x):
    return b(a(x))
```

The a() and b() functions can each be unit-tested because they each do one thing. However, c() cannot be truly unit tested because all of the real work is farmed out to a() and b(). Testing c() will be a test of whether a() and b() can be integrated together.

Integration tests still follow the pattern of comparing expected results to observed results. A sample test_c() is implemented here:

```
from nose.tools import assert_equal

from mod import c

def test_c():
    exp = 6
    obs = c(2)
    assert_equal(exp, obs)
```

Given the lack of clarity in what is defined as a code unit, what is considered an integration test is also a little fuzzy. Integration tests can range from the extremely simple (like the one just shown) to the very complex. A good delimiter, though, is in opposition to the unit tests. If a function or class only combines two or more unit-tested pieces of code, then you need an integration test. If a function implements new behavior that is not otherwise tested, you need a unit test.

The structure of integration tests is very similar to that of unit tests. There is an expected result, which is compared against the observed value. However, what goes in to creating the expected result or setting up the code to run can be considerably more complicated and more involved. Integration tests can also take much longer to run because of how much more work they do. This is a useful classification to keep in mind while writing tests. It helps separate out which tests should be easy to write (unit) and which ones may require more careful consideration (integration).

Integration tests, however, are not the end of the story.

Regression Tests

Regression tests are qualitatively different from both unit and integration tests. Rather than assuming that the test author knows what the expected result should be, regression tests look to the past. The expected result is taken as what was previously computed for the same inputs. Regression tests assume that the past is "correct." They are great for letting developers know when and how a code base has changed. They are not great for letting anyone know why the change occurred. The change between what a code produces now and what it computed before is called a *regression*.

Like integration tests, regression tests tend to be high level. They often operate on an entire code base. They are particularly common and useful for physics simulators.

A common regression test strategy spans multiple code versions. Suppose there is an input file for version X of a simulator. We can run the simulation and then store the output file for later use, typically somewhere accessible online. While version Y is being developed, the test suite will automatically download the output for version X, run the same input file for version Y, and then compare the two output files. If anything is significantly different between them, the test fails.

In the event of a regression test failure, the onus is on the current developers to explain why. Sometimes there are backward-incompatible changes that had to be made. The regression test failure is thus justified, and a new version of the output file should be uploaded as the version to test against. However, if the test fails because the physics is wrong, then the developer should fix the latest version of the code as soon as possible.

Regression tests can and do catch failures that integration and unit tests miss. Regression tests act as an automated short-term memory for a project. Unfortunately, each project will have a slightly different approach to regression testing based on the needs of the software. Testing frameworks provide tools to help with building regression tests but do not offer any sophistication beyond what has already been seen in this chapter.

Depending on the kind of project, regression tests may or may not be needed. They are only truly needed if the project is a simulator. Having a suite of regression tests that cover the range of physical possibilities is vital to ensuring that the simulator still works. In most other cases, you can get away with only having unit and integration tests.

While more test classifications exist for more specialized situations, we have covered what you will need to know for almost every situation in computational physics. In the following sections, we will go over how to write tests more effectively.

Test Generators

Test generators automate the creation of tests. Suppose that along with the function you wished to test, you also had a long list of expected results and the associated arguments to the function for those expected results. Rather than you manually creating a test for each element of the list, the test generator would take the list and manufacture the desired tests. This requires much less work on your part while also providing more thorough testing. The list of expected results and function inputs is sometimes called a *test matrix*.

In nose, test generators are written by turning the test function into a generator with yield statements.[1] In the test function, the assertion for each element of the matrix is yielded, along with the expected value and the function inputs. Corresponding check functions sometimes go along with the test generator to perform the actual work.

For demonstration purposes, take a simple function that adds two numbers together. The function, the check function, and the test generator could all be written as follows:

```
from nose.tools import assert_equal

def add2(x, y):  ❶
    return x + y

def check_add2(exp, x, y):  ❷
    obs = add2(x, y)
    assert_equal(exp, obs)

def test_add2():  ❸
    cases = [  ❹
        (4, 2, 2),
        (5, -5, 10),
        (42, 40, 2),
        (16, 3, 13),
        (-128, 0, -128),
        ]
    for exp, x, y in cases:  ❺
        yield check_add2, exp, x, y
```

❶ The function to test, add2().

❷ The check function performs the equality assertion instead of the test.

❸ The test function is now a test generator.

---

1 See "Generators" on page 109 for a refresher on generators if you need one.

❹ `cases` is a list of tuples that represents the test matrix. The first element of each tuple is the expected result. The following elements are the arguments to `add2()`.

❺ Looping through the test matrix `cases`, we `yield` the check function, the expected value, and the `add2()` arguments. Nose will count each `yield` as a separate full test.

This will produce five tests in `nose`, one for each case. We can therefore efficiently create many tests and minimize the redundant code we need to write. Running *nosetests* will produce the following output:

```
$ nosetests
.....
Ran 5 tests in 0.001s

OK
```

This is a very powerful testing mechanism because adding or removing tests is as easy as modifying the `cases` list. Different testing frameworks implement this idea in different ways. In all frameworks, it makes your life easier. Generating many test cases will hopefully cover more of the code base. The next section will discuss how to determine how many lines of your project are actually being executed by the test suite.

Test Coverage

The term *test coverage* is often used to mean the percentage of the code for which an associated test exists. You can measure this by running the test suite and counting the number of lines of code that were executed and dividing this by the total number of lines in the software project. If you have the `coverage` Python project installed (`pip install coverage`), you can run `nose` and generate coverage statistics simultaneously via the `--with-coverage` switch at the command line:

```
$ nosetests --with-coverage
```

At first glance this metric seems like a useful indicator of code reliability. But while some test coverage is superior to none and broad test coverage is usually superior to narrow coverage, this metric should be viewed critically. All code should ideally have 100% test coverage, but this alone does not guarantee that the code works as intended. Take the following pseudocode for a function `g()` shown here, with two `if-else` statements in its body:

```
def g(x, y):
    if x:
        ...
    else:
```

```
    ...
    if y:
        ...
    else:
        ...
    return ...
```

The following two unit tests for g() have 100% coverage:

```
from nose.tools import assert_equal

from mod import g

def test_g_both_true():
    exp = ...
    obs = g(True, True)
    assert_equal(exp, obs)

def test_g_both_false():
    exp = ...
    obs = g(False, False)
    assert_equal(exp, obs)
```

Every line of g() is executed by these two functions. However, only half of the possible cases are covered. We are not testing when x=True and y=False or when x=False and y=True. In this case, 100% coverage is only 50% of the possible *code path* combinations. In full software projects, 100% coverage is achieved with much less than 50% of the code paths been executed.

Code coverage is an important and often cited measure. However, it is not the pinnacle of testing. It is another tool in your testing toolbox. Use it as needed and understand its limitations.

The next section covers another tool, but one that changes the testing strategy itself.

Test-Driven Development

Test-driven development (TDD) takes the workflow of writing code and writing tests and turns it on its head. TDD is a software development process where you write the tests first. Before you write a single line of a function, you first write the test for that function.

After you write a test, you are then allowed to proceed to write the function that you are testing. However, you are only supposed to implement enough of the function so that the test passes. If the function does not do what is needed, you write another test and then go back and modify the function. You repeat this process of test-then-implement until the function is completely implemented for your current needs.

Developers who practice strict TDD will tell you that it is the best thing since sliced arrays. The central claim to TDD is that at the end of the process you have an implementation that is well tested for your use case, and the process itself is more efficient. You stop when your tests pass and you do not need any more features. You do not spend any time implementing options and features on the off chance that they will prove helpful later. You get what you need when you need it, and no more. TDD is a very powerful idea, though it can be hard to follow religiously.

The most important takeaway from test-driven development is that the moment you start writing code, you should be considering how to test that code. The tests should be written and presented in tandem with the implementation. Testing is too important to be an afterthought.

Whether to pursue classic TDD is a personal decision. This design philosophy was most strongly put forth by Kent Beck in his book *Test-Driven Development: By Example*. The following example illustrates TDD for a standard deviation function, std().

To start, we write a test for computing the standard deviation from a list of numbers as follows:

```
from nose.tools import assert_equal

from mod import std

def test_std1():
    obs = std([0.0, 2.0])
    exp = 1.0
    assert_equal(obs, exp)
```

Next, we write the *minimal* version of std() that will cause test_std1() to pass:

```
def std(vals):
    # surely this is cheating...
    return 1.0
```

As you can see, the minimal version simply returns the expected result for the sole case that we are testing. If we only ever want to take the standard deviation of the numbers 0.0 and 2.0, or 1.0 and 3.0, and so on, then this implementation will work perfectly. If we want to branch out, then we probably need to write more robust code. However, before we can write more code, we first need to add another test or two:

```
def test_std1():
    obs = std([0.0, 2.0])
    exp = 1.0
    assert_equal(obs, exp)

def test_std2():  ❶
    obs = std([])
    exp = 0.0
    assert_equal(obs, exp)
```

```
def test_std3():  ❷
    obs = std([0.0, 4.0])
    exp = 2.0
    assert_equal(obs, exp)
```

❶ Test the fiducial case when we pass in an empty list.

❷ Test a real case where the answer is not one.

A perfectly valid standard deviation function that would correspond to these three tests passing would be as follows:

```
def std(vals):
    # a little better
    if len(vals) == 0:  ❶
        return 0.0
    return vals[-1] / 2.0  ❷
```

❶ Special case the empty list.

❷ By being clever, we can get away without doing real work.

Even though the tests all pass, this is clearly still not a generic standard deviation function. To create a better implementation, TDD states that we again need to expand the test suite:

```
def test_std1():
    obs = std([0.0, 2.0])
    exp = 1.0
    assert_equal(obs, exp)

def test_std2():
    obs = std([])
    exp = 0.0
    assert_equal(obs, exp)

def test_std3():
    obs = std([0.0, 4.0])
    exp = 2.0
    assert_equal(obs, exp)

def test_std4():  ❶
    obs = std([1.0, 3.0])
    exp = 1.0
    assert_equal(obs, exp)

def test_std5():  ❷
    obs = std([1.0, 1.0, 1.0])
    exp = 0.0
    assert_equal(obs, exp)
```

❶ The first value is not zero.

❷ Here, we have more than two values, but all of the values are the same.

At this point, we may as well try to implement a generic standard deviation function. We would spend more time trying to come up with clever approximations to the standard deviation than we would spend actually coding it. Just biting the bullet, we might write the following implementation:

```
def std(vals):
    # finally, some math
    n = len(vals)
    if n == 0:
        return 0.0
    mu = sum(vals) / n
    var = 0.0
    for val in vals:
        var = var + (val - mu)**2
    return (var / n)**0.5
```

It is important to note that we could improve this function by writing further tests. For example, this `std()` ignores the situation where infinity is an element of the values list. There is always more that can be tested. TDD prevents you from going overboard by telling you to stop testing when you have achieved all of your use cases.

Testing Wrap-up

Testing is one of the primary concerns of scientific software developers. It is a technical solution to a philosophical problem. You should now be familiar with the following concepts in testing:

- Tests compare that the result observed from running code is the same as what was expected ahead of time.
- Tests should be written at the same time as the code they are testing is written.
- The person best suited to write a test is the author of the original code.
- Tests are grouped together in a test suite.
- Test frameworks, like `nose`, discover and execute tests for you automatically.
- An edge case is when an input is at the limit of its range.
- A corner case is where two or more edge cases meet.
- Unit tests try to test the smallest pieces of code possible, usually functions and methods.
- Integration tests make sure that code units work together properly.
- Regression tests ensure that everything works the same today as it did yesterday.

- Test generators can be used to efficiently check many cases.
- Test coverage is the percentage of the code base that is executed by the test suite.
- Test-driven development says to write your tests before you write the code that is being tested.

You should now know how to write software and how to follow the best practices that make software both useful and great. In the following chapters we will go over how you can let the world know about the wonderful things that you have done.

PART IV
Getting It Out There

CHAPTER 19
Documentation

> Computational science is a special case of scientific research: the work is easily shared via the Internet since the paper, code, and data are digital and those three aspects are all that is required to reproduce the results, given sufficient computation tools.
>
> —Victoria Stodden, "The Scientific Method in Practice: Reproducibility in the Computational Sciences"

Scientists are nomads. As students, they contribute to a piece of research for no more than four years at a time. As post-docs, their half-life on a project is even shorter. They disappear after three years, maximum. Even once they settle down as faculty or laboratory scientists, their workforce is composed primarily of these fly-by-night individuals. As such, research work in laboratories and universities occurs on a time scale rarely longer than the tenure of a typical PhD student.

In this environment, it is very common for scientists to crank out a piece of code as quickly as possible, squeeze a few publications out of it, and disappear to lands unknown. One victim in all of this is the student or researcher that follows them, seeking to extend their work. Since the first scientist working on a project valued speed over sustainability, the second researcher inherits a piece of code with no documentation. Accordingly, the original work, often termed "legacy code," seems to be understood only by its author. The new contributors to such projects often think to themselves that rewriting the code from scratch would be easier than deciphering the enigmas before them. The cycle, of course, repeats itself.

Why Prioritize Documentation?

Chronic inefficiency permeates this situation, fundamentally disrupting the forward progression of science. In her paper "Better Software, Better Research," Professor Carole Goble relates a favorite tweet on the topic:

One of my favorite #overlyhonestmethods tweets (a hashtag for lab scientists) is Ian Holmes's "You can download our code from the URL supplied. Good luck downloading the only postdoc who can get it to run, though."

Though the original tweet was intended as satire, it's almost too true to be funny. The status quo needs to change. Thankfully, there is hope. The whole culture of science does not adhere to this unfortunate state of affairs out of ill will or malice. It's all a simple misunderstanding—namely, that "Documentation is not worth the time it takes."

This chapter will explain why this statement is so wrong.

Documentation Is Very Valuable

The first false premise behind this statement is that documentation is not valuable. The truth is that documentation is valuable enough to be a top priority, almost irrespective of how much time it takes to generate it. Its value is paramount because:

- The value and extent of your work is clearer if it can be understood by colleagues.
- Documentation provides provenance for your scientific process, for your colleagues and yourself.
- Documentation demonstrates your skill and professionalism.

Other people will interact with your code primarily through its documentation. This is where you communicate the value and intent of your research work. However, the documentation serves as more than an advertisement to your colleagues. It guides the interest of those who might desire to comment on your work or collaborate with you on extensions to it. Somewhat cynically, in this way documentation is superior to modern archival publications, which rarely contain enough detail to fully reproduce work. Rather, they provide enough information to allow informed critique of the methods and serve, frankly, to publicize your efforts as a scientist.

In a similar vein, documentation provides *provenance for your scientific procedure.* That is, documentation is worthwhile because it preserves a record of your thought process. This becomes indispensable as time passes and you inevitably forget how your code works—just in time for a journal editor to ask about the features of the results. Rather than having to frantically reread the code in the hopes of stumbling upon its secrets, you'll have the documentation there to remind you of the equations you were implementing, the links to the journal articles that influenced your algorithm, and everything else that would, were this bench science, certainly be recorded in a laboratory notebook.

Documentation also acts as a demonstration of your skill and professionalism. Stating you have a piece of code is one thing, but without documentation, it will be difficult to demonstrate that this code is a professionally developed, polished piece of work that can be used by others. Furthermore, since most scientists labor under the false assumption that documentation is difficult and time-consuming to write, they will be all the more impressed with your efficiency.

Of course, they're wrong. Documentation is relatively easy; it can even be automated in many cases.

Documentation Is Easier Than You Think

The second false premise behind the idea that documentation isn't worth the effort is that writing documentation takes a lot of time. This is wrong for two reasons:

- Documentation pays for itself with the time it saves in the long run.
- Documentation requires little effort beyond writing the software itself.

Any time you spend on documentation will pay for itself with the time it will save in the long run. New users need either documentation or hand-holding, but hand-holding does not scale. Documentation, on the other hand, scales majestically. Fundamentally, if something is written down, it will never need to be explained again. All questions about how the software works can now be redirected to the user manual. Your brain, then, remains free for something else. Well-documented code is somewhat self-maintaining, because when someone new comes along to use your code, the documentation does the work of guiding them so you don't have to.

Even disregarding future time savings, producing documentation takes little effort beyond writing the software itself. Documentation can be easily streamlined into the programming workflow so that updates aren't a separate task. For every modern programming language, there is a framework for automatically generating a user manual based on well-formed comments in the source code (see "Automation" on page 436). These frameworks minimize the effort on the part of the developer and help to ensure that the documentation is always up to date, since it is version controlled right alongside the code. Additionally, the necessity for comments can be reduced with use of standardized style guides, descriptive variable naming, and concise functions.

Types of Documentation

Documentation comes in myriad forms. Each has its own purpose, benefits, and drawbacks. A single project may have all, some, or none of the following types of documentation. Ideally, they all work together or at least exhibit some separation of concerns. Types of documentation often encountered in research software include:

- Theory manuals
- User and developer guides
- Code comments
- Self-documenting code
- Generated API documentation

We'll look at each of these, beginning with the one academics are typically most familiar with: the theory manual.

Theory Manuals

In the universe of academic and research science, the theory manual most often takes the form of a dissertation describing the theoretical foundations of the code base that existed on the day of the defense. Depending on the origin of the code and the career stage of the lead developer, the theory manual can also take the form of a series of white papers, journal articles, or internal reports. Whatever the case may be, a theory manual has a number of distinctly helpful qualities:

- It captures the scientific goals and provenance of the code.
- It has been peer-reviewed.
- It is archived.
- It can be cited.

However, theory manuals have disadvantages as well. Typically:

- They represent significant effort.
- They are not living documents.
- They do not describe implementation.
- They are not stored alongside the code.

A theory manual is a decidedly necessary and important piece of the documentation menagerie for research software. However, integrating additional documentation into the software development workflow can break the problem into more manageable tasks, allow the documentation to evolve along with the code base, and illuminate implementation decisions.

The theory manual, as its title might suggest, describes the theory, but rarely describes the implementation.

User and Developer Guides

Similar to theory manuals, user guides often accompany mature research software. These documents address more important implementation details and instruction for use of the software. Unless generated automatically, however, they also represent significant effort on the part of the developers and are typically updated only when the developers release a new version of the code.

Readme Files

In many code projects, a plain-text file sits among the source code files. With a name like "readme," it hopes not to be ignored. In most projects, the file is located in the top-level directory and contains all the necessary information for installing, getting started with, and understanding the accompanying code. In other projects, however, a readme file might live in every directory or might be accompanied by other files with more specific goals, like:

- *install*
- *citation*
- *license*
- *release*
- *about*

However, readme files are very common, especially in projects where users or developers are likely to install the code from source. Since readme files are as unique as the developers who write them, their contents are not standardized. However, the following is an example:

```
SQUIRREL, version 1.2 released on 2026-09-20

# About

The Spectral Q and U Imaging Radiation Replicating Experimental Library
(SQUIRREL) is a library for replicating radiation sources with spectral details
and Q and U polarizations of superman bubblegum.

# Installation

The SQUIRREL library relies on other libraries:

- The ACORN library www.acorn.nutz
- The TREEBRANCH database format API

Install those before installing the SQUIRREL library. To install the SQUIRREL
library:
```

```
./configure
make --prefix=/install/path
make install

...
```

Rather than being archived in the university library, in a journal article, or in a printed, bound copy on the shelf of the lead developer, the readme lives alongside the code. It is therefore more easily discoverable by individuals browsing the source code on their own. GitHub, in a nod to the ubiquity of the readme file, renders each readme file on the landing page of the directory containing it.

However, a readme is only one plain-text file, so it can only reasonably hope to communicate the very bare minimum of information about the code base. Techniques that improve readme files include markup formats, installation instructions, minimal examples, and references to additional information.

Comments

A comment is a line in code that is not run or compiled. It is merely there for the benefit of the reader, to help with interpreting code. Comments, ideally, assist us when we face code written by other people or, often, our past selves. As discussed in previous chapters, code comments are denoted syntactically by special characters and are not read when the code is executed.

Code commenting syntax provides a mechanism for inserting metainformation intended to be read by human eyes. In Python, comments can be denoted by a few different special characters. The # precedes comments that occupy one line or less. For longer comments and docstrings, triple quotes or apostrophes are used:

```
def the_function(var):
    """This is a docstring, where a function definition might live"""
    a = 1 + var  # this is a simple comment
    return a
```

However, comments can also pollute code with unnecessary cruft, as in the following example:

```
def decay(index, database):
    # first, retrieve the decay constants from the database
    mylist = database.decay_constants()
    # next, try to access an element of the list
    try:
        d = mylist[index] # gets decay constant at index in the list
    # if the index doesn't exist
    except IndexError:
        # throw an informative error message
        raise Exception("value not found in the list")
    return d
```

In this way, it is decidedly possible to over-document code with clutter. Comments should never simply repeat what the code is doing. Code, written cleanly, will have its own voice.

Nearly all of the comments in the previous example are unnecessary. It is obvious, for example, that database.decay_constants() retrieves decay constants from the database object. Due to good variable naming, the comment adds nothing extra.

Indeed, the need for most comments can be reduced with intelligent naming decisions. For example, if the variable d in the preceding example were instead called decay_constant or lambda, the standard mathematical symbol for the decay constant, the purpose of that line of code would be clear even without the comment. A better version of this function might be:

```
def decay(index, database):
    lambdas = database.decay_constants()
    try:
        lambda_i = lambdas[index]    # gets decay constant at index in the list
    except IndexError:
        raise Exception("value not found in the list")
    return lambda
```

Finally, comments can get out of date if they are not updated along with the code. Even though they're immediately adjacent to the code they describe, they're easy to miss when fixing a bug on the fly. For example, imagine that a change is made elsewhere in the code base such that the database.decay_constants() function starts to return a dictionary, rather than a list.

The keys are all the same as the previous indices, so this doesn't cause a problem for the decay function. It still passes all but one of the tests: the one that checks the exception behavior. That test fails because an IndexError is no longer raised for the wrong index. Instead, because the dictionary analogy to IndexError is KeyError, what is raised is a KeyError. This is not caught by the except clause, and the test fails.

To fix this problem, the developer changes the caught exception to the more general LookupError, which includes both IndexErrors and KeyErrors:

```
def decay(index, database):
    lambdas = database.decay_constants()
    try:
        lambda_i = lambdas[index]    # gets decay constant at index in the list
    except LookupError:
        raise Exception("value not found in the decay constants object")
    return lambda
```

However, when making the change, the developer may never have laid eyes on any other line in this function. So, the comment has remained and states that lambdas is a list. For new users of the code, the comment will lead them to believe that the decay_constants object is a list.

How would you fix this code? Perhaps the whole function is better off without the comment entirely. Can you think of anything else that should be changed in this example? The answers to both of these questions can be found in the concept of *self-documenting code*.

Self-Documenting Code

The only documentation that is compiled and tested for accuracy along with the code *is the code*.

In the exceptional book *Clean Code*, Robert C. Martin discusses many best practices for self-documenting code. Most of his principles of clean, self documenting code revolve around the principle that the code should be understandable and should speak for itself. Transparently written, clean code, after all, hides bugs poorly and frightens away fewer developers. We'll look at a few of those best practices here.

Naming

Chief among best practices is naming, which has already been covered somewhat. A variable, class, or function name, Martin says:

> …should answer all the big questions. It should tell you why it exists, what it does, and how it is used. If a name requires a comment, then the name does not reveal its intent.

In the previous example, among other things that should be changed, the `decay()` function should probably be renamed to `decay_constant()`. For more clarity, one might consider `get_decay_constant()` or `get_lambda()` so that the user can guess that it actually returns the value.

Simple functions

As has been mentioned previously, especially in Chapter 18, functions must be small in order to be understandable and testable. In addition to this, they should *do only one thing*. This rule helps code readability and usability enormously. When hidden consequences are not present in a function, the DRY (don't repeat yourself) principle can be used confidently.

Consistent style

Finally, a key feature in readability is rich syntactic meaning. Programming languages derive their vast power from the density of meaning in their syntax. However, any language can be made rich beyond its defined parameters by use of consistent, standardized style.

When variable and function names are chosen with a particular syntactic style, they will speak volumes to the trained eye. Every language has at least one commonly used style guide that establishes a standard. In Python, that style guide is PEP8 (*https://www.python.org/dev/peps/pep-0008/*).

In addition to dictating the proper number of spaces of indentation in Python code, PEP8 also suggests variable and function naming conventions that inform the developer of the intended purpose and use of those variables and functions. In particular:

```
# packages and modules are short and lowercase
packages
modules

# other objects can be long
ClassesUseCamelCase
ExceptionsAreClassesToo
functions_use_snake_case
CONSTANTS_USE_ALL_CAPS

# variable scope is *suggested* by style convention
_single_leading_underscore_      # internal to module
single_trailing_underscore_      # avoids conflicts with Python keywords
__double_leading_and_trailing__  # these are magic, like __init__
```

The syntactic richness demonstrated here increases the information per character of code and, accordingly, its power.

Docstrings

As discussed in Chapter 5, Python documentation relies on docstrings within functions. As a reminder, a docstring is placed immediately after the function declaration and is the first unassigned string literal. It must occur before any other operations in the function body. To span multiple lines, docstrings are usually enclosed by three pairs of double quotes:

```
def <name>(<args>):
    """<docstring>"""
    <body>
```

Docstrings should be descriptive and concise. They provide an incredibly handy way to convey the intended use of the functions to users. In the docstring, it is often useful to explain the arguments of a function, its behavior, and how you intend it to be used. The docstring itself is available at runtime via Python's built-in `help()` function and is displayed via IPython's ? magic command. The Python automated documentation framework, Sphinx, also captures docstrings. A docstring could be added to the `power()` function as follows:

```
def power(base, x):
    """Computes base^x. Both base and x should be integers,
    floats, or another numeric type.
    """
    return base**x
```

In addition to giving your audience the gift of informative type definitions and variable names, it is often useful to explain a class, its purpose, and its intended contents in a comment near its declaration. Python does this using docstrings as well:

```
class Isotope(object):
    """A class defining the data and behaviors of a radionuclide.
    """
```

Further documentation about Python docstrings can be found in PEP257 (*http://bit.ly/PEP257*). Additionally, docstrings are an excellent example of comments that can be structured for use with automated documentation generators. For more on their importance in the use of Sphinx, read on.

Automation

While taking the time to add comments to code can be tedious, it pays off handsomely when coupled with an automated documentation generation system. That is, if comments are constructed properly, they can be read and interpreted, in the context of the code, to generate clickable, interactive documentation for publication on the Internet.

Tools for automatically creating documentation exist for every language. Table 19-1 shows a few of the most popular offerings. In Java, it's Javadoc (*http://bit.ly/javadoc-tool*); for C and C++, a common tool is Doxygen (*http://doxygen.org/*). For Python, the standard documentation generator is Sphinx (*http://sphinx-doc.org/*).

Table 19-1. Automated documentation frameworks

Name	Description
Doxygen	Supports marked-up comments, created for C++
Javadoc	Supports marked-up comments, created for Java
Pandoc	Supports Markdown, reStructuredText, LaTeX, HTML, and others
Sphinx	Standard Python system; supports reStructuredText

With these tools, well-formed comments in the code are detected and converted into navigable API documentation. For an example of the kind of documentation this can create, browse the documentation for the Python language (version 3) (*http://docs.python.org/3*). In keeping with our focus on Python, we'll look at Sphinx here.

Sphinx

Sphinx was created to automate the generation of the online Python 3 API documentation. It is capable of creating theory manuals, user guides, and API documentation in HTML, LaTeX, ePub, and many other formats. It does this by relying on restructured text files defining the content. With an extension called "autodoc," Sphinx is also capable of using the docstrings in source code to generate an API-documenting final product.

Sphinx is a documentation system primarily for documenting Python code. This section will simply detail getting started with Sphinx and the autodoc extension. For a more detailed tutorial on Sphinx, see the Sphinx documentation (*http://sphinx-doc.org*).

Getting started

Sphinx is packaged along with any scientific Python distribution (like Anaconda or Canopy). The tool itself provides a "quickstart" capability. This section will cover how to use that quickstart capability to build a simple website with *.rst* files and the comments in source code.

As an example, we'll use the object code used to demonstrate classes in Chapter 6. Documentation for this code can be generated in a few simple steps. First, enter the directory containing the source code and create a directory to contain the documentation:

```
~ $ cd book-code/obj
~/book-code/obj $ mkdir doc
```

Next, enter the *doc* directory and execute the Sphinx quickstart utility:

```
~/book-code/obj $ cd doc
~/book-code/obj/doc $ sphinx-quickstart
```

This utility is customized by answers from the user, so be ready to answer a few questions and provide some details about your project. If unsure about a question, just accept the default answer. To prepare for automatic documentation generation, be sure to answer "yes" to the question about autodoc ("autodoc: automatically insert docstrings from modules (y/n)").

This step allows the documentation's arrangement to be customized carefully. It will create a few new files and directories. Typically, these include:

- A *source* directory for holding .*rst* files, which can be used to hold user guides and theory manual content or to import documentation from the code package
- A makefile that can be used to generate the final product (by executing `make html`, in this case)

- A *build* directory to hold the final product (in this case, *.html* files comprising the documentation website)

Once the quickstart step has been completed, you can modify the files in the *source* directory and add to them in order to create the desired structure of the website. The *source* directory will include at least:

- A *conf.py* file, which can be used to customize the documentation and define much of the metadata for your project
- An *index.rst* file, which will be the landing page of the website and can be customized to define the structure of its table of contents

The documentation in the *build* directory is based on files in the *source* directory. To include documentation for a particular module, such as *particle.py*, you can create a corresponding *.rst* file (*particle.rst*) that invokes autodoc on that class. The *index.rst* file must also be modified to include that file. In the end, our *index.rst* file should look like:

```
.. particles documentation master file, created by
   sphinx-quickstart on Sun Jan  1 23:59:59 2999.
   You can adapt this file completely to your liking, but it should at least
   contain the root `toctree` directive.

Welcome to particles's documentation!
=====================================

Contents:

.. toctree::
   :maxdepth: 2



Indices and tables
==================

* :ref:`genindex`
* :ref:`modindex`
* :ref:`search`

API
====
.. toctree::
   :maxdepth: 1

   particle
```

And the *particle.rst* file should look like:

```
.. _particles_particle:

Particle -- :mod:`particles.particle`
=====================================

    .. currentmodule:: particles.particle

    .. automodule:: particles.particle

    All functionality may be found in the ``particle`` package::

        from particles import particle

    The information below represents the complete specification of the classes in
    the particle module.

    Particle Class
    **************

    .. autoclass:: Particle
```

Now, Sphinx has been informed that there is a *particle.py* module in the `particles` package, and within that module is a `Particle` class that has docstrings to be included in the documentation. This will work best if the docstrings are well formed. Read on to find out more about how to format your docstrings for Sphinx.

Comment style

You can get more functionality out of Sphinx by formatting your docstrings in a syntax that it can parse easily. While Sphinx will often pick up comments just before a function declaration, even if it is blank, you can control more of its behavior with specific notation. A reference for this notation is on the Sphinx website (*http://sphinx-doc.org/tutorial.html*), but to give you an idea, here is an example of the Sphinx syntax for documenting a function:

```
.. function:: spin(self, s)

    Set the spin of the particle to the value, s.
```

You can also add more detail, with specific syntax. With the help of this syntax, Sphinx can interpret parts of the comment that are intended to illuminate the parameters or the return value, for instance. In this case the function might have a comment like:

```
.. function:: spin(self, s)

    Set the spin of the particle to the value, s.

    :param s: the new spin value
    :type s: integer or float
    :rtype: None
```

Now, armed with this syntax, take some time with a Python code base of your own. Go back and make appropriate changes to the comments in that code in order to provide Sphinx-style syntax for some of the key functions, classes, and variables. Then, try running `sphinx-quickstart` and the Sphinx autodoc extension to generate documentation accordingly.

Documentation Wrap-up

In this chapter, you have learned how to use comments to communicate the meaning and purpose of your code to future users unfamiliar with its implementation. Proper documentation will have an enormous impact on the usability and reusability of your software, both by others and your future self. Additionally, you have learned how to automate the generation of interactive and comprehensive API documentation based on appropriately styled comments.

Equipped with these skills, you can distribute code to your colleagues and it will serve them as more than just a black box. With proper API documentation, your code becomes a legitimate research product. Of course, even though code is one of the most useful kinds of modern research product, recognition is hard to gain without journal publication as well. For help becoming more effective at publication, proceed to the next chapter.

CHAPTER 20
Publication

> In science one tries to tell people, in such a way as to be understood by everyone, something that no one ever knew before. But in poetry, it's the exact opposite.
>
> —Paul Dirac

> One day, I'll find the right words, and they will be simple.
>
> —Jack Kerouac

Publishing is an integral part of science. Indeed, the quality, frequency, and impact of publication records make or break a career in the physical sciences. Publication can take up an enormous fraction of time in a scientific career. However, with the right tools and workflow, a scientist can reduce the effort spent on mundane details (e.g., formatting, reference management, merging changes from coauthors) in order to spend more time on the important parts (e.g., literature review, data analysis, writing quality). This chapter will emphasize tools that allow the scientist to more efficiently accomplish and automate the former in order to focus on the latter. It will cover:

- An overview of document processing paradigms
- Employing text editors in document editing
- Markup languages for text-based document processing
- Managing references and automating bibliography creation

The first of these topics will be an overview of two competing paradigms in document processing.

Document Processing

Once upon a time, *Homo sapiens* etched our thoughts into stone tablets, on papyrus, and in graphical document processing software. All of these early tools share a commonality. They present the author with a "What You See Is What You Get" (WYSIWYG) paradigm in which formatting and content are inextricably combined. While this paradigm is ideal for artistic texts and documents with simple content, it can distract from the content in a scientific context and can make formatting changes a nontrivial task. Additionally, the binary format of most WYSIWYG documents increases the difficulty of version control, merging changes, and collaborating on a document.

Common document processing programs include:

- Microsoft Word
- Google Docs
- Open Office
- Libre Office

Though these tools have earned an important seat at the table in a business environment, they lack two key features that assist in efficient, reproducible paper-writing workflows. The first shortcoming is that they fail to separate the content (text and pictures) from the formatting (fonts, margins, etc.).

Separation of Content from Formatting

In the context of producing a journal publication, formatting issues are purely a distraction. In a WYSIWYG word processor, the act of choosing a title is polluted by the need to consider font, placement, and spacing. In this way, WSYIWYG editors fail to separate content from formatting. They thus prevent the author from focusing on word choice, clarity of explanation, and logical flow.

Furthermore, since each journal requires submissions to follow unique formatting guidelines, an efficient author avoids formatting concerns until a journal home is chosen. The wise author separates the content from the formatting since that choice may even be reevaluated after a rejection.

For all of these reasons, this chapter recommends a What You See Is What You Mean (WYSIWYM) document processing system for scientific work. Most such systems are plain text–based and rely on markup languages. Some common systems include:

- LaTeX (*http://www.latex-project.org/*)
- DocBook (*http://www.docbook.org/*)
- AsciiDoc (*http://www.methods.co.nz/asciidoc/*)
- PanDoc (*http://johnmacfarlane.net/pandoc/*)

Among these, this chapter recommends the LaTeX context. Due to its powerful interface, beautiful mathematical typesetting, and overwhelming popularity in the physical sciences, LaTeX is a fundamental skill for the effective researcher in the physical sciences.

Plain-text WYSIWYM tools such as these cleanly separate formatting from content. In a LaTeX document, layout specifications and formatting choices can be placed in a completely separate plain-text file than the ones in which the actual content of the paper is written. Because of this clean separation, switching from a document layout required by journal A to the layout required by journal B is done by switching the style files accordingly. The content of the paper is unaffected.

Additionally, this clean separation can enable efficient reference management. In the LaTeX context, this chapter will cover how this is achieved with bibliography files.

One more reason we recommend WYSIWYM editors over WYSIWYG document-processing tools is related to reproducibility: they facilitate tracking changes.

Tracking Changes

At the advent of computing, all information was stored as plain text. Now, information is stored in many complex binary formats. These binary formats are difficult to version control, since the differences between files rarely make logical sense without binary decoding. Many WYSIWYG document processors rely on such binary formats and are therefore difficult to version control in a helpful way.

Of course, an adept user of Microsoft Word will know that changes in that program can be tracked using its internal proprietary track-changes tool. While this is a dramatic improvement and enables concurrent efforts, more transparent and robust versioning can be achieved with version-controlled plain-text files. Since Microsoft's model requires that one person maintain the master copy and conduct the merges manually, concurrent editing by multiple parties can become untenable.

At this point in the book, you should have an appreciation of the effectiveness, efficiency, and provenance of version-controllable plain-text formats. Of course, for the same reasons, *we strongly recommend the use of a plain-text markup language for document processing*. Accordingly, we also recommend choosing a text editor appropriate for editing plain-text markup.

Text Editors

The editor used to write code can also be used to write papers. As mentioned in Chapter 1, text editors abound. Additionally, most text editors are very powerful. Accordingly, it can be a challenge to become proficient in the many features of more than one text editor. When new programmers seek to make an informed decision about which text editor to learn and use, many well-meaning colleagues may try to influence their choice.

However, these well-meaning colleagues should be largely ignored. A review of the features of a few common text editors (e.g., *vi*, *emacs*, *eclipse*, *nano*) should sufficiently illuminate the features and drawbacks of each.

> Another argument for the use of plain-text markup is exactly this array of available text editors. That is, the universality of plain-text formatting and the existence of an array of text editors allows each collaborator to choose a preferred text editor and still contribute to the paper. On the other hand, with WYSIWYG, proprietary formats require that everyone must use the same tool.

Your efficiency with your chosen editor is more important than which text editor you choose. Most have a basic subset of tools (or available plug-ins) to accomplish:

- Syntax highlighting
- Text expansion
- Multiple file buffers
- Side-by-side file editing
- In-editor shell execution

Technical writing in a text editor allows the distractions of formatting to be separated from the content. To achieve this, the elements of a document are simply "marked up" with the special text syntax of a markup language.

Markup Languages

Markup languages provide syntax to annotate plain-text documents with structural information. A *build* step then produces a final document by combining that textual content and structural information with separate files defining styles and formatting. Most markup languages can produce multiple types of document (i.e., letters, articles, presentations) in many output file formats (*.pdf*, *.html*).

The ubiquitous HyperText Markup Language (HTML) may provide a familiar example of this process. Plain-text HTML files define title, heading, paragraph, and link elements of a web page. Layouts, fonts, and colors are separately defined in CSS files. In this way, web designers can focus on style (CSS) while the website owner can focus on the content (HTML).

This chapter will focus on the LaTeX markup language because it is the standard for publication-quality documents in the physical sciences. However, it is not the only available markup language. A few notable markup languages include:

- LaTeX

- Markdown
- reStructuredText
- MathML
- OpenMath

Markdown and reStructuredText both provide a simple, clean, readable syntax and can generate output in many formats. Python and GitHub users will encounter both formats, as reStructuredText is the standard for Python documentation and Markdown is the default markup language on GitHub. Each has syntax for including and rendering snippets of LaTeX. MathML and its counterpart OpenMath are optional substitutes for LaTeX, but lack its powerful extensions and wide adoption.

In markup languages, the term *markup* refers to the syntax for denoting the structure of the content. Content structure, distinct from formatting, enriches plain-text content with meaning. Directives, syntactically defined by the markup language, denote or *mark up* titles, headings, and sections, for example. Similarly, special characters mark up document elements such as tables, links, and other special content. Finally, rather than working in a single huge document, most markup languages enable constructing a document from many subfiles. In this way, complex file types, like images, can remain separate from the textual content. To include an image, the author simply references the image file by providing its location in the filesystem. In this way, the figures in a paper can remain in their native place on the filesystem and in their original file format. They are only pulled into the final document during the build step.

The build step is governed by the processing tool. For HTML, the tool is your browser. For the LaTeX markup language, however, it is the LaTeX program. The next section will delve deeper into LaTeX.

LaTeX

LaTeX (pronounced lay-tekh or lah-tekh) is the standard markup language in the physical sciences. Based on the TeX literate programming language, LaTeX provides a markup syntax customized for the creation of beautiful technical documents.

At a high level, a LaTeX document is made up of distinct constituent parts. The main file is simply a text file with the *.tex* file extension. Other LaTeX-related files may include style files (*.sty*), class files (*.cls*), and bibliography files (*.bib*). However, only the *.tex* file is necessary. That file need only contain four lines in order to constitute a valid LaTeX document. The first line chooses the type of document to create. This is called the LaTeX *document class*.

LaTeX document class

The first required line defines the type of document that should result. Common default options include `article`, `book`, and `letter`. The syntax is:

```
\documentclass{article}
```

This is a typical LaTeX command. It has the format:

```
\commandname[options]{argument}
```

The `documentclass` type should be listed in the curly braces. Options concerning the paper format and the font can be specified in square brackets before the curly braces. However, they are not necessary if the default styles are desired.

Note that many journals provide something called a *class file* and sometimes a *style file*, which contain formatting commands that comply with their requirements. The class file fully defines a LaTeX document *class*. So, for example, the journal publisher Elsevier provides an `elsarticle` document class. In order to convert any article into an Elsevier-compatible format, simply download the *elsarticle.cls* file to the directory containing the *.tex* files, and change the `documentclass` command argument to `elsarticle`. The rest of the document can stay the same.

The next two necessary lines are the commands that begin and end the document *environment*.

LaTeX environments

LaTeX environments are elements of a document. They can contain one another, like Russian dolls, and are denoted with the syntax:

```
\begin{environment} ... \end{environment}
```

`\begin{environment}` and `\end{environment}` are the commands that indicate environments in LaTeX. The top-level environment is the document environment. The document class, packages used, new command definitions, and other metadata appear before the document environment begins. This section is called the *preamble*. Everything after the document environment ends is ignored. For this reason, the `\begin{document}` command and the `\end{document}` command must each appear exactly once:

```
\documentclass{article}
\begin{document}
\end{document}
```

Since all actual content of the document appears within the document environment, between the `\begin{document}` and `\end{document}` commands, the shortest possible valid LaTeX file will include just one more line, a line of content!

```
\documentclass{article}
\begin{document}
Hello World!
\end{document}
```

This is a completely valid LaTeX document. Note that no information about fonts, document layout, margins, page numbers, or any other formatting details need clutter this document for it to be valid. However, it is only plain text right now. To render this text as a PDF, we must build the document.

Building the document

If the preceding content is placed into a document—say, *hello.tex*—a PDF document can be generated with two commands. The first runs the *latex* program, which compiles and renders a *.dvi* file. The second converts the *.dvi* file to the portable document format *.pdf*:

```
$ latex hello.tex      ❶
$ dvipdf hello.dvi     ❷
```

❶ LaTeX uses the *.tex* file to create a *.dvi* file.

❷ The *.dvi* file can be directly converted to *.pdf* with *dvipdf*.

Alternatively, if *pdflatex* is installed on your computer, that command can be used to accomplish both steps at once:

```
$ pdflatex hello.tex
```

As shown in Figure 20-1, the document is complete and contains only the text "Hello World!"

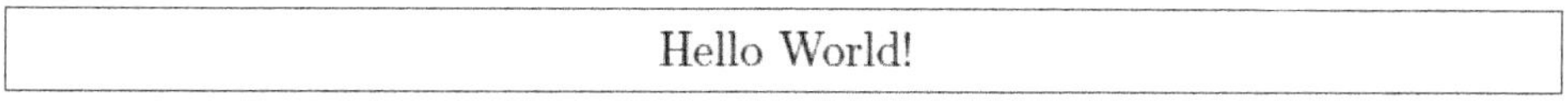

| Hello World! |

Figure 20-1. Hello World!

Now that the simplest possible document has been created with LaTeX, this chapter can move on to using LaTeX to produce publication-quality scientific documents. The first step will be to show how to appropriately mark up metadata elements of the document, such as the author names and title.

LaTeX metadata

Document metadata, such as the title of the document and the name of the author, may appear in many places in the document, depending on the layout. To make these special metadata variables available to the whole document, we define them in a scope outside of the document environment. The preamble holds information that can help to define the document; it typically includes, at the very minimum, a \title{}* and \author{}, but can include other information as well.

When Ada Lovelace, often cited as history's first computer programmer, began to write the first mathematical algorithm, she wrote it in impeccable Victorian handwriting on reams of paper before it was typeset and reprinted by a printing press. This algorithm appeared in the appendices of a detailed technical description of its intended computer, Charles Babbage's Analytical Engine. The document itself, cleverly crafted in the span of nine months, contained nearly all the common features of a modern article in the physical sciences. It was full of mathematical proofs, tables, logical symbols, and diagrams. Had she had LaTeX at her disposal at the time, Ada might have written the document in LaTeX. She would have begun the document with metadata in the preamble as seen here:

```
% notes.tex  ❶
\documentclass[11pt]{article}  ❷

\author{Ada Augusta, Countess of Lovelace}  ❸
\title{Notes By the Translator Upon the Memoir: Sketch of the Analytical Engine Invented by Charles Babbage}  ❹
\date{October, 1842}  ❺
\begin{document}  ❻
\maketitle  ❼
\end{document}  ❽
```

❶ In LaTeX, comments are preceded by a percent symbol.

❷ Ada would like to create an article-type document in 11pt font.

❸ She provides her formal name as the author metadata.

❹ She provides the full title.

❺ Another piece of optional metadata is the date.

❻ The document environment begins.

❼ The \maketitle command is executed. It uses the metadata to make a title.

❽ The document environment ends.

> # Notes By the Translator Upon the Memoir: Sketch of the Analytical Engine Invented by Charles Babbage
>
> Ada Augusta, Countess of Lovelace
>
> October, 1842

Figure 20-2. A Title in LaTeX

Ada's name, as well as the title of the article, should be defined in the preamble. However, they are only rendered into a main heading in the document with the use of the \maketitle command, which takes no arguments and must be executed within the document environment. The document that is produced appears in Figure 20-2.

> **Exercise: Create a Document with Metadata**
>
> 1. Create the *notes.tex* file in the previous code listing.
> 2. Run `latex notes.tex` and `dvipdf notes.tex` to create a *.pdf*.
> 3. View it.
> 4. Remove the value for the date so that it reads \date{}.
> 5. Repeat steps 2 and 3. What changed?

Now that the basics are clear, scientific information can be added to this document. In support of that, the document will need some underlying structure, such as sections and subsections. The next section will show how LaTeX markup can be used to demarcate those structural elements of the document.

Document structure

In the body of the document, the document structure is denoted by commands declaring the titles of each structural element. In the `article` document class, these include sections, subsections, subsubsections, and paragraphs. In a book, the structure includes parts and chapters as well. Ada's foundational notes were lettered A through G. The body of her document, therefore, would have included one \section command for each section:

```
% notes.tex
\documentclass[11pt]{article}
```

```
\author{Ada Augusta, Countess of Lovelace}
\title{Notes By the Translator Upon the Memoir: Sketch of the Analytical Engine
Invented by Charles Babbage}
\date{October, 1842}
\begin{document}
\maketitle

\section{Note A}
\section{Note B}
\section{Note C}
\section{Note D}
\section{Note E}
\section{Note F}
\section{Note G}

\end{document}
```

Since each note is a separate entity, however, it may be wise for Ada to keep them in separate files to simplify editing. In LaTeX, rather than keeping all the sections in one big file, Ada can include other LaTeX files in the master file. If the content of Note A, for example, is held in its own *intro.tex* file, then Ada can include it with the \input{} command. In this way, sections can be moved around during the editing process with ease. Additionally, the content is then stored in files named according to meaning rather than document order:

```
\section{Note A}
\input{intro}

\section{Note B}
\input{storehouse}

...

\section{Note G}
\input{conclusion}
```

Any text and LaTeX syntax in *intro.tex* will be inserted by LaTeX at the line where the command appeared. This multiple-file-inclusion paradigm is very powerful and encourages the reuse of document subparts. For example, the text that acknowledges your grant funding can stay in just one file and can be simply input into each paper.

Now that the document has a structure, we can get to work filling in the text and equations that make up the content of the paper. That will utilize the most important capability in LaTeX: typesetting math.

Typesetting mathematical formulae

LaTeX's support for mathematical typesetting is unquestionably the most important among its features. LaTeX syntax for typesetting mathematical formulae has set the standard for technical documents. Publication-quality mathematical formulae must include beautifully rendered Greek and Latin letters as well as an enormous array of logical, mathematical symbols. Beyond the typical symbols, LaTeX possesses an enormous library of esoteric ones.

Some equations must be rendered inline with paragraph text, while others should be displayed on a separate line. Furthermore, some must be aligned with one another, possess an identifying equation number, or incorporate interleaved text, among other requirements. LaTeX handles all of these situations and more.

To render math symbols or equations inline with a sentence, LaTeX math mode can be denoted with a simple pair of dollar signs ($). Thus, the LaTeX syntax shown here is rendered as in Figure 20-3:

```
The particular function whose integral the Difference Engine was constructed to
tabulate, is $\Delta^7u_x=0$. The purpose which that engine has been specially
intended and adapted to fulfil, is the computation of nautical and astronomical
tables. The integral of $\Delta^7u_x=0$ being    $u_z =
a+bx+cx^2+dx^3+ex^4+fx^5+gx^6$, the constants a, b, c, &c. are represented on the
seven columns of discs, of which the engine consists.
```

Note the dollar signs denoting the beginning and end of each inline mathematical equation. In an equation, mathematical markup can be used. Symbols, like the capital Greek letter delta, are denoted with a backslash. The caret (^) indicates a following superscript, and an underscore (_) means subscript.

> The particular function whose integral the Difference Engine was constructed to tabulate, is $\Delta^7 u_x = 0$. The purpose which that engine has been specially intended and adapted to fulfil, is the computation of nautical and astronomical tables. The integral of $\Delta^7 u_x = 0$ being $u_z = a + bx + cx^2 + dx^3 + ex^4 + fx^5 + gx^6$, the constants a, b, c, etc. are represented on the seven columns of discs, of which the engine consists.

Figure 20-3. Inline equations

Alternatively, to display one or more equations on a line separated from the text, an equation-focused LaTeX environment is used:

```
In fact the engine may be described as being the material expression
of any indefinite function of any degree of generality and complexity,
such as for instance,

\begin{equation} ❶
F(x, y, z, \log x, \sin y, x^p), ❷
\end{equation}

which is, it will be observed, a function of all other possible
functions of any number of quantities.
```

❶   An equation environment denotes an equation separated from the text, nicely centered.

❷   In this environment, mathematical markup can be used.

The equation is thereby drawn out of the text and is automatically given an equation number, as in Figure 20-4.

> The following is a more complicated example of the manner in which the engine would compute a trigonometrical function containing variables. To multiply
>
> $$F(x, y, z, \log x, \sin y, x^p) \qquad (1)$$
>
> which is, it will be observed, a function of all other functions of any number of quantities.

Figure 20-4. The equation environment

LaTeX enables a multitude of such mathematical typesetting conventions common to publications in the physical sciences. For example, multiple equations can be beautifully aligned with one another using the align math environment and ampersands (&) to mark the point of alignment. The American Mathematical Society made this possible by creating a package that it has made available to LaTeX users. To use this aligning environment, Ada will have to load the appropriate package when running LaTeX. That is done in the preamble.

Packages Extend LaTeX Capabilities

In addition to metadata, the preamble often declares the inclusion of any packages that the document relies on. Standard packages include amsmath, amsfonts, and amssymb. These are the American Mathematical Society packages for math layouts, math fonts, and math symbols, respectively. Another common package is graphicx, which allows the inclusion of *.eps* figures.

The align environment is available in the amsmath package, so if Ada wants to use it, she must include that package in her preamble. To enable an extended library of symbols, she might also include the amssymb package. Finally, since the description of the Bernoulli algorithm for the Analytical Engine required enormous, detailed tables spanning many pages, Ada might have also wanted to use the longtable package, which enables flexible tables that break across pages. Here are the lines she'll need to add to her preamble:

```
\usepackage{amsmath}
\usepackage{amssymb}
\usepackage{longtable}
```

If she has added the amsmath package to her preamble, Ada can thus make beautifully aligned equations with the align environment, as in the snippet shown here (rendered in Figure 20-5):

```
The following is a more complicated example of the manner in which the
engine would compute a trigonometrical function containing variables.
To multiply

\begin{align}
&A+A_1 \cos \theta + A_2\cos 2\theta + A_3\cos 3\theta + \cdots    ❶
\intertext{by}     ❷
&B + B_1 \cos \theta.   ❸
\end{align}
```

❶ The ampersand marks the place in this equation that should line up with the next equation.

❷ The progression of mathematical operations can be documented with common interleaved phrases such as "where," "such that," or "which reduces to." To interleave such text in a math environment, the \intertext command is used.

❸ The ampersand in the second equation marks the place in this equation that lines up with the ampersand of the first equation.

The following is a more complicated example of the manner in which the engine would compute a trigonometrical function containing variables. To multiply

$$A + A_1 \cos \theta + A_2 \cos 2\theta + A_3 \cos 3\theta + \qquad (1)$$

by

$$B + B_1 \cos \theta. \qquad (2)$$

Figure 20-5. Aligned LaTeX

As you can see, the equations line up just where the ampersands were placed, but the ampersands do not appear in the rendered document. Of course, this is only a taste of mathematical typesetting capabilities in LaTeX, and equations are only half the battle.

How does LaTeX handle other elements of technical documents, such as tables and figures?

Tables and figures

Tables and figures also often belong in their own files. In addition to the simplicity gained by keeping such elements outside of the text, reusing these elements in other documents becomes much simpler if they are kept in their own files. LaTeX is capable (with the beamer package) of generating presentation-style documents, and these files can be reused in those documents with a simple reference.

By keeping the figures themselves out of the main text file, the author can focus on the elements of the figures that are related to the flow of the document: placement relative to the text, captions, relative size, etc.

In Ada's notes, diagrams of variables related to the algorithm were inserted. These could be created in a LaTeX math environment, but they could also be included as figures. The syntax for including an image is:

```
\begin{figure}[htbp]  ❶
\begin{center}  ❷
\includegraphics[width=0.5\textwidth]{var_diagram}  ❸
\end{center}
\caption{Any set of columns on which numbers are inscribed, represents
merely a general function of the several quantities, until the special
function have been impressed by means of the Operation and
Variable-cards.}  ❹
\label{fig:var_diagram}  ❺
\end{figure}
```

❶ The figure environment begins. Placement options are specified as (h)ere, (t)op, (b)ottom, or on its own (p)age.

❷ The figure should appear horizontally centered on the page.

❸ The name of the image file is indicated and the width option specifies that the figure should be half the width of the text.

❹ A verbose caption is added.

❺ A label is added so that the figure can be referenced later in the document using this name tag.

The result of this syntax is shown in Figure 20-6. In it, the image is brought into the document, numbered, sized, and captioned exactly as was meant.

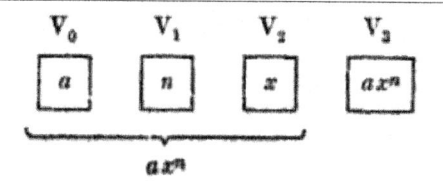

> Figure 1: Any set of columns on which numbers are inscribed, represents merely a general function of the several quantities, until the special function have been impressed by means of the Operation and Variable-cards.

Figure 20-6. Labels in LaTeX

In this example, the figure was labeled with the \label command so that it can be referenced later in the document. This clean, customizable syntax for internal references is a feature of LaTeX that improves efficiency greatly. The next section will show how such internal references work.

Internal references

The LaTeX syntax for referencing document elements such as equations, tables, figures, and sections entirely eliminates the overhead of matching equation and section numbers with their in-text references. The \ref{} command can be embedded in the text to refer to these elements elsewhere in the document if the elements have been labeled with the corresponding \label{} command.

At build time, LaTeX numbers the document elements and expands inline references accordingly. Since tables, figures, and sections are often reordered during the editing process, referring to them by meaningful labels is much more efficient than trying to keep track of meaningless numbers. In Note D (*example.tex*), for instance, Ada presents a complex example and refers to Note B (*storehouse.tex*). Since the ordering of the notes might not have been finalized at the time of writing, referring to Note B by a meaningful name rather than a number—or, in this case, a letter—is preferred. To do this, Ada must use the \label{} command in the *storehouse.tex* file so that the *example.tex* file may refer to it with the \ref{} command:

```
% storehouse.tex ❶
\label{sec:storehouse}❷
That portion of the Analytical Engine here alluded to is called the
storehouse. . .
```

❶ The section on the storehouse is stored in a file called *storehouse.tex*.

❷ That section is called Note B, but Ada remembers it as the storehouse section, and labels it such.

In this example, the label uses the prefix `sec:` to indicate that *storehouse* is a section. This is not necessary. However, it is a common and useful convention. Similarly, figures are prepended with `fig:`, tables with `tab:`, and so on. Ada can now reference Note B from Note D as shown here:

```
% example.tex  ❶
We have represented the solution of these two equations below, with
every detail, in a diagram similar to those used in Note
\ref{sec:storehouse}; ...  ❷
```

❶ Note D is held in *example.tex*.

❷ Ada can reference Note B within this file using the memorable label.

When the document is built with these two files, a "B" will appear automatically where the reference is. The same can be achieved with figure and table labels as well. Now it is clear how to reference figures and sections. However, there is another kind of reference common in publications. Bibliographic references (citations) are also automated, but are handled a bit differently in LaTeX. The next section will explain how.

Bibliographies

An even more powerful referencing feature of LaTeX is its syntax for citation of bibliographic references and its automated formatting of bibliographies. Using BibTeX or BibLaTeX, bibliography management in LaTeX begins with *.bib* files. These contain information about resources cited in the text sufficient to construct a bibliography. Had Ada desired to cite the Scientific Memoir her notes were concerned with, she might have defined that work in a *refs.bib* file as follows:

```
% refs.bib
@article{menabrea_sketch_1842,
        series = {Scientific Memoirs},
        title = {Sketch of The Analytical Engine Invented by Charles Babbage},
        volume = {3},
        journal = {Taylor's Scientific Memoirs},
        author = {Menabrea, L.F.},
        month = oct,
        year = {1842},
        pages = {666--731}
}
```

To cite this work in the body of her text and generate an associated bibliography, Ada must do only three things. First, she uses the `\cite{}` command, along with the key (menabrea_sketch_1842), where she wants the reference to appear:

```
% intro.tex
...
These cards contain within themselves (in a manner explained in the Memoir
```

itself \cite{menabrea_sketch_1842}) the law of development of the particular
function that may be under consideration, and they compel the mechanism to act
accordingly in a certain corresponding order.
...

Second, she must include a command placing the bibliography. Bibliographies appear at the end of a document, so just before the \end{document} command in her main *notes.tex* file, Ada adds two lines:

```
% notes.tex
...

\section{Note G}
\input{conclusion}

\bibliographystyle{plain}
\bibliography{refs}
\end{document}
```

These together define the bibliography style. The choices for this parameter are myriad. The simplest choice is often "plain," as has been used here. However, a one-word change can alter the formatting to comply with Chicago, MLA, or any other bibliography formatting style. The second line names the location(s) of the *.bib* file(s).

The final necessary step is to build the bibliography along with the document. For this, an extra build step is required that employs the bibtex command. In a peculiarity of LaTeX, for the references to appear, you must call latex again twice after issuing the bibtex command. So, at the command line, Ada must type:

```
$ latex notes
$ bibtex notes
$ latex notes
$ latex notes
$ dvipdf notes
```

The result is marvelous. In the text, the \cite command is replaced with "[1]", and on the final page of her document, a bibliography appears as in Figure 20-7.

References

[1] L.F. Menabrea. Sketch of the analytical engine invented by charles babbage. *Taylor's Scientific Memoirs*, 3:666–731, October 1842.

Figure 20-7. Automated bibliography generation

Never again need scientists concern themselves with the punctuation after a title in an MLA-style bibliography—LaTeX has automated this. The only thing LaTeX does not automate about bibliography creation is reading the papers and making the *.bib* file itself. Thankfully, other tools exist to make that process more efficient. The next section introduces these.

Reference management

To generate a *.bib* file easily, consider using a reference manager. Such a tool helps to collect and organize bibliographic references. By helping the researcher automate the collection of metadata about journal articles and other documents, as well as the production of *.bib* files, these tools eliminate the tedious task of typing names, titles, volume numbers, and dates for each reference cited in a paper. It can all be completely automated. A number of open source tools for this task exist. These include, among others:

- BibDesk
- EndNote
- JabRef
- Mendeley
- RefWorks
- Zotero

Reference managers help researchers to organize their sources by storing the metadata associated with them. That metadata can typically be exported as *.bib* files.

Citing Code and Data

One thing that BibTeX lacks is a metadata format appropriate for uniquely referencing code or data, unless it has a digital object identifier (DOI) number associated with it. For truly reproducible publication, you should cite the code and data that produced the analysis using a DOI.

Each commit in your version-controlled code repository has a commit hash number that distinguishes it uniquely from others. For unique identification in a library or bookstore, this book has an ISBN. Analogously, data and software objects can be identified in a persistent way with a DOI number.

It is possible to acquire a DOI for any piece of software using archival services on the Internet. Some are even free and open source.

The use of these reference managers is outside the scope of this chapter. Please go to the individual tools' websites to get started using them.

Publication Wrap-up

Publication is the currency of a scientific career. It is the traditional way in which scientific work is shared with and judged by our peers. For this reason, scientists spend a lot of time producing publication-quality documents. This chapter has sought to provide an overview of the tools available to aid you in this pursuit and to give an introduction to the most ubiquitous, LaTeX. Now that you have read this chapter, you should know that:

- Markup languages separate formatting from content.
- Markup-based text documents are more version-controllable.
- Many markup languages exist, but LaTeX is a particularly powerful tool for scientific publication.

In the context of LaTeX, you should also know how to:

- Produce a simple document
- Give structure to that document
- Add mathematical equations inline
- Display mathematics
- Include figures
- Reference those figures
- Cite bibliographic references
- Automate the creation of a bibliography

With these skills, you are equipped to begin generating lovely publication-quality documents. Many resources are available online to enrich what you have learned here. Two favorites are:

- "The Not So Short Introduction to LaTeX" (*http://bit.ly/tex-short*), by Tobias Oetiker et al.
- Tex-LaTeX Stack Exchange (*http://tex.stackexchange.com/*)

> **IPython Notebook**
>
> As an aside, please note that another option for reproducible document creation that was not mentioned in this chapter (because it is in a class of its own) is the IPython notebook. IPython Notebook is a part of the IPython interpreter that has been used in previous chapters. It is an interface for Python that can incorporate markup languages and code into a reproducible document. With an interface very similar to that of a Mathematica notebook, the IPython (soon, Jupyter) notebook combines plain text, LaTeX, and other markup with code input and output cells. Since the IPython notebook displays tables and plots alongside the code that generated them, a document in this format is especially reproducible.
>
> For more on IPython, Jupyter, and working with the Notebook, see the IPython website (*http://ipython.org/*).

Publication is an essential part of bringing your work to your peers. Another way, however, is direct collaboration. The next chapter will demonstrate how GitHub can make collaboration on papers and software far more efficient.

CHAPTER 21
Collaboration

It was once the case that collaboration involved letters being sent through the mail from scientist to scientist.

Today, collaborations happen via email, conference calls, and journal articles. In addition to these tools, web-based content and task management tools enable scientific collaborations to be made effortlessly across continents, in myriad time zones, and even between scientists who have never met. Indeed, some of the first enormous modern collaborations in the physical sciences spurred the progenitors of the collaboration tools that currently exist (not least of all, the Internet). In the context of computation, issue ticketing systems can be closely tied to version control systems and become powerful tools for peer review.

This chapter will demonstrate how such tools expedite and add peer-review capabilities to collaborative research discussions, writing papers, and developing scientific software. These ticket management systems provide a system for content management alongside version-controlled repositories. Sites like GitHub, Launchpad, and Bitbucket, which provide content management for hosted version-controlled repositories, are essential to modern collaboration.

Additionally, this chapter will describe the interface for *pull requests* that allows collaborators to peer review code. Transparent archiving and opportunity for review do for scientific software what the peer-reviewed journal system does for scientific papers. Scientific code has historically gone unreviewed and unrecognized by the scientific community. However, thanks to these new tools, software is increasingly being seen as a bona fide scientific research product in itself, not unlike a journal article. Without the interfaces for peer review provided by sites like GitHub, this would never be possible.

In Chapter 15, version control was called the "laboratory notebook" of scientific computing. In that paradigm, the tools described in this chapter allow scientists to share, review, and collaborate on laboratory notebooks, both among themselves and with the world.

Scientific collaboration via the post was riddled with inefficiencies, bottlenecks, and obfuscation of provenance. If Lise Meitner, Neils Bohr, Fritz Strassman, and Otto Hahn had had this new kind of system when they were working on the research that would yield the theory of nuclear fission, the process of discovery would have been expedited enormously. In a modern version of their collaboration, their communication would have been instantaneous and the provenance of ideas and effort would have been more transparent to the scientific community. In such a version of their collaboration, perhaps there might have even been sufficient provenance to guarantee a Nobel Prize for Prof. Meitner alongside her colleagues Hahn and Strassman. The next section will discuss how an open source scientific computing project today could rely on ticket managers.

Ticketing Systems

For any research project, computational or otherwise, a ticket management system (sometimes called a *content management system* or *issue tracker*) can vastly simplify collaboration. Web-based ticket management systems allow progress on a project to be tracked and managed at the level of individual tasks by providing a web interface for task assignments, updates, and completion reports.

Almost all web-based services for code repository hosting (e.g., GitHub, Bitbucket, Launchpad) have an associated issue tracker. These provided methods for creating *issues* or *tickets* associated with necessary tasks related to the repository. The resulting dashboard is an annotated, dynamic system for to-do list curation, communication, and peer review.

Tickets are, fundamentally, the first step in the workflow for new features, bug fixes, and other needs within the code base. The next section will give an overview of the workflow associated with such ticketing systems.

Workflow Overview

In the context of a collaborative research effort, community guidelines for using issue trackers must arise organically in a way that appropriately reflects the structure and culture of the collaboration. However, common workflow patterns have emerged in the use of issue trackers in scientific computing projects that share a general structure.

While many types of workflows exist, the workflow in Figure 21-1 is common for a situation when a bug is found.

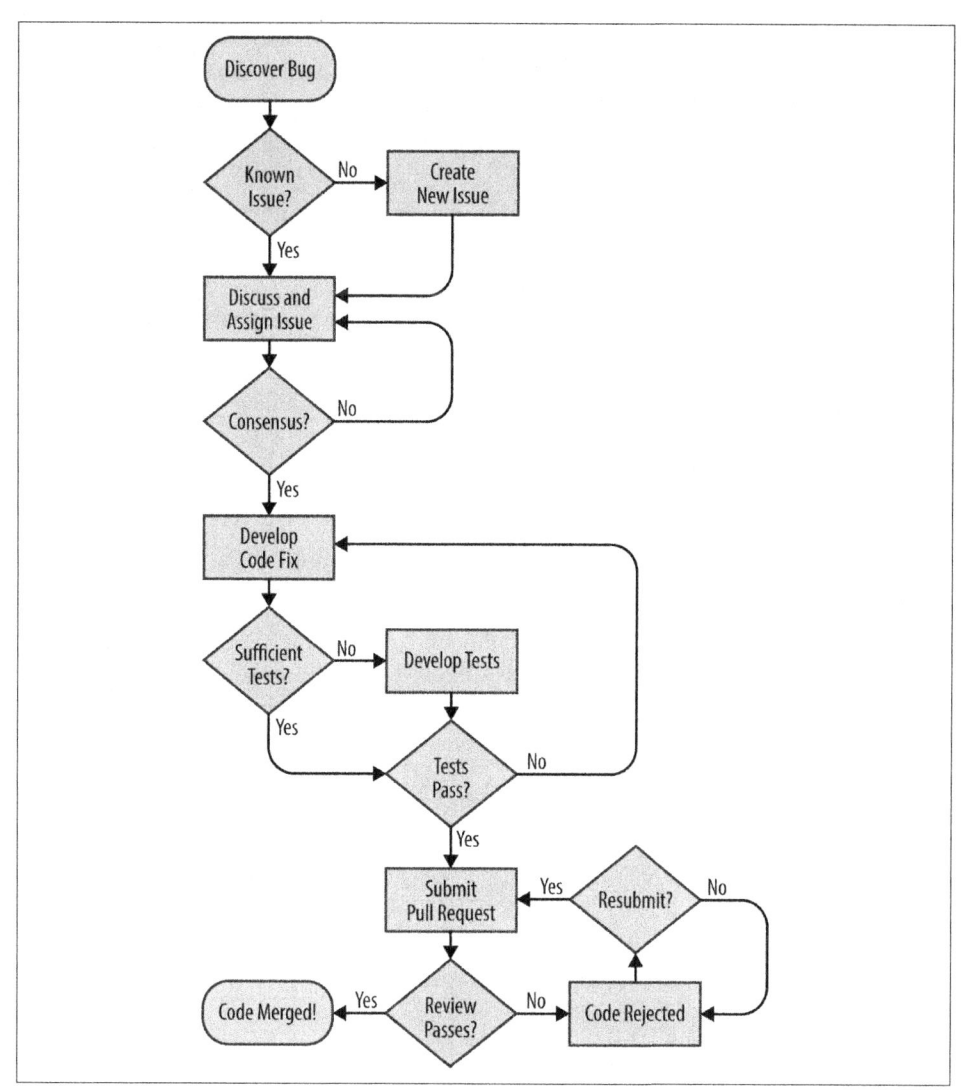

Figure 21-1. A bug resolution workflow

First, before reporting a bug, the user or developer must check the rest of the currently open issues to determine whether it has already been reported. If the bug is not yet known, a new issue can be created describing the bug, and the collaborators can agree upon related goals and subtasks.

When researchers take responsibility for completing issues, those tickets are assigned to them (or they can self-assign it). As a researcher makes progress on the completion of the task, comments and updates can be added to the ticket. If collaborators have questions or comments about the progress of a task, the ensuing discussion can take place directly on the ticket through the web interface. Finally, when a conclusion is reached, code committed to the repository can be referenced in the discussion. A *pull request* holding the new changes is typically submitted, referencing one or more issues. The new code submitted by pull request can then be reviewed, tested on multiple platforms, and otherwise quality checked. When the new code satisfies collaborators, the issue is declared solved or *closed*. It is typically closed by the person who opened it, a project leader, or the person who solved it.

When a new feature is desired, a similar workflow is followed. However, the initial steps can be quite different. Many open source projects have a notion of "enhancement proposals" that are necessary to initiate the process.

Of course, the first step in any of these workflows is to actually create the issue or ticket.

Creating an Issue

Users and developers often find bugs in code. A bug found is better than a bug that goes unnoticed, of course, because only known bugs can be fixed.

"Issues" on GitHub are tickets associated with a particular repository. Issues alert code developers and users to a bug, feature request, or known failure. Primarily, issue tickets exist to specifically designate a place for discussion and updates concerning these topics.

A modern Otto Hahn, when faced with a peculiar result, could begin a discussion with his colleagues in the hopes of solving the problem. Figure 21-2 shows the GitHub issue creation form through which Otto opens an issue and describes the problem.

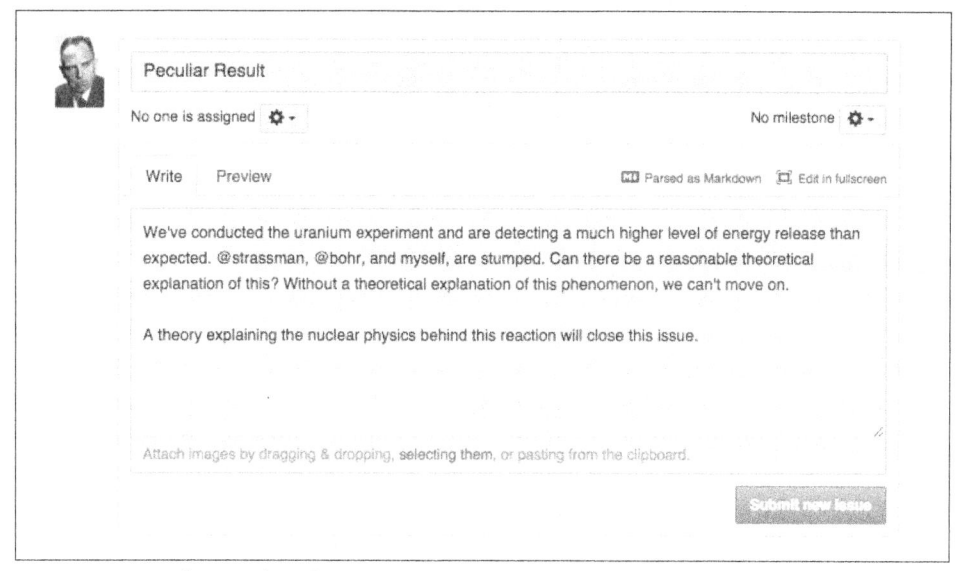

Figure 21-2. Hahn needs a theory

Core features of an issue

The most effective issues have a bit more information than the one in Figure 21-2, however. In particular, issues typically answer a few key questions:

- **What is needed?** A description of the error or feature request.
- **Why is it necessary?** Sufficient information to replicate the error or need.
- **What happens next?** A proposal for a path forward.
- **How will we know this issue is resolved?** A clear end goal.

Without those features, it may be difficult for other collaborators to replicate the issue, understand the need for a change, or move toward a solution. Furthermore, for provenance, even more data about the issue itself can be helpful.

Issue metadata

In addition to this core information, metadata can be added to the issue that helps to organize it and designate its place among others. Different web-based hosting platforms provide different features for defining tickets. Some of the neat features of GitHub issues include tags, user pings, cross-linking with the code base, and commit hooks.

Tags, which are completely customizable for each project, can be used for categorizing and differentiating groups of issues. Many tags may be used for a single issue, and on GitHub, colors can be used creatively to help visually distinguish distinct topics.

Issue tags may be used to categorize the issue along many different axes. For instance, they may indicate the level of importance, the degree of complexity, the type of issue, the component of the code affected, or even the status of the issue.

Table 21-1 gives some examples of the kinds of tags you might apply.

Table 21-1. GitHub tagging examples

Importance	Difficulty	Type of issue	Code component	Issue status
Critical	Expert	Bug	Installation	New
High priority	Challenging	Feature	Input/output	Progress
Medium priority	Mild effort	Documentation	Core	In review
Low priority	Beginner	Test	Visualization	Won't fix

The customizability and power of this metadata are vast. In particular, it can help code developers to decide which issues to tackle next. Together with a tool like HuBoard (*https://huboard.com/*) or waffle.io (*https://waffle.io/*), this metadata can even fuel dashboards for managing projects under sophisticated workflow paradigms (e.g., "agile" or "kanban" systems).

Since so much of this metadata revolves around helping developers to approach and handle tasks, it should make sense that the most important type of metadata in an issue ticket is the assigned developer.

Assigning an Issue

To avoid duplicated effort, an issue can be assigned to a particular developer. Commonly, in the open source world, issues are discussed among developers as soon as they are created. Often, the developer who most clearly has expertise in that area of the code is the one who is assigned to handle it.

An issue can also be left unassigned to indicate that it is unclaimed. Developers who become interested in solving such issues can confidently assign them to themselves.

> ### Grouping Issues into Milestones
> On GitHub, issues can be grouped into *milestones*. Milestones are groups of issues defining broader-reaching goals. Milestones also have due dates. This feature of the GitHub interface can be used as a mechanism for project goal tracking of many kinds, including research group organization and code release management.

Grant-driven research, in particular, is well suited for milestone-based, due date–driven work. Additionally, by bundling all of the issues needed for a desired feature set, milestones are ideal for defining the necessary work remaining for a code release.

Even though she was far away from Otto Hahn, a modern Lise Meitner could have commented on and assigned herself to handle the issue he opened. In Figure 21-3, Lise makes a comment on the GitHub issue. By clicking the "assign yourself" link on the right, she can claim the task.

Figure 21-3. Lise claims this task

Once she has claimed it, she can go ahead and start work on it. As she begins to develop a theory, she may desire to bounce ideas off of her colleagues. Discussing an issue can be done on GitHub as well, so she, her nephew Otto Frisch, and Neils Bohr can discuss their thoughts right alongside the original question.

Discussing an Issue

To discuss an issue on GitHub, just enter a new comment into the comment box associated with the issue. The issue conversation is an appropriate place for:

- Asking and answering clarifying questions about the issue
- Sharing and discussing ideas for an approach
- Requesting and providing updates on the process

Some research groups are tempted to discuss issues via email rather than within the issue tracker. While that strategy seems equivalent, it is not. Discussion directly on the issue page is superior at retaining context, transparency, and provenance.

All that said, very open-ended discussions are typically more appropriate for the email format. Issues are meant to eventually be closed.

Closing an Issue

When the bug is fixed or the new feature implementation is complete, the issue should be closed. The collaboration's cultural norms, expertise distribution, leadership hierarchies, and verification and validation requirements all affect the process by which an issue is deemed complete.

For example, in repositories dedicated to writing a research paper (see Chapter 20), building a research website, or prototyping a quick idea, changes might not require strict quality assurance methods. In those situations, the issue may be closed without much fanfare or oversight at all.

In contrast, the scientific community expects a high level of robustness and quality from scientific software. To assure quality and accuracy, new changes to a scientific software project may need to undergo verification, validation, and peer review. In such a project, closing an issue may therefore involve the effort and consensus of multiple researchers, incorporation of an automated test suite, adherence to a style guide, and appropriate documentation.

Indeed, the level of validation and verification necessary in high-quality software projects typically requires that the issue review culture includes a system of pull requests.

Pull Requests and Code Reviews

Historically, software developers shared, removed, and submitted changes through *patches* passed around via email. The pull request is a hyper-evolved descendant of that technology and, indeed, carries a patch at its core. Pull requests, however, represent an enormous leap forward for collaborative software development. Pull requests are a reasonable, provenance-aware interface for applying peer review to proposed patches.

Chapter 15 demonstrated the power of version control for tracking small changes. Importantly, a patch is just such a small change. Recall from Chapter 1 that the difference between two files can be output to the terminal. Additionally, recall that any output can be redirected to a file instead of the terminal. The resulting file represents the difference between two files. It is called a *patch* because the `patch` command is used to apply that difference to the original file (resulting in the modified file).

Submitting a Pull Request

With the pull-request interface, however, a researcher can submit a change for review in a clean interface that links to actual commits, allows line comments, and persists alongside the code on the GitHub servers.

In Lise Meitner's case, perhaps the project repository might have held a text document outlining the working theory and experimental description for the project. To make changes, Lise first forked the main repository, under the `kaiserwilhelm` username, then cloned it locally:

```
$ git clone git@github.com:lisemeitner/uranium_expmt
```

She might have solved the issue that Otto Hahn opened by creating a branch (see "Listing, Creating, and Deleting Branches (git branch)" on page 365), "newtheory," and editing the text file there:

```
$ git checkout -b newtheory
```

At this point, she might choose to delete some of the text that incorrectly described the theory, and to add lines that outline her theory of fission. After editing the file and committing her changes, she can push that branch up to her fork on GitHub (see "Downloading a Repository (git clone)" on page 375). In the directory containing her local copy of the repository, Lise might perform the following to push her feature branch up to her fork:

```
$ git commit -am "edits the old theory and replaces it with the new theory."
$ git push origin newtheory
```

Once she has pushed the branch up to GitHub, Lise can navigate within a web browser to the dashboard of her repository. There, GitHub provides the option to make a pull request to the master branch of the main *kaiserwilhelm* repository. When that button is clicked, the pull request appears as a new issue in the *kaiserwilhelm* repository, where it should be reviewed by collaborators before being merged into the code base.

Reviewing a Pull Request

Reviewing a pull request is much like reviewing a paper. More accurately, it should be like reviewing a section or paragraph of a paper. Humans are better at reviewing short paragraphs of code rather than hundreds of lines at once—too much for us to hold in our heads at once.[1] For this reason, developers should avoid lengthy or complex pull requests if possible. By addressing changes in an atomistic fashion (one bug fix or fea-

---

1 See Edward R. Tufte's *The Visual Display of Quantitative Information* (Graphics Press).

ture addition at a time), developers reduce the likelihood of introducing a bug that can be missed in the review stage.

At this stage, developers reviewing the pull request may ask a number of questions. Does the code:

- Accomplish the goals?
- Introduce bugs?
- Include sufficient tests?
- Follow the style guide?
- Pass the existing tests?
- Pass new tests?
- Pass the tests on other platforms (Unix, Windows)?

Merging a Pull Request

Once reviewed, the code can be merged. This can be done in one of two ways. On GitHub, within the pull request itself, there is a green button for merging nonconflicting pull requests.

Alternatively, via the command line, a developer can use a combination of `git remote`, `git fetch`, `git merge`, and `git push`. Review Chapter 16 to recall how these commands are used.

Collaboration Wrap-up

Collaboration can be a very complex, time-consuming element of scientific work—especially with old technology. However, readers of this book should now be equipped to collaborate more efficiently using the power of Git and GitHub. In this chapter, you have seen how to create, assign, discuss, and tag issues, as well as how to generate solutions, make pull requests, review code, and incorporate changes efficiently online.

This efficiency should free up time for determining what license is best for distributing your code with. For help with that, keep reading into the next chapter.

CHAPTER 22
Licenses, Ownership, and Copyright

For any software project, the most important file in the project is the license. This file states who owns the work, who is allowed to use the project and under what conditions, and what rights and guarantees are conferred to both the users and the owners. If a license file is not present, it is conspicuous in its absence. Since the license is the most important file in a project, this chapter is the most important one in this book to read and fully understand.

License files should be easy and obvious to find. Most of the time they appear in the top-level directory and go by the name *LICENSE, license.txt,* or another variant. Note that sometimes different parts of a project are provided under different licenses. Some projects also have the dubious practice of being licensed differently depending on how they are used. Be sure to read and understand the license of any software project that you use *before* you use it.

> **Get a Laywer**
>
> This chapter *is not* legal counsel. We are not qualified to help you in a formal dispute. For that, you need to have a lawyer.

This chapter is only intended to provide friendly advice that aims to help you understand the basic concepts in publishing a creative work. Having a good grasp of these fundamentals will help you make informed decisions. If for any reason you do end up needing legal counsel in this area but do not know where to start, you can contact the Electronic Frontier Foundation (*https://www.eff.org/*) (EFF), the Software Freedom Conservancy (*https://sfconservancy.org/*) (SWC), Creative Commons (*http://creativecommons.org/*) (CC), the Free Software Foundation (*http://www.fsf.org/*) (FSF), or Numfocus (*http://numfocus.org/*) (NF); they may be able to help. Each of these organ-

471

izations has experience with the legal aspects of software development and should be able to point you in the right direction, at the very least.

The licenses discussed in detail in this chapter will mostly be open source licenses. This is because without peer review of software and data, scientific code cannot fairly be called reproducible. If something is not reproducible, then it is de facto not science, no matter how deeply it covers a scientific topic. Equal dissemination of knowledge is critical to the scientific method. This is not to diminish the technical prowess of closed source code at all—propietary software is frequently among the most sophisticated. It is just not science in the benefit-for-all-of-humanity-for-all-of-time way. Open source licenses are ideally suited to research software.

This chapter will cover ideas and terms that you are probably already familiar with in their common usage. Here, we seek to improve upon that lay understanding to further the aims of computational physics.

What Is Copyrightable?

Before we talk about licenses in depth, it is important to understand what they cover. In western jurisprudence, from which most copyright law around the world stems, *ideas and concepts are not copyrightable*. However, *expressions* of ideas are copyrightable.

For instance, copyright does not apply to physical laws of nature and mathematical facts. The number pi and the idea that it is the ratio between the area of a circle and the square of its radius is not something that any human can claim ownership of. Humans discovered this knowledge, but humans did not create it nor invent it. Pi just is. Now, if I were to bake a pie with the letter pi cooked into the crust and the digits proudly displayed around the perimeter and then took a picture of my handiwork, I would have copyright over the picture. This tasty expression of pi would be uniquely my own. Anyone claiming otherwise would be wrong.

The same logic applies even outside the world of strictly scientific endeavors. For example, game rules are abstract concepts that are not copyrightable. However, any published version of a game that you read is a copyrighted expression of those rules. The rules for chess, Go, mancala, poker, bridge, basketball, rugby, cricket, *The Settlers of Catan*, and *Dungeons & Dragons* are all not copyrightable. They are just ideas. That said, the rule book that comes with any of these games is a creative and particular expression of the rules and is subject to copyright law.

In software, the breakdown between what is and is not copyrightable is the distinction between the implementation and the interface. The application programming interface (API) is considered to be a set of ideas and therefore not copyrightable. The actual implementation of an API, or how the actual work is performed, is copyrightable. There are many possible implementations for any interface, and so any given implementation is a unique expression. For example, say we wanted a function named std() that computed the standard deviation of a list of values named vals. These name choices and the concept of what the function is supposed to do make up the interface, which is not copyrightable. That said, any code that computes the standard deviation with this interface *is* copyrighted. There is a fundamental distinction both conceptually and legally between how one uses software and how that software is written. Keep this distinction in mind as you read through the following sections.

Right of First Publication

Now that we know what can be copyrighted, we should understand when copyright applies. Most copyright systems feature what is called the *right of first publication*. This is the idea that copyright automatically goes to the first publisher of a creative work. This right is conferred whether or not it is specified by the publisher at the time. Such laws protect publishers from having their work stolen as long as they can demonstrate that they were there first.

This has important implications in the information age, where self-publishing is normal behavior. Anyone who writes a blog post, tweets out "Happy Birthday, Mom!" or puts code on GitHub personally retains the copyright to that work via the right of first publication. Your work is your own unless you give it up.

In software, say you post a piece of code without a license. By rights this code is yours and yours alone, even though it is publicly visible. You own the copyright, and you have not specified how others are allowed to use your code. By default, legally, they are not entitled to use it at all. If your intent was to share your code for reproducibility, provenance, education, or general scientific goodwill, by not having a license you have undermined your intended purpose.

Software licenses are important because they allow you to retain copyright and they state the terms by which other people or organizations are entitled to use or modify your software. That said, it is possible to completely forego all rights.

What Is the Public Domain?

What happens if you do not want to deal with licenses or you do not want to retain copyright? If you just want the code that you produce to be for the unrestricted benefit of all, yielding completely to the better nature of scientific discourse? It is possible to put a work into the *public domain* (PD) with a simple statement along the lines of, "This work has been placed in the public domain."

The public domain is a concept that society as a whole "owns" a work, and it is therefore free and available for anyone and everyone to use and modify. Since everyone owns it, nobody owns it. In most cases, the copyright of an existing work will *expire* after a set number of years (25, 50, 90, etc.), at which point the work will *enter* the public domain. The public domain is what allows anyone to republish the collected works of Mark Twain, Socrates's *Apology*, or Mary Shelley's *Frankenstein*. However, just because copyright will expire naturally does not mean that you have to wait that long. You are free to add your own works to the public domain sooner if you so desire.

That said, the public domain is one of the trickiest parts of international copyright law. Not every country has a public domain that is compatible with the notions expressed here. In some countries it may not be possible for an individual to place a work into the public domain prior to the expiration of copyright. It is important to understand the laws of the country where you live. Wikipedia is often a good first resource. For anything deeper, you should probably seek out legal counsel. If you are at a university, national lab, or private company, your organization will often have resources available for you to use.

If you do not want to put your software in the public domain, but do want it to be free and open, you have to do the work of picking a license.

Choosing a Software License

A license is a legal document that states how software is allowed to be used by its users and what rights the author retains, and serves to protect both the users and the authors. Without such a document, only the original author or publisher has any right to use or modify the code. Having a software license that accurately reflects your needs and the needs of your potential users is extremely important.

A variety of licenses have been created over the years, tailored to different situations. At the broadest level, there are *proprietary licenses* and *free/open source licenses*. Proprietary licenses are usually written by companies that sell software. The Microsoft Windows End-User License Agreement (EULA) is an example of such a document. They typically proclaim the owner of the copyright to be the company, disclaim damages if the software is abused, and promise litigation in the event of piracy. They are often handcrafted by a team of lawyers to minimize the exposure to the company.

Free and open source software licenses, sometimes abbreviated as FOSS, FLOSS, or OSS, are much more relevant to computational physics software, especially if it is primarily for research. Research-grade software typically has the following attributes:

- It does not have immediate and direct commercial interest.
- It must have source code inspectable by peers for review.
- It changes rapidly to fit the needs of the researcher.

Licenses that give users and other developers the freedom to look at and modify code encourage scientific discourse, education, comparison, quality assurance, and participation in the project. Since most researchers do not have the funds available to hire thousands of developers to do all of these activities, an open source license can help establish a community of users and developers. Fellow scientists can pool together to help in these essential tasks that make a software project successful. The exchange is often nonmonetary. One person provides code, and others may help make that code better. To be safe, open source licenses often explicitly include a no-warranty clause. You cannot sue someone for damages because their open source code somehow harmed your machine.

We will not cover proprietary licenses more here. There are a lot of them, since almost every software product has its own license. On the other hand, the open source world has converged on a much smaller number of licenses. However, there are still far too many open source licenses to go into great depth on all of them. Here, we will present only the most important and the most interesting ones. For a more comprehensive review of open source licenses, please see the Open Source Initiative's (*http://bit.ly/osi-license*) (OSI) website or the GNU commentary page (*http://bit.ly/gnu-comm*).

It is highly advisable to use an off-the-shelf open source license. Do not attempt to write your own. Partly this is because you are not a lawyer, and it would be a waste of your time. More importantly, licenses are not generally considered trustworthy until they have been proven in court. This means that one party broke the terms of a license, another party sued, and the license was upheld in a court of law. This is an expensive and time-consuming process. Relatively few open source licenses have gone through this crucible, but almost all of them that have survived the journey. Any license you write will not have this benefit.

The choice of license can have a deep and lasting effect on the community that develops around a code project. Given its importance, picking the right license receives surprisingly little attention from developers. It is a core part of the social aspect of software development. Everyone should know about the license and its implications, not just the law nerds on a project. If you ever need help picking, the excellent ChooseALicense.com will help you along your way.

Let's examine some key licenses now to find out precisely what it means to be open source.

Berkeley Software Distribution (BSD) License

The Berkeley software distribution or *BSD* license is actually a collection of three possible licenses known as the BSD 4-Clause, 3-Clause, and 2-Clause licenses, respectively. Historically, the 4-clause is the oldest and the 2-clause is the most recent. The

4-clause is not recommended anymore, though both the 3- and 2-clause versions are commonly used. Of all of the licenses that we will discuss, either the 3- or 2-clause license is recommend for use in your software projects. These are the licenses best tailored to science and research. Major projects such as NumPy, SciPy, IPython, and the rest of the scientific Python ecosystem use either of these licenses. The text is as follows:

```
Copyright (c) <year>, <owner>
All rights reserved.

Redistribution and use in source and binary forms, with or without
modification, are permitted provided that the following conditions are met:
    * Redistributions of source code must retain the above copyright
      notice, this list of conditions and the following disclaimer.
    * Redistributions in binary form must reproduce the above copyright
      notice, this list of conditions and the following disclaimer in the
      documentation and/or other materials provided with the distribution.
    * Neither the name of the <organization> nor the
      names of its contributors may be used to endorse or promote products
      derived from this software without specific prior written permission.

THIS SOFTWARE IS PROVIDED BY THE COPYRIGHT HOLDERS AND CONTRIBUTORS "AS IS" AND
ANY EXPRESS OR IMPLIED WARRANTIES, INCLUDING, BUT NOT LIMITED TO, THE IMPLIED
WARRANTIES OF MERCHANTABILITY AND FITNESS FOR A PARTICULAR PURPOSE ARE
DISCLAIMED. IN NO EVENT SHALL <COPYRIGHT HOLDER> BE LIABLE FOR ANY
DIRECT, INDIRECT, INCIDENTAL, SPECIAL, EXEMPLARY, OR CONSEQUENTIAL DAMAGES
(INCLUDING, BUT NOT LIMITED TO, PROCUREMENT OF SUBSTITUTE GOODS OR SERVICES;
LOSS OF USE, DATA, OR PROFITS; OR BUSINESS INTERRUPTION) HOWEVER CAUSED AND
ON ANY THEORY OF LIABILITY, WHETHER IN CONTRACT, STRICT LIABILITY, OR TORT
(INCLUDING NEGLIGENCE OR OTHERWISE) ARISING IN ANY WAY OUT OF THE USE OF THIS
SOFTWARE, EVEN IF ADVISED OF THE POSSIBILITY OF SUCH DAMAGE.

Copyright (c) <year>, <owner>
All rights reserved.

Redistribution and use in source and binary forms, with or without
modification, are permitted provided that the following conditions are met:

1. Redistributions of source code must retain the above copyright notice, this
   list of conditions and the following disclaimer.
2. Redistributions in binary form must reproduce the above copyright notice,
   this list of conditions and the following disclaimer in the documentation
   and/or other materials provided with the distribution.

THIS SOFTWARE IS PROVIDED BY THE COPYRIGHT HOLDERS AND CONTRIBUTORS "AS IS" AND
ANY EXPRESS OR IMPLIED WARRANTIES, INCLUDING, BUT NOT LIMITED TO, THE IMPLIED
WARRANTIES OF MERCHANTABILITY AND FITNESS FOR A PARTICULAR PURPOSE ARE
DISCLAIMED. IN NO EVENT SHALL THE COPYRIGHT OWNER OR CONTRIBUTORS BE LIABLE FOR
ANY DIRECT, INDIRECT, INCIDENTAL, SPECIAL, EXEMPLARY, OR CONSEQUENTIAL DAMAGES
(INCLUDING, BUT NOT LIMITED TO, PROCUREMENT OF SUBSTITUTE GOODS OR SERVICES;
LOSS OF USE, DATA, OR PROFITS; OR BUSINESS INTERRUPTION) HOWEVER CAUSED AND
ON ANY THEORY OF LIABILITY, WHETHER IN CONTRACT, STRICT LIABILITY, OR TORT
```

```
(INCLUDING NEGLIGENCE OR OTHERWISE) ARISING IN ANY WAY OUT OF THE USE OF THIS
SOFTWARE, EVEN IF ADVISED OF THE POSSIBILITY OF SUCH DAMAGE.

The views and conclusions contained in the software and documentation are those
of the authors and should not be interpreted as representing official policies,
either expressed or implied, of the <owner> Project.
```

The BSD licenses are known as *permissive licenses*. This is because they allow the further distribution of the code to be done under any license. Only the copyright notice needs to be displayed. Additionally, further versions of the code need make no promises as to what license they will be released under. Modifications to the code may be released under any license that the author of the modification desires. Permissive licenses give a lot of freedom to users and developers other than the original author, while protecting the original author from liability.

For example, suppose person B copies person A's code. A originally released the code under a BSD license. B wants to modify the code and relicense the whole new code base. B is completely free to do so, and does not even have to include a copy of the original BSD license. The only requirement is to include the copyright notice, "Copyright (c) <year>, Person A." This ensures that person A gets credit for the work, but without a lot of headache.

The freedom to modify and relicense is a major reason why BSD is the recommended license for scientific computing. It leaves the most number of options open to future scientists. The MIT license (*http://opensource.org/licenses/MIT*) is considered to be equivalent to the BSD 2-clause and is a perfectly reasonable substitute. Up next is one of BSD's main competitors.

GNU General Public License (GPL)

The GNU General Public License (GPL) is again a collection of three distinct licenses: GPLv1, GPLv2, and GPLv3. Additionally, there are v2.1 and v3 GNU Lesser General Public Licenses (LGPLs). These are compatible with the corresponding GPLs of the same major version but are closer in spirit to BSD and MIT licenses. All of the GPL options are promoted by the FSF for both GNU and non-GNU software projects. GPLv1 is out of date and should not be used. There remains debate over whether v3 is an improvement over v2.

Linux is almost certainly the largest project that uses GPLv2, and it will continue to do so until the end of time. The GNU Compiler Collection (GCC) is likely the largest project to use GPLv3. The texts of both GPLv2 (*http://bit.ly/gpl-2*) and GPLv3 (*http://bit.ly/gnu-pl3*) are too long to include here. However, the following preamble should be added to the top of every GPLv3 file:

```
<one line to give the program's name and a brief idea of what it does.>
Copyright (C) <year>  <owner>
```

```
This program is free software: you can redistribute it and/or modify
it under the terms of the GNU General Public License as published by
the Free Software Foundation, either version 3 of the License, or
(at your option) any later version.

This program is distributed in the hope that it will be useful,
but WITHOUT ANY WARRANTY; without even the implied warranty of
MERCHANTABILITY or FITNESS FOR A PARTICULAR PURPOSE.  See the
GNU General Public License for more details.

You should have received a copy of the GNU General Public License
along with this program.  If not, see <http://www.gnu.org/licenses/>.
```

Unlike BSD, the GPL licenses are not permissive. They instead exemplify a class of licenses known as *copyleft* by their punning proponents, or *viral* by their disinterested detractors. In a copyleft license, modifications to the code must be licensed in the same (or a similar) way as the original code. For open source code, this means that any third party that forks your code also has to make the fork open source. In principle, this is a good idea. Over time, it builds up an ecosystem of programs that are all open source and work well with one another. In practice, however, sometimes this requirement limits the freedom of the developers to the point where an improvement, modification, or fork will not even be written in the first place.

In general, GPL licenses are a good and reasonable choice for software projects. However, the potential disincentives to contribution can be a barrier to entry for small and medium-sized projects. Since physics programs—even when enormously popular—are never large, GPL is not recommended for default use. There may still be situations where it is the best choice, though. It's important to always carefully consider your options.

Another important concept for this suite of licenses is *GPL compatibility*. The FSF defines compatibility (*http://bit.ly/gnu-compat*) as whether different portions of a code base can be released under different licenses, where one license is the GPL. A license need not be copyleft itself to be compatible with the GPL. Notably, the BSD 3- and 2-clause licenses are both GPL compatible.

Even though the GPL may not be ideally suited to scientific endeavors, it is still a wildly successful series of licenses that have been proven in court. In general, if you want GPL-style copyleft but also want something less restrictive in terms of redistribution, the LGPL offers a middle path that should be treated as a serious alternative.

Permissive and copyleft licenses are the main players in open source software. However, there are also licenses that may be used for generic creative works and are not restricted to software that may be appropriate.

Creative Commons (CC)

The Creative Commons (*http://creativecommons.org/*) (CC) license suite is an alternative that applies not only to software but, more broadly, to all creative works. This includes poetry, prose, film, audio, and more. The CC suite is now in its fourth instantiation, sometimes called v4. The goal of CC licenses is to make the ideas of sharing and collaboration evidenced in open source software more broadly applicable to other endeavors. For example, all content on Wikipedia is licensed under CC. This has been a remarkably successful suite of licenses in its own right.

Creative Commons licenses are all quite lengthy, so it is not possible to print them here. Their length comes partly from the fact that the licenses are designed to keep the same spirit in every country in the world. As you might imagine, this can make the documents quite wordy. They are a tremendous feat of legal engineering.

The CC licenses are distinguished from one another through one or more of the following four modifiers:

BY *(Attribution)*
: Other people must credit you when they use your work. This applies to all CC licenses.

SA *(ShareAlike)*
: Other people must reshare the work under the same license if they use or modify your work.

ND *(NoDerivatives)*
: Other people may share your work but are not allowed to modify it in any way.

NC *(NonCommercial)*
: The work may be modified and shared as long as the result is not sold commercially.

The six licenses that Creative Commons supplies are thus known under the following monikers: CC BY, CC BY-SA, CC BY-ND, CC BY-NC, CC BY-NC-SA, and CC BY-NC-ND. Naturally, ND and NC cannot be applied to the same license.

The CC BY license (*http://bit.ly/by-license*) fills the same role as the BSD license. It has permissive terms that allow modification of the work as long as credit is given where credit is due. The CC BY-SA license (*http://bit.ly/cc-by-sa-4*) fills the same role as the GPL or the LGPL: it adds copyleft provisions to the attribution clause. Both of these are great choices for a scientific computing project, depending on whether you need copyleft or not.

The ND and NC modifiers, on the other hand, are considered extremely harmful to scientific software. They may have a place in other domains, such as the arts (notably, the webcomic xkcd (*http://xkcd.com/*) is released under a CC BY-NC license, which is why you do not see any witty stick figures appearing in this book). However, for physics programs, the inability to modify the software effectively renders the open source aspects useless. This is true even with just commercial applications restricted. Such license terms hamstring the development of vibrant science and software communities. Regard any use of either ND or NC with extreme prejudice. Luckily, examples of their use are very rare in computational science.

Lastly, Creative Commons supplies a public domain substitute called CC0 (*http://bit.ly/cc-zero-pd*) (pronounced *see-see-zero*). In many ways, licensing code as CC0 is better than placing it in the public domain because of the consistency problems with the public domain. CC0 applies everywhere, including in countries where analogous public domain laws do not exist. Moreover, in countries with a robust public domain, the CC0 license effectively reduces to the public domain. Of all of the licenses that are discussed here, CC0 is the freest. No major code projects of the size of Linux use this license, though it is slowly gaining traction.

In summary, CC BY, CC BY-SA, and CC0 are all reasonable choices for a scientific software project. The others should be avoided due to their implications being antithetical to evolving a scientific community.

The next section discusses licenses that you should not use, but which highlight interesting aspects of open source software.

Other Licenses

There are many other licenses that we are not able to cover in depth. A large number of these fall into the broad categories of either being permissive or copyleft, though with their own unique wording. Some of these have been defended in court. Unlike the previously discussed licenses, most of these were initially written to apply to a single code project.

For instance, the Apache License (*http://bit.ly/apache-v2*) was written for the Apache web server and is supported by the Apache Software Foundation. It has a lot of use outside of Apache itself. Unlike other licenses, it contains clauses pertaining to patents.

The Python Software Foundation License (*http://python.org/psf/license*) is a license that applies to the CPython interpreter and some other Python Software Foundation code. It occasionally sees some use in other parts of the Python ecosystem but is not commonly used outside of that. It is a permissive license that is GPL compatible.

On the more problematic side is the JSON License (*http://www.json.org/license.html*). This is easily one of the most commonly broken and flagrantly abused licenses in the world. The full text of the license reads:

```
Copyright (c) 2002 JSON.org

Permission is hereby granted, free of charge, to any person
obtaining a copy of this software and associated documentation
files (the "Software"), to deal in the Software without restriction,
including without limitation the rights to use, copy, modify, merge,
publish, distribute, sub-license, and/or sell copies of the Software,
and to permit persons to whom the Software is furnished to do so,
subject to the following conditions:

The above copyright notice and this permission notice shall be
included in all copies or substantial portions of the Software.

The Software shall be used for Good, not Evil.

THE SOFTWARE IS PROVIDED "AS IS", WITHOUT WARRANTY OF ANY KIND,
EXPRESS OR IMPLIED, INCLUDING BUT NOT LIMITED TO THE WARRANTIES
OF MERCHANTABILITY, FITNESS FOR A PARTICULAR PURPOSE AND
NONINFRINGEMENT. IN NO EVENT SHALL THE AUTHORS OR COPYRIGHT
HOLDERS BE LIABLE FOR ANY CLAIM, DAMAGES OR OTHER LIABILITY,
WHETHER IN AN ACTION OF CONTRACT, TORT OR OTHERWISE, ARISING
FROM, OUT OF OR IN CONNECTION WITH THE SOFTWARE OR THE USE OR
OTHER DEALINGS IN THE SOFTWARE.
```

The provision for Good and against Evil is considered indefensible due to the subjective nature of these terms. While the broad strokes of good and evil can be largely understood and agreed upon, the specifics, even in a strictly utilitarian context, can get murky quickly. How any of this applies to software at all is even less well understood. The JSON License brings into relief the supposition that other open source software allows for "more evil."

Even though the pursuit of evil is antithetical to most open source software developers and scientists, explicitly disallowing it is not reasonable. It is an unenforceable claim. Suppose someone really wanted to do bad things and felt the need to use JSON in order to accomplish the nefarious tasks. Would not the first evil act simply be to break the JSON license? It ends up being prescriptive, not preventative. Suffice it to say that no one has ever been successfully prosecuted for breaking the "Good, not Evil" clause. This license serves as a cautionary tale as to why you should never write your own license—it probably will not turn out like you thought.

Lastly, the FLASH code (*http://flash.uchicago.edu/site/*) from the University of Chicago has a particularly interesting license. FLASH is a plasma and high-energy density physics program. FLASH is open source, though it is not directly redistributable by the user. Furthermore, users need to sign a document registering themselves with the Flash Center prior to being given access to the code. Item 7 of the FLASH license is unique in its academic nature:

```
7. Use Feedback. The Center requests that all users of
   the FLASH Code notify the Center about all publications that
```

```
incorporate results based on the use of the code, or modified
versions of the code or its components.  All such information
can be sent to info@flash.uchicago.edu.
```

Here we see an attempt to gather statistics about the impact factor of scientific software being enforced through the license itself. Given the 850+ publications (*http://bit.ly/flash-pubs*) found on the FLASH website, this strategy has been very successful. The other nonstandard provisions of the FLASH license are all justified in their own right, but it is not recommended that you adopt licenses such as the FLASH one (indeed, the authors are not aware of another implementation of this kind of license). Your code is likely not in a situation like that of the FLASH code. Using one of the stock licenses from the previous sections is much easier for scientific programs. Uniqueness is not an asset for software licenses.

Now that we have seen many license types, including permissive, copyleft, and more exotic ones, it is reasonable to wonder how an existing project might transfer from one license to another.

Changing the License

Occasionally developers will decide that it is time to change the license that their code is distributed under. This can happened because of newfound legal implications in the license, because the broader ecosystem has moved away from a license, or, more commonly, because it is felt that the project will be easier to sustain under an alternative license. One prominent and successful example of a computational physics project that has undergone relicensing is yt; you can read about the campaign on the project blog (*http://bit.ly/yt-relicense*).

Even under the best conditions, changing a license is a long and painful process. It has a high likelihood of failure even if all of the current developers are on board with the decision.

Relicensing an open source project requires that you obtain consent from all previous contributors to make the switch. Consent can be given either actively (sign a waiver, send an email, etc.) or passively (if we do not hear from you in three months, we will assume you do not object). If consent is not given by a former or current developer, either the code that the developer worked on must remain with its original license, or all of that person's code must be rewritten to conform with the new license. Both of these options are their own headaches.

If enough people do not agree with the relicensing activity, it becomes too much of a hassle to actually relicense the code. In such cases it is easier to start from scratch than to try to force the issue with the existing code base. The scorched-earth strategy happens more frequently than many people would like to admit. Many of the worst horror stories come from trying to switch between copyleft and permissive licenses, or vice versa.

That said, relicensing is not impossible. It is just hard. When done successfully it will take at least three or four months. The steps to proceed through a relicense are as follows:

1. Privately and tentatively, present the idea of the new license to current core developers. Make the case for the new license as a possibility and get a feel for the level of support. Solicit feedback from this team first. If the current core team is not interested, stop.

2. Publicly present the option for hypothetically changing the license based on the previous feedback to all core developers. Solicit opinions from the developers and gauge the interest. If all members of the core team are happy with the change and most of the current developers are either for it or do not care, proceed. Otherwise, stop.

3. Publicly present the option for changing the license to the users. If the project will lose a significant minority of users due to the change, stop. Users are your most precious resource, and changing the license is not worth the setback.

4. Promise to change the license only on the following and further releases. Do not apply the relicense retroactively to previous releases.

5. Obtain written consent from all current and past developers that the relicense is acceptable. An email suffices. Put a healthy time limit on the opportunity to submit the consent form. This allows for the opportunity to raise objections. A month is typically enough. Still allow the relicensing to fail at this point.

6. Relicense the code on the next release after waiting an appropriate amount of time.

There are many places where even a single person who has not worked on the project in a decade can block this entire process. It also is not a technical concern that deals with how the code operates. There are no right or wrong answers. This is a recipe for hurt feelings and battered egos. You should enter into such an effort treading as softly and as slowly as possible.

Even with the license's chief position in a software project, you also need to understand its limitations.

Copyright Is Not Everything

Copyright and licensing are not all that there is to what is broadly known as *intellectual property*. *Patents* cover temporary monopolies over ideas and processes. The original idea behind patents was to allow creators, inventors, and innovators a brief opportunity to bring their ideas to market. Patenting software has recently come under intense scrutiny due to "patent trolls" who hoard patents but do not attempt to produce anything but lawsuits. *Trademarks* also fall into the realm of intellectual property. They are recognizable symbols of a business or organization that uniquely identify that group in a specific domain. Trademarks must be continually used or they lapse and may be taken up by another group. Some people believe that it's incorrect to use the umbrella term "intellectual property," as copyright, patents, and trademarks are all distinct legal areas. Each of these requires a lawyer that specializes in that area. Still, for us plebs, the term has stuck.

Intellectual property is not the only instrument of control over software. In fact, most intellectual property does not truly apply to computational physics. Other mechanisms are much more effective and more present in many physics software projects.

Export control is a particularly strong system. This is when the government steps in and forbids the transfer of source code to any other country or foreign national without explicit permission. Typically even citizens who have access must sign a document stating that they promise not to violate the terms of the export restrictions. Export control is particularly powerful because it applies not only to software, but also to data and ideas. As an extreme and unreasonable example, the government could restrict you from telling anyone the value of pi on the basis of export control. When a program is export controlled, it cannot just be put on the Internet. The rules for sharing become much more complicated. Violating export control typically comes with a long jail sentence. It is important to take it *very* seriously. In general, this is not something you ever want to deal with personally.

Physics programs are subject to export control far more frequently than other software. The typical reason that a physics program becomes export controlled is because it solves a class of problems that make it easier to build various kinds of highly restricted weaponry. Furthermore, some contain or produce data that is considered sensitive or secret and therefore may not be shared. Mathematics and computer science programs are sometimes export controlled due to their applications to cryptography. Be aware of and abide by your government's laws regarding export control.

Lastly, in the United States there is the Health Insurance Portability and Accountability Act of 1996, or *HIPAA*. Since this deals with medicine, it comes up only occasionally in physics software. Software that deals with human patients must be properly anonymized to ensure the public's right to privacy. This is known as *HIPAA compliance*. The Department of Health and Human Services is responsible for overseeing such compliance. Other countries have similar laws.

These alternative structures can be extraordinarily effective at limiting what can be done with software. They are not always in the control of the author, either. While patents and trademarks and other concerns in intellectual property may not apply, HIPAA rules and privacy violations can be extraordinarily incriminating. And every computational physicist needs to be cognizant of export control rules at all times. Momentary lapses in good judgment are not allowed with respect to export control.

Licensing Wrap-up

You have now seen a wide swath of the legal issues that surround software development and how they apply to computational physics. You should be familiar with the following ideas:

- The license document is the most important file in your project.
- Scientific software should be free and open source.
- The permissive BSD licenses are recommended for most computational science projects.
- The copyleft GPL licenses are also reasonable choices.
- The public domain is a great alternative to licensing your code, but it does not apply everywhere. Use CC0 as a substitute for the public domain.
- Creative Commons licenses can apply to more than just software.
- You should not write your own license.
- Relicensing a project can be very difficult.
- It is important to be aware of the export control laws in your country.

Now that you've made it to the end of the book, the next chapter dives into some parting thoughts about computational physics.

CHAPTER 23
Further Musings on Computational Physics

At last, you have arrived! You are now ready to go forth into the wide world of computational physics. No matter where your specialty may take you, you now have the skills, abilities, and understanding to perform and reproduce great feats of scientific computing. For some of you, this book is all you will need to succeed. For others, this is only the beginning.

Where to Go from Here

What is so beautiful about the skills that you have learned in this book is that they empower you to go anywhere. Computational physics has taken us from deep within the Earth's crust to the farthest reaches of the universe, from pole to pole, all around the world, and everything in between. Even asking "Where to?" can seem daunting.

The answer is that you should go where your interests lie. If you have a special part of physics that you already call home, research what computational projects are out there already. Then try a project out as a user. Join the mailing list. Ask the developers if they need help with anything, and try contributing back. A good project will be very welcoming to new users and contributors.

If there is nothing out there that does what you want, you don't like the languages that the existing projects are written in, or you don't agree with their licenses, try starting your own project—one that suits your needs. This is not scary, and with the existence of repository hosting websites like GitHub, it has become very easy. Starting a new project is a great way to hone your software architecture skills while also clarifying what part of physics you find most interesting.

The following is a list of projects that might pique your interest. We have grouped them according to subdomain. Most, but not all, of these have strong Python and physics components. Here they are, in their own words:

- Astronomy and astrophysics
    - yt (*http://yt-project.org/*): A Python package for analyzing and visualizing volumetric, multi-resolution data from astrophysical simulations, radio telescopes, and a burgeoning interdisciplinary community.
    - Astropy (*http://www.astropy.org/*): A community effort to develop a single core package for astronomy in Python and foster interoperability between Python astronomy packages.
    - SunPy (*http://sunpy.org/*): The community-developed, free and open source solar data analysis environment for Python.
- Geophysics, geography, and climate
    - UV-CDAT (*http://uvcdat.llnl.gov/*): A powerful and complete frontend to a rich set of visual data exploration and analysis capabilities well suited for climate data analysis problems.
    - Iris (*http://scitools.org.uk/iris/*): A Python library for meteorology and climatology. The Iris library implements a data model to create a data abstraction layer that isolates analysis and visualization code from data format specifics.
    - ArcPy (*http://bit.ly/py-arcgis*): Python for ArcGIS.
    - Shapely (*http://bit.ly/shapely-docs*): A Python package for set-theoretic analysis and manipulation of planar features using functions from the well-known and widely deployed GEOS library.
- Nuclear engineering
    - PyNE (*http://pyne.io/*): A suite of tools to aid in computational nuclear science and engineering. PyNE seeks to provide native implementations of common nuclear algorithms, as well as Python bindings and I/O support for other industry-standard nuclear codes.
    - OpenMC (*http://mit-crpg.github.io/openmc/*): A Monte Carlo particle transport simulation code focused on neutron criticality calculations. It is capable of simulating 3D models based on constructive solid geometry with second-order surfaces. The particle interaction data is based on ACE format cross sections, also used in the MCNP and Serpent Monte Carlo codes.
    - Cyclus (*http://fuelcycle.org/*): The next-generation agent-based nuclear fuel cycle simulator, providing flexibility to users and developers through a dynamic resource exchange solver and plug-in, user-developed agent framework.
- Physics
    - QuTiP (*http://qutip.org/*): Open source software for simulating the dynamics of open quantum systems.

- Trackpy (*http://soft-matter.github.io/trackpy/*): A Python package providing tools for particle tracking.
- Mathematics
  - FiPy (*http://www.ctcms.nist.gov/fipy/*): An object-oriented, partial differential equation (PDE) solver, written in Python, based on a standard finite volume (FV) approach.
  - SfePy (*http://sfepy.org/doc-devel*): Software for solving systems of coupled partial differential equations (PDEs) by the finite element method in 1D, 2D, and 3D.
  - NLPy (*http://nlpy.sourceforge.net/*): A Python package for numerical optimization. It aims to provide a toolbox for solving linear and nonlinear programming problems that is both easy to use and extensible. It is applicable to problems that are smooth, have no derivatives, or have integer data.
  - NetworkX (*https://networkx.github.io/*): A Python-language software package for the creation, manipulation, and study of the structure, dynamics, and functions of complex networks.
  - SymPy (*http://sympy.org*): A Python library for symbolic mathematics. It aims to become a full-featured computer algebra system (CAS) while keeping the code as simple as possible in order to be comprehensible and easily extensible.
  - Sage (*http://www.sagemath.org/*): A free, open source mathematics software system licensed under the GPL.
- Scientific Python
  - IPython (*http://ipython.org/*): A rich architecture for interactive computing.
  - NumPy (*http://www.numpy.org/*): The fundamental package for scientific computing with Python.
  - SciPy (*http://www.scipy.org/*): The SciPy library is one of the core packages that make up the SciPy stack. It provides many user-friendly and efficient numerical routines such as routines for numerical integration and optimization.
  - Pandas (*http://pandas.pydata.org/*): An open source, BSD-licensed library providing high-performance, easy-to-use data structures and data analysis tools for the Python programming language.
  - matplotlib (*http://matplotlib.org/*): A Python 2D plotting library that produces publication-quality figures in a variety of hardcopy formats and interactive environments across platforms.
  - PyTables (*http://pytables.github.io/*): A package for managing hierarchical datasets, designed to efficiently and easily cope with extremely large amounts of data.

- h5py (*http://www.h5py.org/*): A Pythonic interface to the HDF5 binary data format. It lets you store huge amounts of numerical data, and easily manipulate that data from NumPy.
- PyROOT (*http://bit.ly/pyROOT*): A Python extension module that allows the user to interact with any ROOT class from the Python interpreter.
- lmfit (*http://lmfit.github.io/lmfit-py/*): A high-level interface to non-linear optimization and curve fitting problems for Python. Lmfit builds on the Levenberg-Marquardt algorithm of `scipy.optimize.leastsq()`, but also supports most of the optimization methods from `scipy.optimize`.
- scikit-image (*http://scikit-image.org/*): A collection of algorithms for image processing.

- Enthought Tool Suite (ETS) (*http://code.enthought.com/projects/*): A collection of components developed by Enthought and its partners, which can be used every day to construct custom scientific applications.
  - Mayavi (*http://bit.ly/3d-mayavi*): An application and library for interactive scientific data visualization and 3D plotting in Python.
  - ParaView (*http://www.paraview.org/*): An open source, multiplatform data analysis and visualization application. ParaView users can quickly build visualizations to analyze their data using qualitative and quantitative techniques. The data exploration can be done interactively in 3D or programmatically through ParaView's batch processing capabilities.
  - VisIt (*http://bit.ly/visit-app*): An open source, interactive, and scalable visualization, animation, and analysis tool. From Unix, Windows, or Mac workstations, users can interactively visualize and analyze data ranging in scale from small desktop-sized projects to large leadership-class computing facility simulation campaigns.
  - Vispy (*http://vispy.org/*): A high-performance interactive 2D/3D data visualization library. Vispy leverages the computational power of modern graphics processing units (GPUs) through the OpenGL library to display very large datasets.
  - Cython (*http://cython.org/*): An optimizing static compiler for both the Python programming language and the extended Cython programming language. It makes writing C extensions for Python as easy as Python itself.

- Pedagogy and community
  - Software Carpentry (*http://software-carpentry.org*): A nonprofit membership organization devoted to improving basic computing skills among researchers in science, engineering, medicine, and other disciplines. Its main goal is to teach researchers software development skills to help them do more, in less time and with less pain.

— NumFOCUS (*http://numfocus.org*): A nonprofit foundation that supports and promotes world-class, innovative, open source scientific software. NumFOCUS aims to ensure that money is available to keep projects in the scientific Python stack funded and available.

— Mozilla Science Lab (*http://mozillascience.org/*): A Mozilla Foundation initiative that helps a global network of researchers, tool developers, librarians, and publishers collaborate to further science on the Web.

— PyLadies (*http://www.pyladies.com/*): An international mentorship group with a focus on helping more women become active participants and leaders in the Python open source community.

— Open Hatch (*https://openhatch.org/*): A nonprofit dedicated to matching prospective free software contributors with communities, tools, and education.

Now, go forth and make waves!

Glossary

absolute path
 The full, unambiguous path from the root directory at the top of the tree, all the way to the file or directory being indicated.

API
 Application programming interfaces are the public-facing functions, methods, and data with which users and developers interact.

assertion
 An assertion, in software, is an operation that compares two values. If the assertion returns a false Boolean value, a runtime exception is thrown.

assignment
 Assignment statements apply values to variable names. Most modern languages use = to denote assignment. Notably, R uses left and right arrows (<- and ->).

attributes
 Any variable may have attributes, or *attrs*, which live on the variable. Attributes are also sometimes known as *members*. They may be accessed using the binary . operator.

awk
 A shell program for manipulation and analysis of column-based text files.

bash
 The Bourne Again SHell, which combines features from previous shells and seeks to provide a scripting language for interacting with an operating system through a terminal emulator.

binary
 An operator that takes exactly two variables or expressions.

call stack
 Also called the execution stack, this is the list of function calls that have been made but not completed. At a certain point of execution, such as a crash, the call stack makes up the traceback.

class
 A class defines a collection of functions and data. Classes also define constructors, which describe how to create the objects that those functions and data are associated with.

command-line interface
 The command-line interface, or CLI, provides access to the shell. Commands can be entered into the command-line prompt to navigate the filesystem, run programs, and manipulate files.

compilation
 A step in building software when a compiler converts source code into binary machine code.

compiled language
A programming language in which source code is first converted to binary machine code and then executed. This two-step process requires that all statements that are to be executed be known before the compilation step.

concatenate
A fancy word meaning "to append."

configuration
A step in building software or executing an analysis pipeline when platform-dependent variables are detected and used to customize a makefile or installation script.

container
A data structure that holds other variables. For example, a tuple, list, or dictionary.

continuous integration
A software development strategy in which new code is tested and built regularly. Usually, new code must pass all tests in order to be accepted and cross-platform installation issues are checked once or more per day.

CPU-bound
Describes an operation that is limited by the speed of the processor.

csh
The C SHell—an early shell program, based on sh, for interacting with an operating system through a terminal emulator.

current working directory
When you're using the terminal, the current working directory is the one that you are in. That is, it is the location from which all commands are executed. It is output as the result of the `pwd` command.

docstring
A specific syntax for documentation in Python. It is the first unassigned string literal in a function body and is usually enclosed by three double quotes.

DRY
The "don't repeat yourself" principle states that any piece of code should be defined once and only once and have a single meaning. This promotes code reuse.

dynamic language
A programming language where variable types are not declared before they are used.

exception
In software, an exception is a way of alerting the user to runtime errors in code behavior. An exception can be thrown from some place in the code and, optionally, caught elsewhere in the code. If it is thrown but not caught before reaching global scope, it halts code execution and prints a (hopefully) informative error message.

executable
A runnable program.

general-purpose language
A programming language that is meant to cover a broad range of domains and computational problems.

lobal scope
The namespace of the current module.

glue language
A programming language that can easily interact with multiple other programming languages and their libraries.

grep
A command-line program for regular expression–based pattern matching.

high-level language
A programming language that provides useful, common abstractions over lower-level languages. Such languages are typically more concise and easier to use than their counterparts. Examples include Python, R, and MATLAB.

inheritance
When one class is a subclass of another, the subclass is said to be inheriting from the superclass. Indeed, data and behavioral attributes of the parent (super)class are passed down to the child (sub)class.

installation
The step in building software where executables, libraries, and include files associated with a program are put in an accessible place in the filesystem.

integration test
A type of test that exercises more than a few units (functions) in a code base. Integration tests often check that for simple inputs, the code arrives at a final expected answer as an ultimate output.

interpreted language
A programming language in which statements in the language are executed at runtime by sequentially being fed into a special interpreter loop. There is no need for a separate compile step in order to execute code because of the interpreter. High-level languages are often interpreted, though not always!

issue trackers
Also known as issue ticketing systems or bug trackers, these systems help to streamline a project. Most often, they provide an interface for organizing the process of identifying, describing, collaborating on, and solving software bugs and new features.

ksh
The Korn SHell is an early shell program that is backward-compatible with sh, but extends its ability to interact with an operating system through a terminal emulator.

linking
The step in building software in which an executable is attached to an external library on which it depends.

Linux
An open source operating system kernel first developed by Linus Torvalds. There are many flavors of Linux operating systems available. The most popular of these is Ubuntu.

local scope
The namespace of the current function.

memory-bound
Describes an operation whose speed is limited by how much memory (typically RAM) there is available.

metacharacter
A character that has a special meaning aside from its literal meaning.

metadata
Data about data.

no-op
A no-operation statement, function, class, or other language construct that exists as a placeholder for when a value is required but nothing should be executed.

object
In object-oriented programming, objects are instances of a class. They are entities that possess data and methods specific to their class.

object orientation
A computational paradigm that associates data and method attributes to classes of objects.

redirection
When the output of one program is diverted into the input of another program or file. Typically, this relies on the > syntax.

regression test
A test that serves to guarantee the preservation of expected behavior through changes in the code base. A suite of unit tests can serve as regression tests when they are run after changes are made.

regular expressions
A language of metacharacters for pattern matching.

relative path
This is a string describing the path from the current directory to the file or directory being indicated.

REPL
A read-eval-print loop is a standard mechanism for text-based interactive computing. Users type in commands that are read in by the interpreter and evaluated while the user waits, and the results are printed to the screen. This process repeats until the user closes the interpreter session.

root
This word has two meanings in Unix parlance. In the context of filesystems, the "root" directory is the one at the top of the directory tree, indicated by /. In the context of user permissions, "root" is the administrator, the top-level user of the machine.

scope
Scope defines what variables are available inside of different constructs in a programming language. Scoping rules determine how variables are looked up and vary from language to language.

sed
A command-line program for regular expression–based pattern matching and substitution.

sequence
A term used to describe any data structure that imposes an integer ordering on its values. This is roughly equivalent to the mathematical construct by the same name.

sh
The Bourne SHell, a venerable and popular Unix shell that first appeared in Version 7 Unix.

singleton
A singleton is a class that only has one instance in memory throughout the lifetime of a program. The term "singleton" may also apply to the variable itself. Singletons are not necessarily constants.

string
A string is a character or list of characters. It is a data type appropriate for data such as names and paths.

symbolic link
A filesystem object that points to another filesystem object (the "target").

tcsh
The TENEX C SHell. Like most shells, this shell seeks to provide a scripting language for interacting with an operating system through a terminal emulator.

terminal emulator
A graphical user interface that allows the user access to a text prompt and applications such as the command-line interface.

ternary
An operator that takes exactly three variables or expressions.

test framework
A system for collecting and running unit tests. Examples include nose, jUnit, xUnit, and GTest.

traceback
Also called the stack trace, backtrace, or stack backtrace. The traceback is a report of the active function calls at a certain point during execution.

unary
An operator that takes only one variable or expression.

unit test
A test that operates on a single unit of code. Typically, functions and methods are the atomic units of code that are tested in this case.

Unix

Any computer operating system platform that derives from the original AT&T Unix, developed at the Bell Labs research center.

variable type

A variable's type defines the internal properties of the value, how it is stored, and how other parts of a language may use the variable. Variable types may be primitive and built into the language or defined by users and developers. A variable may be checked to verify whether it "is a" certain type. Certain types may be converted to other types.

version control

Version control is a method by which a repository is created for holding versions of a set of files. Version control systems include Git, Mercurial, SVN, CVS, and more. These systems allow storage, recall, and distribution of sets of files under development. Plain-text files, such as code and markup, are well suited to version control.

Bibliography

- Albrecht, Michael, Patrick Donnelly, Peter Bui, and Douglas Thain. "Makeflow: A Portable Abstraction for Data Intensive Computing on Clusters, Clouds, and Grids." *Proceedings of the 1st ACM SIGMOD Workshop on Scalable Workflow Execution Engines and Technologies* (2012): 1–13. ACM Press. doi: 10.1145/2443416.2443417.

- Alted, F. "Why Modern CPUs Are Starving and What Can Be Done about It." *Computing in Science & Engineering* 12 (2010): 68–71. doi:10.1109/MCSE.2010.51.

- Barendregt, Henk P., and Erik Barendsen. "Introduction to Lambda Calculus." *Nieuw archief voor wisenkunde* 4 (1984): 337–372.

- Beck, Kent. *Test-Driven Development: By Examples.* Boston, MA: Addison-Wesley, 2002.

- Chacon, Scott, and Ben Straub. *Pro Git.* 2nd ed. New York: Apress, 2014. http://git-scm.com/book/en/v2.

- Collette, Andrew. *Python and HDF5: Unlocking Scientific Data.* Sebastopol, CA: O'Reilly Media, 2013.

- Donoho, David L., Arian Maleki, Inam Ur Rahman, Morteza Shahram, and Victoria Stodden. "Reproducible Research in Computational Harmonic Analysis." *Computing in Science & Engineering* (11): 8–18. doi:10.1109/MCSE.2009.15.

- Drepper, Ulrich. "What Every Programmer Should Know About Memory." September 21, 2007. http://lwn.net/Articles/250967/.

- Duarte, Gustavo. "What Your Computer Does While You Wait." November 30, 2008. *http://bit.ly/comp-wait*.

- Feathers, Michael. *Working Effectively with Legacy Code*. Upper Saddle River, NJ: Prentice Hall, 2004.

- Fenner, Martin. "One-Click Science Marketing." *Nature Materials* 11 (2012): 261–63. doi:10.1038/nmat3283.

- Friedl, Jeffrey E.F. *Mastering Regular Expressions*. Sebastopol, CA: O'Reilly Media, 2002.

- Gamma, Eric, John Vlissides, Ralph Johnson, and Richard Helm. *Design Patterns: Elements of Reusable Object-Oriented Software*. Reading, MA: Addison-Wesley, 1995.

- Goble, Carole. "Better Software, Better Research." *IEEE Internet Computing* 18 (2014): 4–8.

- Goldberg, David. "What Every Computer Scientist Should Know About Floating-Point Arithmetic." *ACM Computing Surveys* 23 (1991): 5–48.

- Inman, Matthew. "Why the Mantis Shrimp is my new favorite animal." *http://theoatmeal.com/comics/mantis_shrimp*.

- Irving, Damien. "Authorea: The Future of Scientific Writing?" April 20, 2014. Dr Climate. Accessed September 29. *http://bit.ly/authorea*.

- Knight, Steven. "Building Software with SCons." *Computing in Science & Engineering* 7 (2005): 79–88.

- Knuth, Donald Ervin. *Computers & Typesetting, Volume A: The TeXbook*. Reading, MA: Addison-Wesley, 1986. *http://bit.ly/texbook*.

- Lamport, Leslie. *LaTeX: A Document Preparation System*. 2nd ed. Reading, MA: Addison-Wesley, 1994.

- Martin, Ken, and Bill Hoffman. *Mastering CMake*. Clifton Park, NY: Kitware, 2010.

- Martin, Robert C. *Clean Code: A Handbook of Agile Software Craftsmanship.* Upper Saddle River, NJ: Prentice Hall, 2008.

- McKinney, Wes. *Python for Data Analysis: Data Wrangling with Pandas, NumPy, and IPython.* Sebastopol, CA: O'Reilly Media, 2012.

- Merali, Zeeya. "Computational Science: …Error." *Nature* 467 (2010): 775–77. doi: 10.1038/467775a.

- MissMJ. "Standard Model of Elementary Particles" June 27, 2006. *http://bit.ly/particle-model*.

- Mozilla Developer Network. "Regular Expressions." *http://bit.ly/mdn-regex*.

- Newham, Cameron. *Learning the bash Shell.* 3rd ed. Sebastopol, CA: O'Reilly Media, 2005.

- NumPy developers. "NumPy." *http://www.numpy.org/index.html*.

- Oetiker, Tobias, Hubert Partl, Irene Hyna, and Elisabeth Schlegl. "The Not So Short Introduction to LaTeX2e." Version 5.04, October 29, 2014. *http://bit.ly/tex-short*.

- Pérez, Fernando, and Brian E. Granger. "IPython: A System for Interactive Scientific Computing." *Computing in Science and Engineering* 9 (2007): 21–29. doi: 10.1109/MCSE.2007.53.

- Perl 5 Porters. "Perl 5 version 20.1 documentation - perlre." *http://perldoc.perl.org/perlre.html*.

- Prechelt, Lutz. "An Empirical Comparison of Seven Programming Languages." *Computer* 33 (2000): 23–29.

- Preshing, Jeff. "Hash Collision Probabilities." May 4, 2011. *http://bit.ly/hash-prob*.

- Press, William H., Saul A. Teukolsky, William T. Vetterling, and Brian P. Flannery. *Numerical Recipes 3rd Edition: The Art of Scientific Computing.* Cambridge: Cambridge University Press, 2007.

- The SciPy community. "NumPy Reference Manual- Universal functions (ufuncs)." *http://bit.ly/ufunc*.

- Stodden, Victoria. "The Scientific Method in Practice: Reproducibility in the Computational Sciences." MIT Sloan School Working Paper 4773-10, February 2010. *http://bit.ly/repro-sci*.

- Terry, Matthew. SciPy 2013 Yoink Lightning Talk. *http://bit.ly/scipy-talk*.

- Tufte, Edward R. *The Visual Display of Quantitative Information*. 2nd ed. Cheshire, CT: Graphics Press, 2001.

- Van Rossum, Guido, Barry Warsaw, and Nick Coghlan. "PEP8 — Style Guide for Python Code." Python Software Foundation, 2001. *http://bit.ly/pep-8*.

- Verborgh, Ruben. *Using OpenRefine*. Birmingham: Packt Publishing, 2013.

- Wikipedia. "Comparison of revision control software." *http://bit.ly/revcontrol*.

- Wilson, Greg, D. A. Aruliah, C. Titus Brown, Neil P. Chue Hong, et al. "Best Practices for Scientific Computing." *PLoS Biol* 12 (2014): e1001745. doi:10.1371/journal.pbio.1001745.

- Zach, Richard. "LATEX for Philosophers." November 7, 2013. *http://bit.ly/tex-phil*.

Index

Symbols
" (double quote), 49, 54
""" (triple double quotes), 55
(comment character), 41, 432
$ (dollar sign), 1
% (modulo operator), 54
' (single quote), 49, 54
''' (triple single quotes), 55
( ) (parentheses operator), 55
* (wildcard character), 20, 180
+ (addition operator), 53
, (comma), 70
-- (double minus sign), 23
-r (recursive flag), 19
-rf (recursive force flag), 19
. (dot), 4, 55, 62, 183
.. (double dot), 62
/ (forward slash), 3
<> (angle brackets), 8
= (assignment operator), 48
= (equals sign), 23, 42
== (equality operator), 79
? (question mark), 183
@ sign, 112, 139
[ ] (square brackets), 8, 66, 186
[[]] (double square brackets), 8
\ (backslash character), 13, 54, 185
\ (escape character), 13, 54, 184
\n (newline character), 54
\r (carriage return character), 54
\t (tab character), 54
_ (underscore character), 13, 42
__ (double underscore character), 120
__init__( ) method, 127, 141
__new__( ) method, 141
{} (curly braces), 56, 71, 186
~ (tilde character), 5
| (pipe command), 25, 186
> (arrow character), 25
>> (double arrow), 25
>>> (Python prompt), 40

A
abs( ) function, 121
absolute imports, 61
absolute path, 3, 34, 493
accumulate( ) method, 223
add( ) method, 73
addition operator (+), 53
algorithms
    data size and, 259
    implementing, 161
    in- and out-of-core, 249
    non-parallel, 285-290
    parallel, 282
    zipping, 252
alias command, 36
alternation, 186
Amdahl's law, 281
American Mathematical Society, 452
Anaconda, installing, xxiii
analysis and visualization
    analysis
        data-driven, 162
        model-driven, 160
        tools for, 159
    cleaning and munging data, 155-159
    data preparation

      automated approach to, 148
      experimental data, 149
      metadata, 151
      simulation data, 150
      steps of, 145
   importance of, 145
   loading data
      Blaze, 155
      NumPy, 152
      Pandas, 153
      PyTables, 153
      tools available, 151
   visualization
      best practices, 162, 176
      Bokeh, 172
      Gnuplot, 164
      Inkscape, 174
      matplotlib, 167-172
      tool selection, 175
      tools for, 164
analysis pipeline
   automating creation of, 333
   building/installing software
      automation tools, 343
      compilation, 345
      dependency configuration, 345
      makefile configuration, 343
      overview of, 341
      platform configuration, 344
      system and user configuration, 344
   compilation, 334
   configuration, 333
   execution, 334
   installation, 334, 346
   linking, 334
   make utility
      automatic updating by, 337
      benefits of, 334
      enabling, 337
      makefiles, 337
      running, 337
      special targets, 340
      target definition, 338
      vs. bash scripts, 336
   overview of tasks, 335
   strategies for, 333
angle brackets (<>), 8
anonymous functions, 108
antipatterns, 142

APIs (application programming interfaces), 66, 472, 493
append( ) method, 67
approximations, 42
apropos command, 24
arange( ) function, 202, 221
arguments
   adding, 22
   keyword, 99
   naming, 128
   optional, 8
   positional, 98
   variable numbers of, 101
arithmetic operations, 211, 226, 280
array data languages, 201
array( ) function, 202
arrays
   adding dimensions to, 214
   Boolean arrays, 217
   broadcasting, 212
   comparison operators, 218
   copying a slice, 210
   creating, 202, 207
   data types (dtypes), 204
   fancy indexing, 215
   fixed size of, 204
   forcing data types in, 207
   HDF5 data format, 240
   immutability of dtypes, 221
   manipulating attributes, 204
   organizing with B-trees, 270
   purpose of, 201
   record arrays, 220
   slicing and views, 208
   structured arrays, 220
   support for in programming languages, 201
   temporary, 211
   transforming, 223
arrow character (>), 25
as keyword, 59
ASCII characters, 49
assertions, 405, 493
assignment, 42, 493
assignment operator (=), 48
astronomy and astrophysics projects, 487
at( ) method, 223
atof( ) function, 238
atoi( ) function, 238
atomic types, 238

attributes, 493
  listing, 119
  manipulating in arrays, 204
  ndarray attributes, 203
  string, 55
Automake/Autoconf, 343
automatic resizing, 261
awk, 493
  adding action with, 196
  benefits of, 195
  example of, 195
  overview of, 188
  vs. sed and grep, 195

B

B-trees
  array organization with, 270
  best application of, 257, 269
  libraries supporting, 271
  rotation of, 270
  rules for, 270
  structure of, 269
  vs. binary search trees, 270
backslash character (\), 13, 54, 185
backtrace (see traceback report)
Bash (Bourne Again SHell), 493
Bash scripts, 36, 335
Basic regular expressions (BRE), 182
.bashrc file, 33
Berkeley Software Distribution (BSD), 475
bibliographies, 456
BibTex/BibLaTex, 456
bidirectional/bijective hash maps, 263
Big Data, 229
big O notation, 259
bin directory, 3
binary operators, 46, 493
binary package managers, 311, 316
binary search trees, 270
  (see also B-trees)
Binstar, 318
birthday paradox, 261
Bitbucket, 371
bitwise operators, 219
blank lines, deleting, 193
Blaze, 155
blist package, 271
Bokeh, 172
Boolean arrays, 217

Boolean values, 45
boolean variable type, 45
Bourne Shell (sh), 496
branches, 365
break statement, 87
broadcasting, 212
buckets, 262
build system automation tools, 343

C

call stacks, 395, 493
capturing parentheses, 192
carriage return character (\r), 54
cat command, 13
CC0 (Creative Commons Zero) license, 480
cd (change directory) command, 7
central processing unit (CPU), 235, 244
change mode (chmod) command, 29
change ownership (chown) command, 28
changes, tracking in publication process, 443
char (character) type, 49
character sets, 186
child threads, 291
chmod (change mode) command, 29
chown (change ownership) command, 28
chunking, 245
citations, 456
classes
  attribute types associated with, 124
  class files, 446
  class inheritance, 133, 495
  class keyword, 123
  class variables, 124
  constructors, 127
  decorators, 139
  defining, 123
  definition of term, 493
  duck typing, 133
  function of, 117
  instance variables, 126
  main ideas of, 118
  metaclasses, 140
  methods, 129
  polymorphism, 135-138
  purpose of, 123
  static methods, 132
CLI (command-line interface), 493
  additional resources, 38
  basics of, 1

Index | 505

computing environment, 31-36
   getting help, 21-26
   HDF5 utility commands, 254
   managing files and directories, 11-21
   metacharacters on, 179-187
   navigating the shell, 1-11
   permissions and sharing, 26-31
   scripting with bash, 36
   SSL (secure socket layer) protocol, 30
climate projects, 488
close( ) method, 234
cloud computing, 310, 325
CMake, 343
code
   backing up online, 371
   citing, 458
   code reuse, 95
   code reviews, 468
   creating documentation for, 427-440
   deploying, 309-329
   direct collaboration, 461-470
   legacy code, 427
   legal issues surrounding, 471-485
   scalability of, 281
   scaling up, 281
   self-documenting, 434
   style guides for, 434
   text editors for, 443
   writing clean, 414, 434
code reuse, 104
collaboration
   overview of, 461
   pull requests, 468
   ticketing systems
      assigning issues, 466
      benefits of, 462
      closing issues, 468
      discussing issues, 467
      issue creation, 464
      workflow, 462
collisions, 261
columns
   manipulating data in, 195
   working with in data frames, 268
comma (,), 70
comment character (#), 41, 432
commits, making, 374, 458
communicator objects, 301
community and pedagogy projects, 490

COMM_WORLD communicator, 302
comparison operators, 78, 218
compatibility, 478
compilation, 493
compiled languages, 39, 316, 334, 341, 494
complex patterns, 192
composable operators, 48
compound dtypes, 220
comprehensions, 90
compression, 252
computational physics
   astronomy and astrophysics projects, 487
   basic project steps, xix-xxi
   benefits of regular expressions for, 178
   geophysics, geography, and climate projects, 488
   mathematics, 489
   nuclear engineering projects, 488
   pedagogy and community projects, 490
   physics projects, 488
   Python open source community, 491
   Python operators for, 46
   relationship of physics and computation, xvii
   scientific Python projects, 489
computational scale, 280
computer architecture, 235
computers, original use of term, 177
computing environment
   configuring/customizing, 33
   investigating with echo program, 31
   nicknaming commands, 36
   running programs, 34
   saving variables, 33
computing systems
   distributed computing, 283
   high-performance, 283
   high-throughput, 283, 327
   simulated, 319
concatenation, 13, 53, 494
Conda package manager, xxiii, 316
conditionals
   if-elif-else statements, 81
   if-else statements, 80, 82
   positive vs. negative, 81
   syntax of, 77
configuration, 494
constants, scientific, 131
constructors, 127

containers, 494
    deployment with, 321-325
    lists, 66
    purpose of, 65
    sets, 71
    tuples, 70
    types of, 65
content management systems (see ticketing systems)
contents, listing, 6
context management, 234
contiguous datasets, 245
continuous integration, 494
copyleft licenses, 478
copyright (see legal issues)
corner cases, 410
countdown( ) generator, 110
cp (copy) command, 17
cProfile, 396
CPU-bound, 249, 291, 494
CPython, 291
create_array( ) method, 240
create_carray( ) method, 247
create_group( ) method, 240
create_table( ) method, 240
Creative Commons (CC), 471, 479
cross references, creating, 29, 455
csh (C SHell), 494
CSV (comma separated values), 152
Ctrl-c command, 14
Ctrl-d command, 14
curly braces ({}), 56, 71, 186
current working directory, 4, 494

D

daemon threads, 291
dash dash (--), 23
data
    chunking, 245
    citing, 458
    concealing with masks, 218
    converting between formats, 155
    CSV format, 152
    experimental data, 149
    HDF5 format, 153
    improving access to, 244, 252, 259
    manipulating columns of, 195
    missing, 158
    mutable vs. immutable types, 65

processing raw, 177
saving/loading in Python, 231
simulation data, 150
size limitations, 259
structuring, 155
time series data, 149
various formats for, 153
wrangling, 155
data analysis and visualization
    analysis, 159-162
    cleaning and munging, 155-159
    importance of, 145
    loading data, 151-155
    preparing data, 145-151
    visualization, 162-175
data frames
    benefits of, 263, 269
    best application of, 257, 266
    creating/working with, 267
    handling of missing data, 265
    makeup of, 263
    series, 264
    structure of, 266
    vs. other structures, 266
    vs. tables, 263
data structures
    arrays, 201-228
    B-trees, 269
    data frames, 263-269
    hash tables, 258-263
    k-d trees, 272-277
    overview of, 257
data-driven analysis, 162
datasets
    B-trees and, 271
    chunking, 245
    compressing, 252
    contiguous, 245
    HDF5 format, 240
    inspecting via command line, 254
    inspecting via graphical interfaces, 255
debugging
    encountering bugs, 386
    functions and, 95
    importance of, 386
    interactive debugging, 389
    linting, 401
    overview of, 385
    pdb interactive debugger

Index | 507

    continuing the execution, 394
    features of, 390
    getting help, 392
    importing, 390
    querying variables, 393
    running functions and methods, 394
    setting breakpoints, 395
    setting the state, 393
    setting the trace, 391
    stepping forward, 392
  print statements, 387
  profiling, 396-401
  reporting bugs with issue trackers, 463
decision making (see flow control and logic)
decorators
  class, 139
  function, 112-116
def keyword, 96, 123
default constructors, 127
degree of parallelism, 280
del (deletion) operator, 48
deployment
  best practices for, 313
  challenges of, 309
  cloud-based, 325
  develop-then-use cycle, 309
  documentation, 427-440
  goals for, 309
  overview of, 329
  packaging
    binary package managers, 311, 316
    Conda package manager, 316
    containers, 321-325
    cross-platform package managers, 316
    distributable files, 311
    package managers, 311
    pip packaging tool, 312-316
    source-based distributions, 311
    virtual machines, 319
    virtualizations, 311
  supercomputer-based, 327
design patterns, 142
develop-then-use cycle, 309
developer guides, 431
dictionaries, 73, 89, 261
digital object identifier (DOI), 458
dimensions, adding to arrays, 214
dir( ) function, 119
directories

bin, 3
changing, 7
current working, 4, 494
deleting, 18
flags and wildcards, 20
home, 5
lib, 3
listing files and subdirectories, 6
making, 18
manipulating, 11-21
printing working, 4
root, 3, 496
searching multiple, 181
trees, 3
discard( ) method, 73
distributed computing, 283
distutils module, 312
Docker, 321-325
docstrings, 98, 122, 435, 494
document processing
  common programs, 441, 442
  separating content from formatting, 442
  tracking changes, 443
  WYSIWYG systems, 442
documentation
  automated creation of, 436
  avoiding over-documentation, 433
  benefits of, 429
  choices for, 429
  comments, 432
  docstrings, 435
  importance of, 427
  naming, 434
  readme files, 431
  self-documenting code, 434
  theory manuals, 430
  user/developer guides, 431
  value of, 428
dollar sign ($), 1
dot (.), 4, 55, 62, 183
double arrow (>>), 25
double dot (..), 62
double minus sign (--), 23
double quote ("), 49, 54
double square brackets ([[]]), 8
double underscore character (__), 120
double-space formatting, 193
Doxygen, 436
DRY (don't repeat yourself) principle, 95, 494

dtypes (data types), 204, 220
duck typing, 65, 133
dunder (see double underscore)
dynamic languages, 39, 44, 494

E

echo program, 31
edge cases, 409
Electronic Frontier Foundation (EFF), 471
element-wise operations, 211
emacs text editor, 15
embarrassingly parallel problems, 282
empty files, 13
empty( ) function, 202, 221
encapsulation, 118, 123
encodings, 50
End-User License Agreement (EULA), 474
env command, 33
environment variables, 32
equality operator (==), 79
equals sign (=), 23, 42
error messages, 43, 84
escape character (\), 13, 54, 184
Evaluated Nuclear Data File (ENDF), 150
exception handling, 82
exceptions, 84, 405, 494
executables, 1, 494
execution pathways, 77
execution stacks (see call stacks)
exit( ) function, 40
experimental data, 149
explicit indexing, 51
explicit relative imports, 62
explicit typing, 134
export command, 32
export control, 484
expressions, 48
extend( ) method, 67
Extended regular expressions (ERE), 182
extension modules, 57

F

"fail early and often" credo, 44
False variable, 45
fancy indexing, 215
figures, publishing in LaTeX, 454
file handle object, 231
file space, 3
filenames, 13

files
    accessing remote, 30
    appending contents of, 25
    benefits of regular expressions for, 177
    class files, 446
    closing, 232
    copying and renaming, 17
    creating
        choices for, 11
        GUIs for, 12
        text editors, 13
        touch command, 13
    creating links in, 455
    creating links to, 29
    cross-referencing, 29
    deleting, 18
    finding and replacing patterns in, 190-195
    finding filenames with patterns, 182-187
    finding patterns in, 188
    flags and wildcards, 20
    formats for, 230
    granting access to, 26
    handling in HDF5 format, 239-242
    handling in Python, 230-235
    hidden, 34
    inspecting head and tail, 10
    license files, 471
    listing, 6
    listing with simple patterns, 180
    manipulating, 11-21
    modes for, 233, 239
    operating on multiple, 179-187
    overwriting contents of, 25
    reading, 233
    readme files, 431
    redirecting, 15, 25
    reducing size of, 252
    safety code, 234
    setting ownership of, 28
    setting permissions on, 29
    sharing, 26, 41
    style files, 446
    user-specific configuration in, 34
    version control of
        citations, 458
        local, 349-369
        remote, 371-383
    viewing permissions on, 26
    writing to, 233

filters, 92, 253
find and replace function, 190
find command, 182-187
fixed points, 107
flags
    contextual meanings for, 23
    using wildcards with, 20
FLASH license, 481
Flexible Image Transport System (FITS), 150
float( ) function, 238
floating-point arithmetic, 42, 161
FLOPS (floating-point operations per second), 280
flow control and logic
    conditionals
        if-elif-else statements, 81
        if-else statements, 80, 82
        syntax of, 77
    exceptions, 82
    importance of, 77
    loops
        comprehensions, 90
        for loops, 88
        formats for, 85
        while loops, 86
    types of, 77
for loops, 88
forks, 297
format( ) method, 56
forward slash (/), 3
FOSS/FLOSS licenses, 474
fragments command, 35
Free Software Foundation (FSF), 471
from keyword, 62
from-import statement, 58
frozenset data type, 73
full stop (.) (see dot)
functional programming, 109
functions
    anonymous, 108
    basis of, 95
    best practices for, 434
    constructors, 127
    decorators, 112-116
    defining, 96
    dunder vs. built-in, 121
    generators, 109
    in NumPy, 226
    keyword arguments, 99
    lambdas, 108
    modifying, 112
    multiple return values, 103
    object orientation and, 118
    purpose of, 95
    recursion, 107
    scope, 104
    string, 55
    universal functions (ufuncs), 223
    variable number of arguments, 101
    vs. methods, 129
    working with, 95

G

general-purpose languages, 39, 201, 494
generators, 109
geometry problems, 277
geophysics and geography projects, 488
Git
    additional resources on, 384
    checking repo status (git status), 357
    common commands, 353
    configuring, 354
    conflict resolution, 369
    creating repositories (git init), 355
    discarding revisions (git revert), 364
    getting help, 352
    installing, 352
    merging branches (git merge), 367
    saving a snapshot (git commit), 358
    staging files (git add), 357
    switching branches (git checkout), 366
    unstaging (reverting) files (git reset), 363
    version control tasks, 355
    viewing file differences (git diff), 362
    viewing repo history (git log), 361
    working with branches (git branch), 365
Git Bash, 1
GitHub
    account creation, 372
    adding new remotes, 376
    checking repo status (git status), 382
    conflict resolution, 381
    customizable tags in, 465
    declaring a remote (git remote), 373
    downloading repositories (git clone), 375
    features of, 372
    fetching and merging (git pull), 380
    fetching remote contents (git fetch), 379

forking a repository, 377
merging remote contents (git merge), 380
milestones in, 466
repository creation, 373
repository hosting, 371
sending commits (git push), 374
ticketing system in, 462-468
global interpreter lock (GIL), 291
global scope, 104, 494
global substitution, 191
globals( ) function, 252
glue languages, 39, 494
GNU General Public License (GPL), 477
GNU Portable Threads, 307
Gnuplot, 164
Google Code, 371
Google search, 188
GPUs (graphics cards), 237, 280
gradient( ) function, 212
graphicx, 452
grep, 494
  finding patterns in files with, 188
  history of, 178
  overview of, 187
  search options, 189
  vs. sed, 195
Grid Engine, 328
group access, 26

H

h5dump command-line tool, 254
h5ls command-line tool, 254
h5py, 230
halting problem, 87
hard disk drives (HDDs), 235
hard links, 29
hash tables, 73
  benefits of, 259, 263
  best application of, 257, 277
  bidirectional/bijective hash maps, 263
  collisions, 261
  example, 258, 260
  inner workings of, 258
  popularity of, 258
  Python hash( ) function, 73, 258
  resizing, 259
hashability, 72
HDF5
  benefits of, 229, 235, 237

chunking, 245
compression, 252
converting into NumPy arrays, 153
features of, 238
file manipulation in, 239-242
hierarchical structure of, 238, 242
in- and out-of-core operations, 249
inspecting files
  via command line, 254
  via graphical interface, 255
memory mapping, 242
querying, 252
tables, 240
utility tools, 254
hdfview graphical tool, 255
head command, 10, 25
help
  apropos command, 24
  combining simple utilities, 25
  man (manual) program, 21
help( ) function, 40
heterogeneous data, 71
heterogeneous problems, 282
Hierarchical Data Format 5 (see HDF5)
high-latency tasks, 292
high-level languages, 39, 95, 312, 494
high-performance computing (HPC), 283
high-throughput computing (HTC), 283, 327
HIPAA (Health Insurance Portability and Accountability Act), 484
HMTL plots, 172
home directory, 5
homogeneous data, 71
HyperText Markup Language (HTML), 444
hypervisors, 319

I

identity operator (is), 79
if statement, 78
if-elif-else statements, 81
if-else statements, 82
immutable data types, 65, 73
implicit indexing, 51
implicit relative importing, 61
import statement, 58
imprecise types, 42
in-core operations, 249
indentations, 79
indexing

duck typing and, 66
fancy indexing, 215
of data frames, 263
of hash tables, 262
of lists, 67
techniques for, 50
infinite loops, 86
Infrastructure-as-a-Service (IaaS), 325
inheritance, 495
    class interface and, 133
    definition of term, 118
    graphic representation of, 138
    in forks, 297
    multiple inheritance, 138
    polymorphism and, 135
    vs. polymorphism, 138
initial state, 127
initialization, 127
Inkscape, 174
installation, 495
instance variables, 126, 131
instantiation, 133
int( ) function, 238
integers, 42
integration tests, 414, 495
integration, continuous, 494
interfaces, 66
interior tests, 409
internal references, 455
interpreted languages, 39, 291, 495
IPython, 40
IPython Notebook, 460
IPython Parallel, 307
is (identity operator), 79
isdigit( ) method, 56
isinstance( ) function, 134
issue trackers, 461, 495
items( ) method, 89
iteration, 86

J
JSON License, 480

K
k-d trees
    best application of, 257, 272, 277
    documentation on, 277
    example, 273
    KDTree class, 274
    organization of, 273
    vs. binary search trees, 272
k-dimensional trees (see k-d trees)
kernel, 319
kernprof, 400
key function, 109
key-value mapping, 258, 272
keys( ) method, 89
keyword arguments, 99
keywords, searching for, 24
ksh (Korn SHell), 495
KVM, 320

L
lambdas, 108
LaTeX
    basics of, 335
    benefits of, 442
    building documents, 447
    constituent parts of, 445
    document class, 446
    document structure, 449
    environments, 446
    extension packages, 452
    internal references, 455
    math formulae, 450
    metadata, 447
    preamble, 446
    tables and figures, 454
legacy code, 427
legal issues
    Berkeley Software Distribution (BSD), 475
    changing licenses, 482
    choosing a software license, 474
    copyrights, 472
    Creative Commons (CC), 479
    export control, 484
    GNU General Public License (GPL), 477
    HIPAA compliance, 484
    license files, 471
    obtaining council, 471
    other licenses, 480
    patents, 483
    public domain, 473
    right of first publication, 473
    trademarks, 484
less program, 23
lib directory, 3
licenses (see legal issues)

line numbers, adding, 193
link (ln) command, 29
linking, 455, 495
linspace( ) function, 203
linting, 401
Linux, 495
Linux Containers (LXC), 321
list( ) conversion function, 68
lists, 66-70
literal characters, 179
literal types, 42, 49, 54
literals, escaping, 185
ln (link) command, 29
ln -s command, 29
loadtxt( ) function, 152
local scope, 104, 495
locals( ) function, 252
logic (see flow control and logic)
logical operators, 78
logspace( ) function, 203
loops
    comprehensions, 90
    customizing with generators, 109
    for loops, 88
    formats for, 85
    infinite, 86
    nonterminating, 86
    while loops, 86
low-level languages, 66
lower( ) method, 56
ls (list) command, 6, 26, 181

M

magic objects, 120
make utility
    automatic updating by, 337
    benefits of, 334
    enabling, 337
    makefile configuration, 343
    makefiles, 337
    running, 337
    special targets, 340
    target definition, 338
    vs. bash scripts, 336
man (manual) program, 21
markup languages
    bibliographies, 456
    choices of, 444
    LaTeX

        building documents, 447
        constituent parts of, 445
        document class, 446
        document structure, 449
        environments, 446
        extension packages, 452
        internal references, 455
        math formulae, 450
        metadata, 447
        preamble, 446
        tables and figures, 454
    process of using, 444
    reference managers, 458
masks, 217
math formulae, typesetting, 450
mathematics projects, 489
matplotlib, 167-172
max( ) function, 101
memory mapping, 242
memory-bound, 249, 495
metacharacters, 495
    basic rules for, 179
    escaping, 184
    globally finding filenames with patterns, 182-187
    listing files with simple patterns, 180
    special vs. literal meaning of, 185
    usefulness of, 179
    vs. literal characters, 179
    wildcard character, 180
metaclasses, 140
metadata, 495
    importance of including, 151
    in issue tracking, 465
    in LaTeX, 447
    reference managers and, 458
    updating with touch command, 13
metaprogramming, 139
methods
    instance variables and, 131
    listing, 119
    requirements for, 128
    static methods, 132
    string methods, 55
    vs. functions, 129
Microsoft Word, 443
milestones, 466
Miniconda, installing, xxiii, 317
minus minus sign (--), 23

missing data
    dealing with in Pandas, 158
    handling of in data frames, 265
mkdir (make directory) command, 18
model-driven analysis, 160
module scope, 104
modules
    aliasing imports, 59
    aliasing variables on import, 59
    basics of, 57
    benefits of, 57
    distutils module, 312
    extension modules, 57
    importing, 58
    importing variables from, 58
    in Python standard library, 62
    math module, 226
    multiprocessing, 297
    packages, 60
    pstats module, 396
    regular expression module, 177, 197
    scipy.constants module, 131
    third-party, 63
    threading module, 291
modulo operator (%), 54
more program, 23
MPI (Message-Passing Interface)
    appropriate use of, 301
    basics of, 301
    benefits of, 301
    role in supercomputing, 300
    scalability of, 306
    specifications for, 300
mpi4py package, 301
msysGit, xxiv
multidimensional slicing, 209
multiline strings, 55
multiple inheritance, 138
multiprocessing
    appropriate use of, 300
    benefits of, 296
    implementation of, 297
    scalability of, 299
    vs. threading, 297
multitasking, 296
munging, 155
mutability, 65
mv (move) command, 17

N

N-body problem, 284, 292
N-dimensional array class, 202
NaN (Not a Number) values, 265
nano text editor, 15
natural naming, 240, 434
negative conditionals, 81
negative indices, 51
nesting, 87, 114
newaxis variable, 214
newline character (\n), 54
next function( ), 109
nil value, 45
no-op (no-operation) statements, 495
nodes, 283
non-parallel programs, 285-290
None variable, 45
nonterminating loops, 86
normalization, 91
nose testing framework, 404
notebook web-based browser, 41
NotImplemented, 45
np (see NumPy)
nuclear engineering projects, 488
null( ) function, 96
NULL/null value, 45
numexpr library, 251
NumFOCUS (NF), 471
NumPy, 152
    arithmetic, 211
    array class in, 202, 302
    array creation, 202
    attribute modification, 204
    benefits of, 201
    bitwise operators, 219
    broadcasting, 212
    documentation on, 203
    dtypes (data types), 204
    fancy indexing, 215
    linspace( ) and logspace( ) functions, 203
    masking, 217
    miscellaneous functions, 226
    ndarray attributes, 203
    newaxis variable, 214
    np.dot( ) function, 213
    np.reshape( ) function, 204
    slicing and views in, 208
    structured arrays, 220, 267
    universal functions, 223

where( ) function, 219

O
object orientation, 495
    additional resources, 142
    applications for, 117
    basics of, 117
    benefits of, 126
    concept of, 118
    design patterns, 142
    features of, 118
    main ideas of, 118
    polymorphism in, 135
    reductionism and, 118
objects, 119-123, 495
ones( ) function, 202, 221
online reference manuals, 21
open addressing, 262
open source community, 491
open( ) function, 231
open-source licenses, 474
OpenMP, 307
options, 22
"or" syntax, 186
OSS licenses, 474
out-of-core operations, 249
outer( ) method, 223
output, sending into files, 25
ownership (see legal issues)

P
packages
    definition of, 57
    installing, xxiv
    managing, xxiii, 311
        (see also deployment)
    Python syntax for, 60
Pandas, 153, 263
parallelism
    basics of, 279
    benefits and drawbacks of, 279
    challenges of, 283
    cross-platform, 307
    event-driven, 307
    libraries for, 291, 297, 302
    low-level, 307
    maximum degree of, 280
    measurement of, 280
    multiprocessing, 296-306
    N-body example problem, 284
    problem classification, 282
    scale and scalability, 280
    threads, 290-296
    vs. sequential programs, 285-290
    vs. web-based parallelism, 307
parameterization, 97
parent classes, 137
parentheses operator ( ), 55
particle physics, 123
pass statement, 96
patches, 468
patents, 483
PATH environment variable, 35
path-generating shortcuts, 10
paths, 3, 34
pdb interactive debugger
    continuing the execution, 394
    features of, 390
    getting help, 392
    importing, 390
    querying variables, 393
    running functions and methods, 394
    setting breakpoints, 395
    setting the state, 393
    setting the trace, 391
    stepping forward, 392
pedagogy and community projects, 490
PEP8 Style Guide, 120, 434
Perl-compatible regular expressions (PCRE), 182
permissions and sharing
    basics of, 26
    connecting to other computers, 30
    linking files/programs, 29
    seeing permissions, 26
    setting file ownership, 28
    setting permissions, 29
permissive licenses, 477
physics
    particle physics, 123
    physics-based fields, xviii
    projects in, 488
pickling, 301
pip packaging tool, 312-316
pipe command (|), 25, 186
pipeline (see analysis pipeline)
plain text markup, 444
Platform-as-a-Service (PaaS), 325

plotting libraries
    Gnuplot, 164
    matplotlib, 167-172
point-to-point communication, 302
polymorphism
    graphic representation of, 138
    inheritance and, 135
    multiple inheritance, 138
    overview of, 119
    subclasses, 136
    superclasses, 137
    vs. inheritance, 138
positional arguments, 98
positive conditionals, 81
preamble, 446
precise types, 42
precompiled languages, 341
print statements, 387
print working directory (pwd) command, 4
profiling, 396-401
programs
    chaining together, 25
    creating links to, 29
    interrupting, 14
    locating built-ins in bash shell, 24
    never-terminating, 14
    running, 34
    writing in parallel, 290-306
    writing in serial, 285-290
projects
    astronomy and astrophysics, 487
    basics steps of, xix-xxi
    geophysics, geography, and climate, 488
    legal issues surrounding, 471-485
    mathematics, 489
    nuclear engineering, 488
    pedagogy and community, 490
    physics, 488
    Python open source community, 491
    scientific Phyton, 489
prompt, 1, 40
proprietary licenses, 474
pstats module, 396
ptdump command-line tool, 254
public domain, 473
publication
    document processing
        common programs, 442
        overview of, 441
        separating content from formatting, 442
        tracking changes, 443
        WYSIWYG systems, 442
    legal issues surrounding, 471-485
    markup languages
        bibliographies, 456
        choices of, 444
        LaTeX, 445-456
        process of using, 444
        reference managers, 458
    right of first publication, 473
    self-publishing, 473
    text editors, 443
pull requests, 468
pwd (print working directory) command, 4
pyflakes linting tool, 401
PyPy project, 291
PyTables
    compression routines in, 253
    dataset classes in, 238
    loading data with, 153
    obtaining, 239
    querying, 252
    vs. h5py, 230
Python
    B-tree support libraries, 271
    benefits of, 39, 63
    built-in data containers, 65-76
    comment characters, 41
    dictionary resizing, 261
    drawbacks of, 39
    duck typing and, 65
    dunder vs. built-in functions, 121
    exiting, 40
    expressions, 48
    file handling in, 230-235
    getting help, 40
    hash( ) function in, 73, 258
    installing, xxiii
    math module, 226
    modules
        aliasing imports, 59
        aliasing variables on import, 59
        basics of, 57
        importing, 58
        importing variables from, 58
        in standard library, 62
        packages, 60
    mutability and, 65

numexpr library, 251
open source community, 491
operators in, 46
pandas package, 263
PEP8 Style Guide, 434
Python prompt, 40
reference counting in, 68
regular expression module in, 177, 197
running, 40
scientific Python, 489
special variables
    Boolean values, 45
    None, 45
    NotImplemented, 45
standard library, 57, 62
statements, 49
strings
    basics of, 49
    concatenation, 53
    indexing, 50
    literals, 54
    string methods, 55
threading in, 290
variables, 42
whitespace separation in, 79
Python 2, 50
Python 3.3, 50
Python Packaging Authority (PyPA), 312
Python Software Foundation License, 480

Q

querying
    HDF5 data format, 252
    k-d trees, 275
question mark (?), 183

R

raise keyword, 84
random-access memory (RAM), 235, 280
ranks, 301
read-only mode, 233
read/write permission, 26, 29
readme files, 431
record arrays, 220
recursion, 107
recursive flag (-r), 19
recursive force flag (-rf), 19
redirection, 13, 25, 495
reduce( ) method, 223

reduceat( ) method, 223
reductionism, 118
reference counting, 68
reference managers, 458
reference manuals, 21
references, internal, 455
regex (see regular expressions)
regression tests, 416, 495
regular expressions, 496
    applications for, 177, 179
    awk
        adding action with, 196
        benefits of, 195
        example of, 195
        overview of, 188
        vs. sed and grep, 195
    benefits of, 177
    grep
        finding patterns in files with, 188
        overview of, 187
        search options, 189
    history of, 178
    matching examples, 194
    metacharacters
        basic rules for, 179
        escaping, 184
        globally finding filenames with patterns, 182-187
        listing files with simple patterns, 180
        special vs. literal meaning of, 185
        usefulness of, 179
        vs. literal characters, 179
        wildcard character, 180
    sed
        adding double-space formatting, 193
        adding line numbers, 193
        complex patterns, 192
        deleting blank lines, 193
        multiple replacement tasks, 191
        overview of, 188
        saving output to new files, 191
        syntax for, 190
        vs. grep, 190
    text matching with, 178
    varied implementations of, 182
relative paths, 4, 34, 496
REPLs (read-eval-print loops), 40, 496
repositories
    checking status of, 357

creating, 355, 373
downloading, 375
forking, 377
hosting, 371
sending commits to, 374
reproducibility, 148, 309, 349, 472
resizing hash tables, 259
resources, sharing, 29, 327
return keyword, 97
right of first publication, 473
rm (remove) command, 18
root directory, 3, 496
root users, 496
RunSnakeRun, 397

S

scalability, 281
scale
    definition of term, 280
    measurement of, 280
    strong scaling, 281
    weak scaling, 281
scaling up, 281
schedulers, 328
scientific constants, 131
scientific plotting (see plotting libraries)
scientific Python, 489
scipy.constants module, 131
SCons, 343
scope, 104, 496
scp (secure copy) command, 31
scripts, creating, 36
search and replace function, 190
search function, 188
    (see also grep)
sed, 496
    adding double-space formatting, 193
    adding line numbers, 193
    complex patterns, 192
    deleting blank lines, 193
    multiple replacement tasks, 191
    overview of, 188
    saving output to new files, 191
    syntax for, 190
    vs. awk, 195
    vs. grep, 190
self argument, 128
self-documenting code, 434
self-publishing, 473

separate chaining, 262
sequence, 50, 496
sequential programs, 285-290
series, 263
sets, 71
sh (Bourne SHell), 496
shared resources, 327
shell
    basics of, 1
    benefits of, 3
    changing directories, 7
    characteristic of, 2
    escape characters in, 184
    file inspection (head and tail), 10
    home directory (~), 5
    listing contents, 6
    navigating, 1-11
    paths and pwd, 3
    types of, 2
shell variables, 32
simulation data, 150
single quote ('), 49, 54
singletons, 44, 496
Slashdot effect, 326
slices, 51, 208
SnakeViz, 399
software (see analysis pipeline; deployment; programs)
Software Freedom Conservancy (SWC), 471
Software-as-a-Service (SaaS), 325
solid state drives (SSDs), 235
sort command, 23
sorted( ) function, 109
source code, 40
source command, 34
source-based package managers, 311
SourceForge, 371
spaces, vs. tabs, 80
spawning, 291, 297
special variables, 44
speedup (s), 281
Sphinx, 436
square brackets ([ ]), 8, 66, 186
SSH (Secure SHell), 30
ssh command, 31
stack backtrace (see traceback report)
stack trace (see traceback report)
"starving CPU" problem, 244
state, 65

statements, 49
static methods, 132
statistical calculations, 263
StopIteration error, 110
strings, 496
    basics of, 49
    concatenation, 53
    converting other types to, 49
    defining string literals, 49
    escape characters for, 54
    indexing, 50
    multiline strings, 55
    prefixes for, 55
    purpose of, 42
    string methods, 55
    working with string literals, 54
strip( ) method, 56
strong scaling, 281
structured arrays, 220
style files, 446
style guides, 434
subclasses, 136
subdirectories
    listing, 6
    packages, 60
    searching multiple, 181
subpackages, 60
substitution, 190, 191
sum( ) function, 226
superclasses, 137
supercomputers, 300, 310, 327
swapcase( ) method, 56
switches, 22
symbolic links, 29, 496
syntactic style, 434

T

tab character (\t), 54
tables
    HDF5 data format, 240
    in NumPy, 220
    publishing in LaTeX, 454
    vs. data frames, 263
tabs, vs. spaces, 80
tcsh (TENEX C SHell), 496
temporary arrays, 211
terminal emulators, 1, 496
ternary operators, 46, 496
test frameworks, 404, 496

test suites, 403
test-driven development (TDD), 419
testing
    as core principal, 404
    benefits of, 404
    concept of, 403
    corner cases, 410
    edge cases, 409
    for equivalence vs. equality, 407
    importance of, 404
    integration tests, 414
    interior tests, 409
    placement of, 405
    regression tests, 416
    running tests, 409
    selecting targets for, 406
    test coverage, 418
    test fixtures, 412
    test generators, 417
    test matrix, 417
    timing of, 405
    unit tests, 412
text editors, 13, 15, 443
text matching, 178
    (see also regular expressions)
theory manuals, 430
third-party modules, 63
threads
    appropriate use of, 290, 292
    benefits of, 291
    child threads, 291
    communication and spawning in, 291
    daemon threads, 291
    drawbacks of, 291
    module for, 291
    N-body problem solution using, 292
    speed limitations on, 291
    vs. multiprocessing, 297
ticketing systems
    assigning issues, 466
    benefits of, 462
    closing issues, 468
    discussing issues, 467
    issue creation, 464
    workflow, 462
tilde character (~), 5
time series data, 149
TORQUE, 328
touch command, 13

traceback report, 395, 496
tracking changes, 443
trademarks, 484
triple double quotes ("""), 55
triple single quotes ('''), 55
True variable, 45
tuples, 70
Twisted, 307
2-body problem, 284
type( ) function, 43
TypeError message, 45

U

unary operators, 46, 496
underscore character (_), 13, 42
Unicode, 49
unit tests, 412, 496
universal functions (ufuncs), 223
Unix, 497
update( ) method, 75
upper( ) method, 56
user access, 26
user guides, 431
UTF-8 encoding, 50

V

ValueError message, 43
values( ) method, 89
values, multiple return, 103
variables
    assigning names, 42, 434
    Boolean, 45
    class-level variables, 124
    environment, 32
    in Python, 42
    instance variables, 126, 131
    specifying, 23
    types of, 42, 497
variables, in Python, 58
vector graphics, 174
version control, 497
    basics of, 349
    benefits of, 349
    example of, 350
    importance to reproducibility, 349

local, 349-369
remote, 371-383
tool types, 351
tools for, 350
with Git, 352-369, 371
view( ) method, 210
views, 209
vim text editor, 15, 190
viral licenses, 478
virtual machines (VM), 319
VirtualBox, 320
virtualization, 310, 321
visualization (see analysis and visualization)
ViTables database viewer, 255
VMware, 320

W

weak scaling, 281
where( ) function, 219, 252
while loops, 86
whitespace separation, 79
whitespace syntax, 80
whitespace, removing, 56
wildcard character (*), 20, 180
write (w) mode, 233
WYSIWYG (What You See Is What You Get), 441

X

XenServer, 320

Y

yes program, 14
yield keyword, 109

Z

zero, 45
zero-indexed languages, 50
ZeroMQ, 307
zeros( ) function, 202, 221
zipping, 252
zlib library, 253
Zope Object Database (ZODB), 271

About the Authors

Anthony Scopatz is a computational physicist and longtime Python developer. Anthony holds a BS in Physics from UC Santa Barbara and a Ph.D. in Mechanical/Nuclear Engineering from UT Austin. A former Enthought employee, he did his postdoctoral studies at the Flash Center at the University of Chicago, in the Astrophysics Department. He is currently a staff scientist in the Department of Engineering Physics at the University of Wisconsin–Madison. Anthony's research interests revolve around essential physics modeling of the nuclear fuel cycle, and information theory and entropy. Anthony is proudly a fellow of the Python Software Foundation and has published and spoken at numerous conferences on a variety of science and software development topics.

Kathryn Huff is a fellow with the Berkeley Institute for Data Science and a postdoctoral scholar with the Nuclear Science and Security Consortium at the University of California Berkeley. In 2013, she received her Ph.D. in Nuclear Engineering from the University of Wisconsin–Madison. She also holds a BS in Physics from the University of Chicago. She has participated in varied research in areas including experimental cosmological astrophysics, experimental non-equilibrium granular material phase dynamics, computational nuclear fuel cycle analysis, and computational reactor accident neutronics. At Wisconsin, she was a founder of scientific computing group The Hacker Within, and she has been an instructor for Software Carpentry since 2011. Among other professional services, she is currently a division officer in the American Nuclear Society and has served two consecutive years as the Technical Program Co-Chair of the Scientific Computing with Python (SciPy) conference.

Colophon

The animal on the cover of *Effective Computation in Physics* is a bobtail squid (of the order Sepiolida). Bobtail squids are part of a group of cephalopods that are closely related to cuttlefish, but do not have a cuttlebone. They have eight arms and two tentacles and are generally quite small (usually between 1 and 8 centimeters).

Bobtail squid can be found in the shallow coastal waters of the Pacific Ocean as well as in some areas of the Indian Ocean and off the Cape Peninsula of South Africa. In some parts of the world, they are known as the "dumpling squid" or "stubby squid" because of their rounded bodies. Like cuttlefish, they can swim either by using the fins on the outside of their bodies or by using jet propulsion.

Because they live in shallow waters, the bobtail squid has developed a symbiotic relationship with the bioluminescent bacteria *Vibrio fischeri*, which provide camouflage in return for food and habitat. The bacteria live in a special organ inside the squid's mantle (body cavity) and are fed a sugar and amino acid mixture. The bacteria then emit enough light to match what hits the top of the squid's body, hiding its silhouette from predators swimming below. This symbiosis begins almost immediately after the squid hatches and even induces the morphological changes that lead to maturity.

About 70 species of bobtail squid are known, but taxonomy within the Cephalopoda class is controversial. The number could change in the future as more species and evolutionary evidence are discovered.

Many of the animals on O'Reilly covers are endangered; all of them are important to the world. To learn more about how you can help, go to *animals.oreilly.com*.

The cover image is from a loose plate of unknown origin. The cover fonts are URW Typewriter and Guardian Sans. The text font is Adobe Minion Pro; the heading font is Adobe Myriad Condensed; and the code font is Dalton Maag's Ubuntu Mono.

Get even more for your money.

Join the O'Reilly Community, and register the O'Reilly books you own. It's free, and you'll get:

- $4.99 ebook upgrade offer
- 40% upgrade offer on O'Reilly print books
- Membership discounts on books and events
- Free lifetime updates to ebooks and videos
- Multiple ebook formats, DRM FREE
- Participation in the O'Reilly community
- Newsletters
- Account management
- 100% Satisfaction Guarantee

Signing up is easy:

1. Go to: oreilly.com/go/register
2. Create an O'Reilly login.
3. Provide your address.
4. Register your books.

Note: English-language books only

To order books online:
oreilly.com/store

For questions about products or an order:
orders@oreilly.com

To sign up to get topic-specific email announcements and/or news about upcoming books, conferences, special offers, and new technologies:
elists@oreilly.com

For technical questions about book content:
booktech@oreilly.com

To submit new book proposals to our editors:
proposals@oreilly.com

O'Reilly books are available in multiple DRM-free ebook formats. For more information:
oreilly.com/ebooks

O'REILLY®

©2014 O'Reilly Media, Inc. O'Reilly logo is a registered trademark of O'Reilly Media, Inc. 14373